Sustainable Radio Frequency Identification Solutions

Edited by

Cristina Turcu

Sustainable Radio Frequency Identification Solutions
Edited by Cristina Turcu

Republished by InTech
Janeza Trdine 9, 51000 Rijeka, Croatia

Additional hard copies can be obtained from orders@intechopen.com

Sustainable Radio Frequency Identification Solutions, Edited by Cristina Turcu
p. cm.
ISBN 978-953-7619-74-9

For EU product safety concerns:
IN TECH d.o.o., Prolaz Marije Krucifikse Kozulić 3, 51000 Rijeka, Croatia,
info@intechopen.com or visit our website at intechopen.com.

Meet the editor

Cristina Turcu is an Associate Professor of Software Engineering and Artificial Intelligence at Stefan cel Mare University of Suceava, Romania. She received her Diploma (M.Sc.) in Automatics and Computers and her Doctorate (Ph.D.) in Automatics, in 1991 and 2000, respectively, both from the Gheorghe Asachi Technical University of Iasi, Romania where she has been Head of Computer Department since 2004. Her research interests include software engineering, RFID applications for the end-consumer, and intelligent systems. Dr. Turcu is an Editor of four books and has served on various program committees of conferences in computing and RFID systems. She also has served as a reviewer for numerous referred journals and conferences. She is the Editor in Chief of the International Journal of Radio Frequency Identification & Wireless Sensor Networks. Dr. Turcu has published over 80 publications in books or book chapters, refereed journals, technical reports, and refereed conference/workshop/seminar proceedings.

Preface

Radio frequency identification (RFID) is a fascinating, fast developing and multidisciplinary domain with emerging technologies and applications. It is characterized by a variety of research topics, analytical methods, models, protocols, design principles and processing software. With a relatively large range of applications, RFID enjoys extensive investor confidence and is poised for growth. To put it differently, significant technological advances and falling RFID tag and readers prices enable more and more organisations to adopt RFID solutions. RFID has been successfully applied in manufacturing, retail, supply chain, livestock, asset and people tracking, etc. and also has received much attention from the medical care industry (e.g. asset tracking in operating rooms and emergency rooms, patient and baby tracking solutions, etc.). Revolutionary experiments have been also conducted on the International Space Station.

A number of RFID applications proposed or already used in technical and scientific fields are described in this book. *Sustainable Radio Frequency Identification Solutions* comprises 19 chapters written by RFID experts from all over the world. In investigating RFID solutions experts reveal some of the real-life issues and challenges in implementing RFID.

In the first chapter of this book, the authors identify and discuss major challenges of RFID infrastructure for large scale supply chains involving small and medium enterprises. Their analysis is based on two national projects in Australia. Then, they discuss a virtualisation model which incorporates an existing RFID framework based on the EPC Network to be applied in large-scale supply chains.

Chapter 2 analyzes the CJ-Global Logistics Service business model with Blue Ocean strategy to show how a company in the Red Ocean reinforces its competitive advantage to move towards a less competitive new market space by utilizing information technologies.

In chapter 3 several authors share the challenges of using RFID tags with embedded temperature sensors in cold chain logistics for perishable food and pharmaceuticals products.

Chapter 4 analyzes the main issues of RFID-enabled traceability along the supply chain: existing methods, applications and future challenges. Thus, the authors point out the characteristics of RFID data and the requirements for RFID enabled traceability. They introduce data types, storage schemes and system frameworks proposed in the existing literatures. Then, the authors discuss tracing methods based on the traceability system architecture. The chapter also contains an overview of current applications in real settings of both discrete and continuous production. The authors also discuss challenges that are preventing companies from adopting RFID for their traceability solutions.

Chapter 5 presents a number of traceability methods that can be used to improve traceability in continuous processes with a special focus on RFID. The authors exemplify how RFID may be applied and combined with other methods to improve traceability in continuous processes with two examples from continuous refinement process of iron ore and from wood refinement process. Also, additional RFID applications to improve traceability in continuous processes are described. Various benefits of using RFID to improve traceability in continuous processes are identified and discussed from well-described RFID applications. Also, special challenges using RFID in continuous processes are identified; they are followed by the presentation of several solutions to be adopted in solving or avoiding such challenges. Finally, the authors describe some ideas for future research projects and ongoing research projects.

Chapter 6 presents a case study of an RFID-based system for the identification and tracking of pilgrims. The author describes a developed prototype employing a passive RFID wristband tag for identification of pilgrims in holy areas during Hajj. Thus, RFID technology was chosen assist the authorities in the Hajj season.

Chapter 7 approaches the implementation protocol using RFID technology and biometric identifiers in the context of Irish bovine traceability. The use of retinal images as a biometric, to verify identity would provide a system check that would be virtually fraud-proof. Thus, the described system would be able to identify cattle whose ear tag has been tampered with, providing a mechanism for source and identity verification of Irish beef products.

In chapter 8 the author identifies and analyzes several airports issues and describes ways to use RFID for improving passenger and baggage handling at airports. Finally, the benefits of applying the RFID technology for several airport operations are presented.

Chapter 9 discusses the implementation of RFID integrated with biometric sensor to improve boarding school management of students and facilities. The technologies used to develop monitoring systems and some methods to enhance the security level of the system are presented. Finally, based on the application scenario described, the authors describe the developed monitoring system and discuss the flow of the overall system.

In chapter 10 the authors closely examine the evolution of technology, architectures, and approaches to solution building in the RFID field. They examine those components that have benefited from standards that now have become the building blocks for solutions that when combined with business processes unleash a greater business value. Furthermore, the authors look at the options to templatize parts of the process to accelerate the integration and reuse of these components and to ultimately deliver a greater ROI in a shorter period of time.

Chapter 11 focuses on RFID applications for sanitary environments. The main characteristics of sanitary environments are described and a major focus lies on scenarios where the use of RFID technology is indispensable. Then, some standard application scenarios for item tracking and for people tracking are presented. Also, potential problems deriving from the introduction of radio technology into an environment with sharp safety requirements are approached; the authors discuss possible interferences with other devices and refer to several law recommendations. Finally, significant potential improvements deriving from the refinement of the technology are considered and discussed.

The broad objective of chapter 12 is to show how RFID technology can be used to reduce medical mistakes, to improve patient safety and to enhance the quality of medical service in hospitals. After a brief introduction of the eHealth domain and the survey of several healthcare issues, this chapter focuses on how RFID technology can be efficiently used in the healthcare domain; thus, some concrete hospital issues and case studies are presented. An overview of a medical staff and patients tracking application architecture is given. The authors show how to use this application to improve the quality of hospital services. This chapter also identifies several problems in the healthcare system that must be solved prior to the introduction of the RFID technology.

Chapter 13 deals with an application based on RFID technology which directs a person to a desired destination in an unfamiliar environment. Related studies on current signage solutions are described. The proposed system architecture and the employed technology are presented. The hardware and software aspects of the implementation of the system are described by suggesting cost-effective hardware solutions and briefing the used algorithm. A deployment scenario is also presented. Finally, potential future improvements are discussed.

The major aim of chapter 14 is to highlight the contributions of RFID systems modeling. Conventional and emerging RFID applications in healthcare are reviewed. This chapter introduces the agent based modeling technique, invoked to investigate system performance in an application for RFID-enabled patient tracking within a hospital emergency department. The agent-based model simulation results are outlined. Finally, the authors discuss implementation strategies for an RFID real-time location system for patient tracking, reflecting a service oriented architecture approach.

Chapter 15 focuses on the progress achieved in the development of RFID writers/readers used in the collection of damaged information, as well as in information sharing and damaged information collection systems. Thus, this chapter presents an RFID writer/reader functioning as an information resource tool in rescue activities in various disaster areas; consequently, it allows rescue crews to instantaneously collect vital information in a non-contact fashion and then inform the authorities outside the afflicted areas about the conditions in the investigated regions.

Chapter 16 deals with low-cost identification applications in traffic vehicular environments. Firstly, a state of the art of the applications developed in a worldwide framework is outlined. Secondly, a summary of the ERI standards and the architecture of the developed application for several technologies (RFID, Bluetooth and WiFi) are presented. Finally, the tests performed on the developed system are explained and the results for each test are analyzed.

Chapter 17 focuses on ways of integrating data across disparate sensing systems over both time and space in order to design smart environments. This chapter presents several research problems and their solutions developed for building hazard aware spaces. A method for deploying sensors remotely with a robot is developed. Different modalities of controlling the robot (e.g. voice, gestures, a mouse or a keyboard) and several aspects related to fusing multiple simultaneous commands are considered. After analyzing the tradeoffs of localization approaches, the authors explore the effectiveness of passive RFID technology in localization procedures. They have proposed a methodology for building a sensor model for accurate RFID-based localization in addition to the standard acoustic time-of-flight ranging and stereo vision localization methods. Finally, they have concentrated on proactive approaches to camera control, spectral image analysis and human alert mechanisms; they have briefly outlined a few challenges in application scenarios similar to the hazard aware space.

Chapter 18 proposes an investigation of RFID context-aware systems. Thus it surveys and classifies existing RFID context-aware systems. The authors concentrate upon defining and clarifying concepts related to context-awareness. Moreover, several design issues related to context-aware systems are investigated. The chapter comprises a section investigating the use of RFID in a museum exhibition guide system that provides the most suitable services to be offered to visitors based on the context information. The described system proves highly efficient in guided tours because it never ignores the location of its users and because it is always ready to perform its duties in accordance with every user's instructions and preferences.

The final chapter of this book goes beyond identification matters and also addresses aspects referring to the representation of real world objects and actors in radio frequency identification. The chapter introduces a novel RF tag architecture called Object Tag (OTag), to represent the real world objects and actors in RF tags. Finally, possible applications are derived using the OTag architecture. OTag is introduced to improve the existing intelligent transportation systems and it enables a large array of novel applications.

I hope that this book will give to the readers many solutions and ideas in design and implementation of another efficient RFID architectures and applications. Many thanks go to all authors who contributed to this book and shared their knowledge, and without whom this book would not have been possible. Finally, my appreciation goes to Intech team that is behind the scenes, but without its hard work this book would never have appeared.

Editor

Cristina TURCU

Stefan cel Mare University of Suceava

Romania

Contents

1

RFID Infrastructure for Large Scale Supply Chains Involving Small and Medium Enterprises

John P.T. Mo and William Lorchirachoonkul
RMIT University
Australia

1. Introduction

A "manufacturing based supply chain" containing a broad spectrum of partners, including manufacturers, third party logistic providers, distributors and retailers, all of whom, have mixed modes of value adding processes (Chauhan & Proth, 2005). An important success factor for this type of supply chains is the ability for the partners to identify and track products moving through the supply chain, using the latest information technologies (IT). However, these value enhancements would vary substantially according to the nature of the partners' business requirement. As such, the flexibility of managing the IT infrastructure for supporting value adding activities is of the utmost importance.

Radio Frequency Identification (RFID) is a wireless communication technology for precisely identifying objects. It uses radio-frequency waves to identifying information between tagged objects and readers without line of sight, thus enabling automatic tracking and tracing. Passive RFID can track products in supply chains from the supplier to the distribution centre, warehouse, and point of sale. RFID technology is increasingly used in supply chain management (SCM). However, like most emerging technologies there are different vendors, standards, systems, applications, appliances and processes that exist today. Therefore, potential increase in productivity is heavily offset by huge capital expenses on assets which may be obsolete in a few months' time. As a consequence, most small and medium enterprises (SMEs) simply wait until such time that they feel the technology is mature and stable enough to be integrated into their existing operations.

Traditionally, supply chain network is built on existing IT infrastructure of the companies involved in the supply chain. Since major companies have IT policies that naturally channel data in a global fixed network structure, the supply chain information backbone is built based on the assumption that communication will be directed to the Internet. Typical examples are the two Australian national scale projects that were supported by the Australian government to investigate the properties and limitations of electronic product code (EPC) technology for fast moving consumer goods (FMCG) supply chains (Mo et al, 2009b). In Europe, the project "Building Radio frequency IDentification solutions for the Global Environment" (BRIDGE) was supported by the European Community sixth framework programme to resolve the barriers to the implementation of the EPCglobal Network in Europe (Soppera et al, 2007). The BRIDGE project aimed to develop easy-to-use technological solutions for the European business communities, including the SMEs. These projects show that the application of EPC technology can lead to significant benefits in system efficiency and reliability. Ironically, large organisations, which already embrace RFID technology, will often need to communicate with SMEs that may not choose to implement RFID due to resource constraints. Under these circumstances, there is no other physical way in that the RFID tags can be read or updated in some parts of the supply chain, the entire information link is disrupted.

One of the ways to resolve such issue is by the method of virtualisation. Virtualization implies the utilisation of IT and communication technology by the business organizations to manage their key operations with stakeholders, such as customers, suppliers and employees (Demchenko, 2004). Virtualisation is to encapsulate a physical device to behave as multiple virtual devices where the resources of the physical device are shared among the virtual devices. Most of the existing frameworks are based upon physical RFID tags being read by RFID readers which reside over a gate or doorway. As products containing RFID tags move across the RFID readers, the location of the products is confirmed. This feature may be very useful for traceability in a warehouse operation environment, but it may not be necessary when products move across two trading parties. Since the objective of the RFID reader is to verify that the product has arrived at the receiving party, the sending party does not require extensive and details of information, for example, which door or gate their product are received. The main objective is to verify that it has been received. Therefore, if the RFID readers themselves can be emulated or have virtual machine build-in, the implementation and maintenance process would be more efficient and the overall costs of ownership reduced.

In this chapter, we review the lessons learnt from the two national projects in Australia. We examine the activities in the projects by modelling and simulating them in a discrete event simulation environment. The outcome is an estimation of the benefits that can be derived from the national projects if it was continued for a much longer period. We then discuss a virtualisation model which incorporates existing RFID framework based on the EPC Network for application in large scale supply chains. The model allows alternative technologies to verify the position of the RFID tags and then updates the existing RFID framework to ensure that the traceability of the RFID tags remain intact, even if there is an interruption in data along sections in the supply chain. With these new developments, a mixed mode value added supply chain will be able to include any members in the supply chain, large and small, without the concern of information infrastructure incompatibility.

2. RFID pilot projects

In a typical deployment, RFID tags are affixed to individual products, and readers installed at various checkpoints capturing tag-reading events. Companies then transport these tagged items along the supply chain, leaving trails of tag-reading events behind. These events enable the companies to record product movement in the supply chain. Unfortunately, passive RFID technology has limitations, both in terms of physical constraints placed on signal transmissions and how much data a tag is able to store. To access more data about a product (such as its production date, batch number, or package size), organizations could store most of that information externally in a central data warehouse and use the RFID tag as a key. All organizations would have to agree on a common storage format and continuously upload their data to the central data warehouse. This approach has several

severe drawbacks, for example competing. Competing enterprises are often reluctant to put their data in a shared central database.
To explore the functionality requirements associated with large-scale RFID applications, two Australian RFID national projects were developed. The projects were conducted in real industrial environments integrating the experimental investigation with normal business operations between industry partners in two cities, Melbourne and Sydney.

2.1 The national EPC network demonstrator project

The EPCglobal Network standard provides a promising architecture for tracking and tracing objects over the Internet (EPCglobal, 2006). However, few real-world large-scale application examples were reported. Consequently, practitioners lack guidance and only have their own limited deployment experience from which to learn. To gain first-hand, practical experience the National EPC Network Demonstrator Project (NDP) aimed to identify the business benefits of sharing information securely using the EPC Network, providing authentication to interacting parties, and enhancing the ability to track and trace movement of goods within the entire supply chain involving transactions among multiple enterprises. The key concepts in the NDP include:

- using the EPCglobal Network standard,
- providing authentication to interacting parties, and
- enhancing organizations' ability to track and trace product movement within the entire supply chain and for transactions among multiple enterprises.

This project was the first in the world to demonstrate the full stack of the EPC network architecture enabling inter-organisational transactions and supply chain management. When a given tag was detected, instead of having each company storing this information and communicating to the next partner, the EPCglobal model defined one authoritative registry of numbers that could be queried for links to access detail information from local servers. Hence, all items to be identified were allocated a unique global EPC by GS1. Items of interests included products (identified by a serial GTIN - SGTIN), pallets (identified by a Global Returnable Asset Identifier - GRAI), and unit loads (identified by a Serial Shipping Container Codes - SSCC).
The NDP was a large scale project, involving 13 consortium members. To ensure a good chance of success, the consortium simplified the material flow process by limiting to 9 product items. The system designs were incorporated in 15 use cases developed by the industry partners of the consortium. Use cases are the description of how the business processes work with the system (software). Several innovative process designs were developed to support data integrity of the system (Mo, 2008a).
In order to share information securely among the partners, the NDP web site was set up on a global server. Partners could access the web site using a username and password pair control. Once logged in, the product information, containment (content), history (track and trace) information can be accessed using the EPC as the search key (Figure 1). This data sharing capability was the biggest advantage of the NDP which demonstrated the data transparency about the traded items. Detailed transaction data such as location and time about an item (e.g. a shipment on pallet) were immediately available to other partners once the information was logged to the global EPC information server. The timely information improved the efficiency of the supply chain.

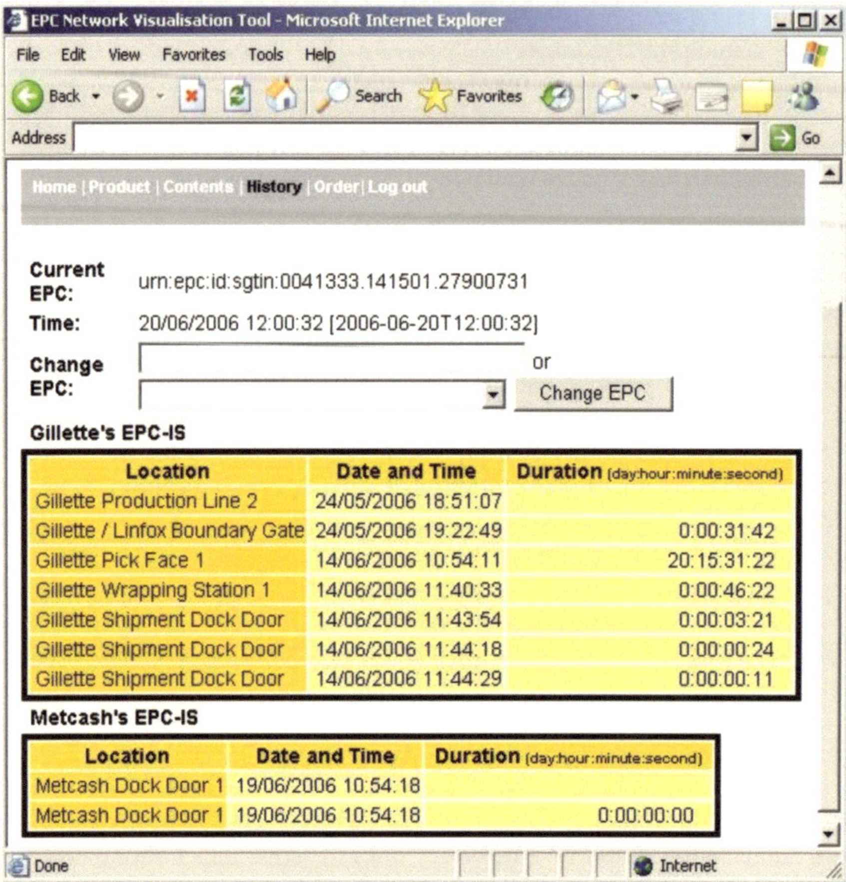

Fig. 1. Track and trace for item urn:epc:id:sgtin:0041333.141501.27900731

2.2 The NDP extension

Following the NDP, some of the partners continued the research and re-grouped as a second consortium working on an extension project "National EPC Network Demonstration Business Information Integration" (NDP Extension) (Mo et al, 2009a). The NDP Extension aimed to address the industry's expectation to see RFID building into real processes and integrating the data with business information system. The NDP Extension concentrated on assets, in this case pallets. Asset management was one of the potential benefit areas identified in the NDP.

Management of assets throughout the supply chain was an onerous and surprisingly complex task. As assets moved between trading partners, irregularities were hard to spot at the time. Errors in quantities sent and received was not spotted until the statement was available, which could be several weeks later. The discrepancy then had to be negotiated between trading partners, each of whom had their own version of events and supporting paperwork that may be in conflict. Since there were more than 10 million pallets in circulation throughout Australia, the effort spent managing assets was extremely costly.

The NDP Extension provided an industrial environment to test the EPC Network technology for solving the asset management problem. Six sites, three in New South Wales and three in Victoria were installed with the EPC hardware and network infrastructure. The project had the advantage to use the latest RFID hardware so a high read rate in the system was anticipated.

Some problems were reported. Initially test runs were unsatisfactory because the readers were unstable in detecting the RFID tags passing through their RF fields. As more tests were done, it was found that there were quality problems of the tags. Some tags were coded incorrectly. Some tags had manufactured problem that they were inherently weaker tags. Since the consortium used 3 different brands of readers, there were also site and reader specific issues. These problems were rectified by a number of remedial measures. Each reader was tested thoroughly as a standalone unit as well as when it was connected to the EPC Network. Tags were individually checked before used.

The consortium was able to perform 4 paperless test runs of pallet despatch and delivery transactions for each site. The test runs proved that business information integration for supporting paperless delivery of pallets could be practised provided that 100% read rate was achieved in the transaction. Process models were developed for each of the transaction routes and benefits were reported as shown in Table 1 (Gajzer, 2007).

Company	Labour cost $/hr	Standard Time min/ delivery	EPC Process Time min/ delivery	Increased Efficiency %
CHEP Erskine Park	$32	35	23	34.3%
Masterfoods	$28	35	30	14.3%
Linfox	$25	38	36	5.3%
Acco	$23	35	30	14.3%
Westgate Logistics	$31	45	35	22.2%
Average				18.1%

Table 1. EPC process efficiency gain

The efficiency gain and hence cost reduction due to the elimination of data entry, verification and reconciliation processes was significant, especially for the pallet supplier. Furthermore, improvement in inventory accuracy as well as improvement in quality area, such as accuracy and transparency of information and real time processing had great impact on the other logistics operations such as planning and forecasting.

In addition, the project developed a business scenario in which the EPC Network could be provided to the general public as a subscribed managed service. The normal business model was that for any company to implement EPC Network, the company had to invest significant amount of resources at the start. The investment includes cost of IT infrastructure, tagging process, training, fault tolerant systems and others. This was the capital expenditure model. The new subscribed service had the potential to change it as an operating expenditure.

2.3 Key features of NDP and its extension

The distance between transaction locations was a significant factor, causing undesirable delays in the system's ability to complete transactions or provide visibility for the items in transition. The various NDP partners jointly developed the system to support these transactions. The system must maintain consistency across physical and organizational boundaries. The partners adopted a six-layer model (Figure 2), from the EPCglobal architecture, which specified the top three layers. The partners developed the operational elements that span layers 4, 5 and 6:

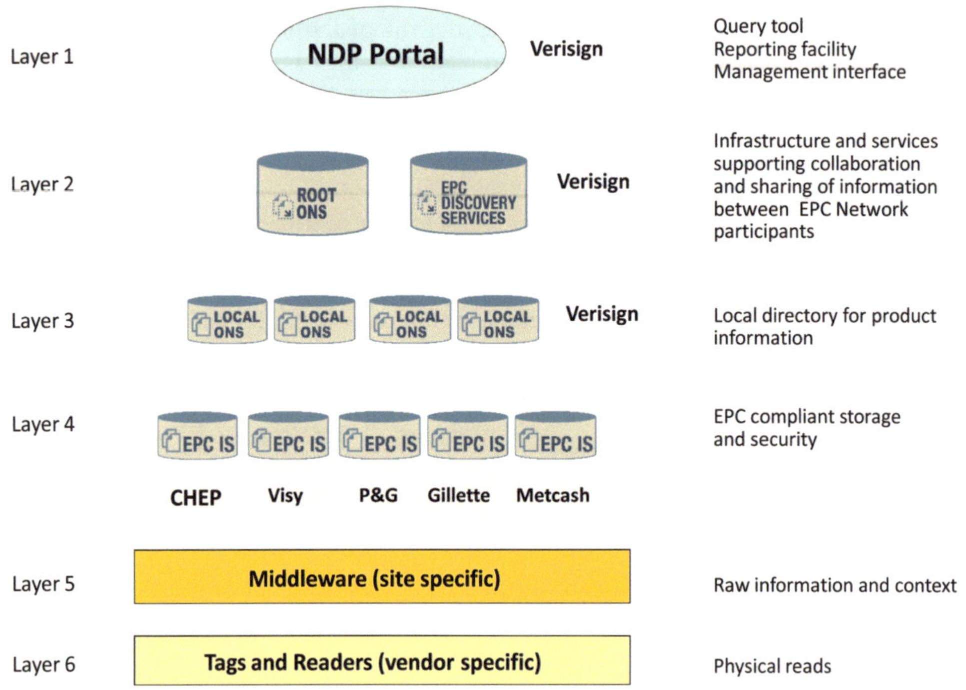

Fig. 2. The NDP layered architecture.

- *NDP portal.* The consortium developed the portal to display the data collected and stored in the system and to allow companies to query the movement of goods through the supply chain.
- *Root objects name service (ONS) and discovery service.* The ONS identifies unique numbers for manufacturers, whereas the discovery service points to a particular EPC information service (EPC-IS) in which companies can obtain detailed information on a specific item.
- *Local ONS.* Each industry partner maintains its own repository of product-specific data. The local ONS provides a pointer to the local database.
- *EPC-IS.* At each site, the industry partners maintain details about site-specific product data that other participants can query.
- *Middleware.* An onsite software component converts multiple reads to one read, adds contextual information (such as the reading location and timestamp), and formats data for storage in an EPC-IS.
- *Tags and readers.* Physical reads capture RF signals in the form of hexadecimal numbers and transmit them to the server for the middleware layer to filter.

In Figure 2, VeriSign implemented the first three layers. While the rest of the EPC Network were implemented by partners involved in the NDP. Adopting a common architecture solved half the visibility problems. However, partners still had their own internal systems and processes that weren't compatible for information exchange.

3. Analysis of the national projects

EPC technology is designed on the premise that a supply chain network is built on the existing IT infrastructure of the companies involved in the supply chain (Kelepouris et al, 2007). However, if one of the participants is not EPC compliant, the information link for the entire supply chain can be disrupted. Capital expenditures are limited. It would be difficult to convince these companies to invest in new technology infrastructures such as EPC. The common concern is that the technology is unstable and not mature enough to remain an industry standard. With rapid advances in technological progress, the current technology may soon be obsolete. Moreover, the costs of training and implementing such technologies are high, thus putting further pressure on already narrow profit margins. To assist decision making for companies in a supply chain, the NDP and its Extension are analysed by two methods.

3.1 Modelling and simulation

This research has been analysed by digital simulation. The computational logic of the simulation model is based on the work flow and information flow of the NDP processes (Mo, 2008b). At the beginning of each day, Metcash DC checks inventory to decide whether to place an order. If the inventory level is less than a pre-defined level, Metcash DC orders replenishment up to certain level. Figure 3 shows part of the work flow in the NDP simulation model.

Metcash's purchase order initiates most activities at P&G, Gillette CHEP and Visy. Only Gillette Manufacturing is independent from Metcash's purchase. Gillette Manufacturing's activities are based on the sales forecast and the production quantity will be based on the purchase order quantity rather than inventory. The NDP simulation model is further separated into five sub models. Each sub model represents a consortium member and has several functions for different work.

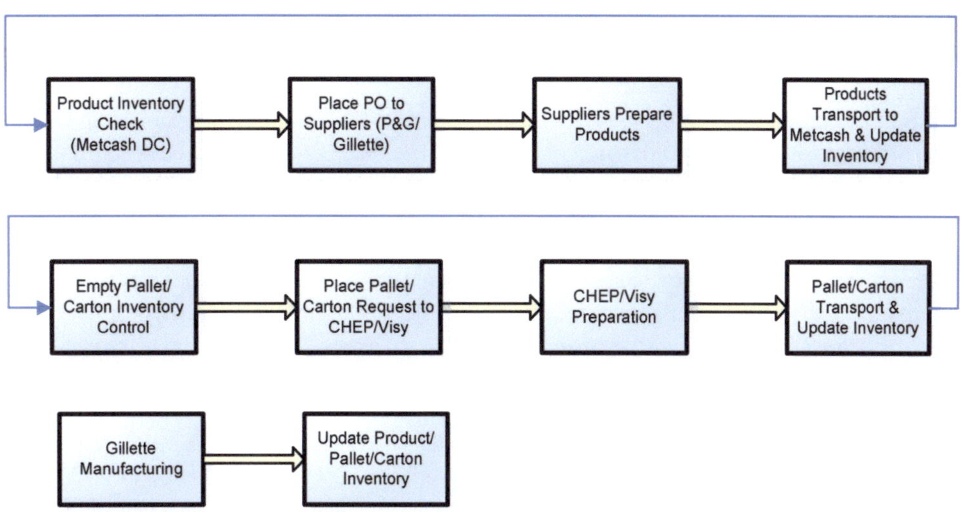

Fig. 3. Basic work flow in NDP simulation model

By running the simulation model, it was found that a saving of $192,158 on overall labour cost can be achieved if the NDP continues for a year. Other costs such as overhead and materials costs are not included. Therefore, it only represents a tiny section of the potential savings of the supply chain. Given that there are thousands of products handled by the consortium companies, the savings will be significant if the EPC technology is rolled out to the whole supply chain.

3.2 Network compatibility

A stable and reliable network and IT infrastructures of value added supply chain is vital for both of the national projects in Australia. To ensure such stability and reliability, each partner maintain their own EPC information server (EPC-IS), behind their firewall (Figure 4). It's common for large scale operators to host their network equipments and IT infrastructures at a data centre. These provide additional layer of redundancy for back up data. Should the power shortage occur, secondary and tertiary back up power supply will be activated? Data centres will also provide regulated temperature and humidity control, such that these equipments work at optimal operating condition. Such facilities are expensive, and may not be an affordable option for smaller organisations.

In order to maintain the speed and reliability of the information transfer between the RFID reader, the middleware and the EPC-IS, the quality of the link between sites must be of high standard. However, it may not be cost effective to to establish such wide area network (WAN) connection. In some remote sites, where the land value is lower for new warehouse to be built, the link may not be possible due to lack of infrastructures. The Australian and European projects used connections in established areas, where the network has been operating in an ideal optimised condition. In addition, the RFID gates were considered a fixed asset in the two national projects. Each member in the consortium used their own RFID readers which were assigned to their specified location. However, in the 3PL environment, where a single warehouse location can be used to represent different suppliers and customers, the RFID readers cannot be modelled as fixed assets and are only updated by a single EPC-IS which belongs to a single entity. After the tags are read and filtered by the middleware, it will still need to be managed by an intelligent decision system that ultimately routes the information to the correct EPC-IS. The information is then translated to a meaningful business function and process.

As shown in Figure 4, an EPC Network supports access to EPC portals. Queries to the ONS server which ultimately points to an EPC-IS that stores the database behind each organisation's firewall. Similarly a decision system is also needed to do similar query as those of the EPC portals. A set of queries can be sent to multiple EPC-IS at the same time to query an order that consists of multiple tags. However, if one of the links is broken (due to system response, connectivity, security or other reasons), that request may not be executed. If the result of that particular query is required to trigger an event or change of ownership, failure of transaction may occur. Moreover, a gap exists between those supply chain members who are fully EPC compliance and those who do not have such network infrastructure in place. Although EPCglobal has published a universal EPC standard, individual countries are still free to impose further restrictions to the frequency band and power ratings that EPC devices can be used on their land. EPCglobal has been widely used amongst European and western countries. However, there are other RFID frameworks being utilised around the world. These include China's National Product Codes (NPC) (more information can be found at http://www.chinapt.com/CptNpc/En/Npc_e.aspx) and Japan's Ubiquitous ID (UID) (more information can be found at http://www.uidcenter.org/english/introduction.html), which is supported by 352 Japanese companies including Sony, Toyota, and Mitsubishi. These networks operate under different radio frequencies and have different structure to those of EPCglobal. As China is now the world largest exporter, it is important to acknowledge their views on the technology. In order for the tags to be read successfully across the supply chain, a system that is flexible to handle a wide variety of operating parameters (within the universal standards) is mandatory.

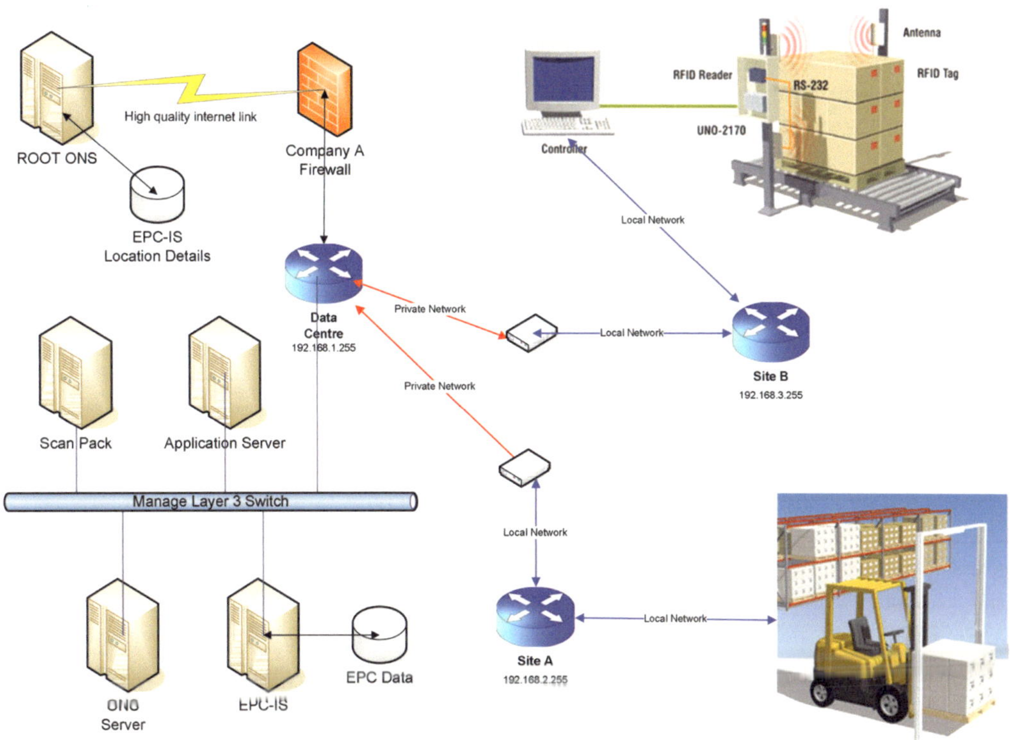

Fig. 4. EPC Based System Architecture

4. Virtual network infrastructure

Virtualised IT environments have been developed to reduce costs and improve reliability (Liu et al, 2008). Virtualization implies the use of IT and communication technology by organizations in managing their interactions and key business operations with stakeholders, such as customers, suppliers and employees (Zarour et al, 2005). In a virtualized environment, the virtual devices are seen externally as handling different functions. Such virtualisation techniques have been used in medium to large scale information technology infrastructures, where vendor such as VMware, Citrix Xen Server and Microsoft Hyper V use a single physical server to host many virtual servers that perform a range of tasks (Kim, 2008). It uses the concept of Hardware Abstract Layer (HAL) which is presented to the operating system in each virtual server as its usual hardware platform. Therefore, a virtual machine can be moved to and from different physical servers with different physical configurations and vendors. By simplifying the hardware and hence reducing the cost of new system implementation, it is then possible to increase the redundancy of the system or reduce power consumption with less physical hosts.

Virtualisation of supply chain systems depend on the use of the Internet. In a network environment, hierarchical virtualization of a network provides a flexible, granular, protection resource. Using the HAL concept, the network services are handled by a number of virtual servers. In the event that a physical server hosting the virtual servers fails, another physical server with the virtual systems (potentially another server already operating on the network) can stand in as an alternative physical host. This improves the robustness of the system in operational functions and data backup. Since supply chain participation is a loosely coupled relationship, increased mobility of information infrastructure will significantly improve the efficiency of supply chain system set-ups and encourage cost effective partnering relationships.

To develop a virtualised RFID infrastructure, it is necessary to use alternative technologies that function as RFID in the absence of RFID. Since RFID is relative expensive to established, establish, we cannot assume that all SME can afford to implement and maintain such technologies. SME who do not subscribe to RFID network, are not visible to the rest of the RFID subscriber, hence, creating a broken chain to the global supply chain. For this reason, we cannot afford to have a complete reliance on a single technology to drive a global supply chain network.

4.1 GPS

The issue of high infrastructure cost for RFID implementation has attracted serious questions with respect to its suitability for the supply chain industries, as many are SMEs. (Pedro & Reyes, 2007). Alternative technologies are being investigated, one of which is global positioning system (GPS). Since late 1990's, it has been a common trend for third party logistics (3PL) companies to incorporate GPS technologies via mobile data networks such as GSM or GPRS to track vehicles and drivers (Zito et al, 1995). The key function for such technology is to provide accountability and traceability to the end customer, especially when billing on an hourly basis (taxi truck, over dimension freight). It is also used in reverse, for checking subcontractor invoices and accounting for the time charged (Hamilton, 1993).

GPS has been utilised in most Fleet Management System (FMS) for some years (Visser, 1991). Its usage is largely confined to business management such as travel time control (Quiroga & Bullock, 1998). In most FMS applications, users can specify an address location and then set a permitter around the area as shown in Figure 5. The GPS coordinates are transmitted at set time intervals from the personal digital assistant (PDA) which is mounted inside the vehicle. If the vehicle is in an area that mobile reception is not available, the coordinates and timestamps are stored inside the PDA, until such time that mobile reception is re-established (Chiang & Huang, 2008).

4.2 Geo-fence

According to a study by the Pennsylvania University RFID Study Group (2006), the major issue in the use of GPS for goods tracking in supply chains is the difficulty of system interoperability. The only key data being transmitted over the mobile network is the GPS data. This usually consists of longitude, latitude, current speed, headings, and altitude

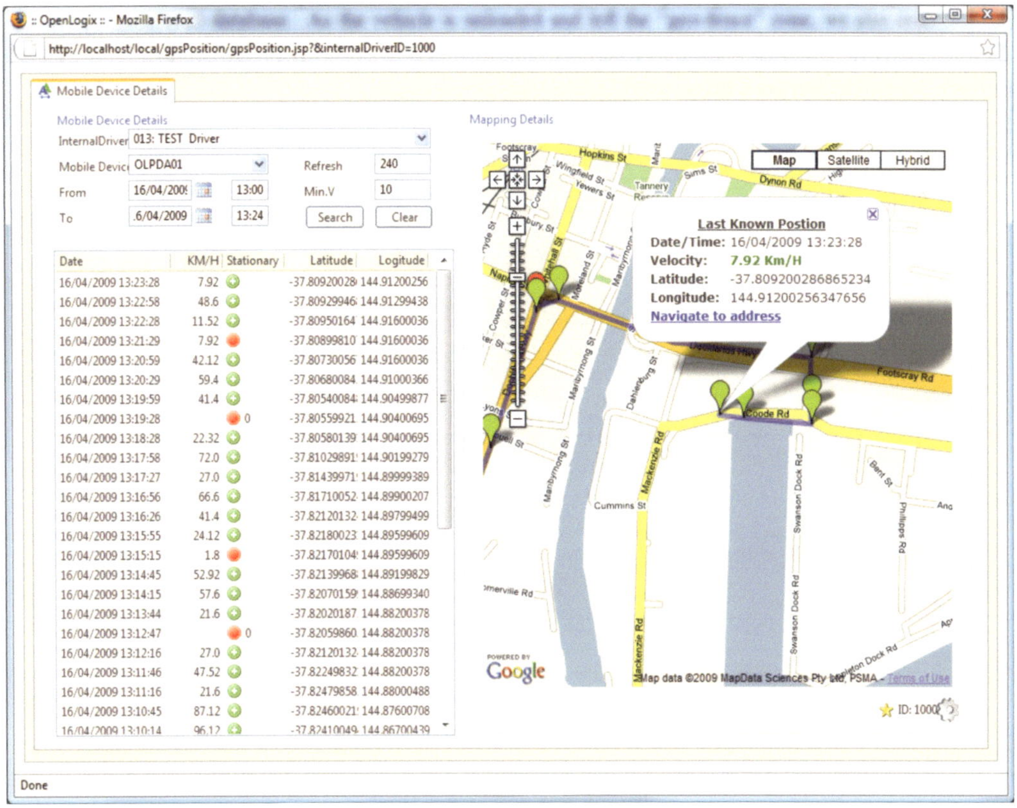

Fig. 5. Vehicle tracking using GPS

(Behzadan et al, 2008). Most GPS solutions would incorporate mapping database which allow users to locate the vehicles' locations. Since the mapping database is proprietary to the GPS software, it is inaccessible to any other applications. This makes it impossible to provide accurate interpretation of the GPS data in the map. Therefore, a system architecture that is flexible enough to adapt to device incompatibility is required for integration with RFID implementation.

A mobile server can be configured as an agent to filter out any invalid signal and feed useful information to the track and trace service. This can trigger a pre-defined process when the event profile is detected. Theoretically, for goods track and trace in logistics, the mobile server keeps track of GPS coordinates against the defined virtual geo-fence. When a vehicle enters the valid geo-fence zone, an arrival timestamp is stored inside the FMS. An event will be triggered that leads to a set of predefined tasks being put to action, such as an update on the job status or on an EPC-IS database. As the vehicle is unloaded and leaves the virtual geo-fence, the departure timestamp can be stored and vehicle loading time calculated.

The geo-fence allows emulated RFID data to be used at locations where installation of the RFID infrastructure is not possible. In Figure 6, a company's enterprise resources planning (ERP) system has been integrated with EPC-IS to provide track and trace functionality for its goods delivery. This tracing capability works well within the company's boundaries. Event management of the local system will update the consignment with all EPCs of goods and the truck that is assigned to this route. Once the truck leaves the depot, the connection of the consignment to the EPC Network is lost. In a virtual geo-fence environment, the task of tracing the truck can then be moved to the FMS module.

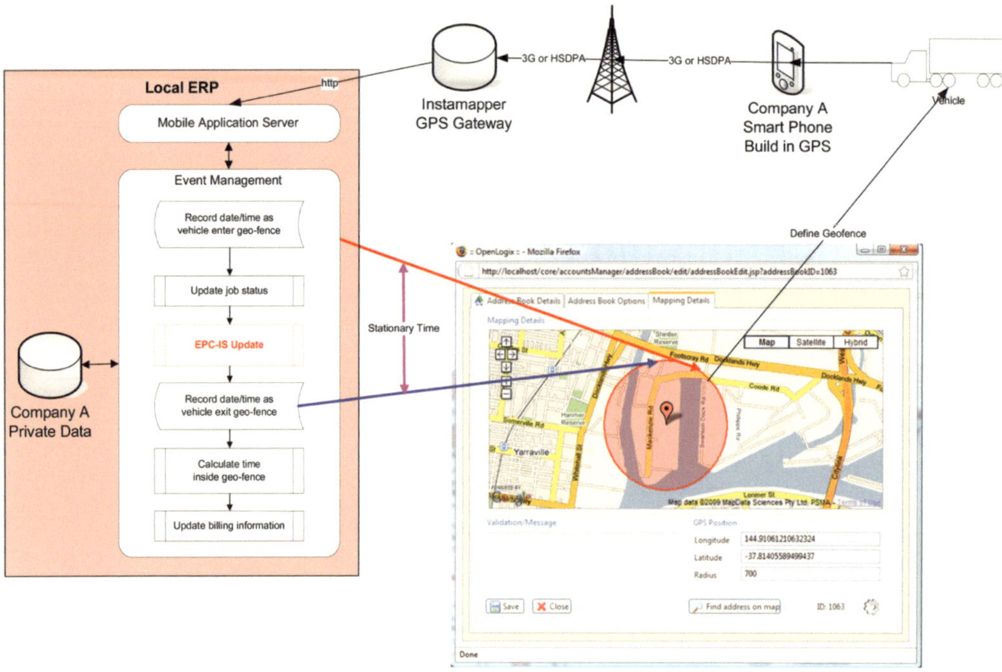

Fig. 6. Geo-fence data integrated with ERP

It is noted that, the actual arrival timestamp will be the time when the system detects the vehicle's entry into the geo-fence. This is dependent on the size of the geo-fence (i.e. the logical radius) and GPS update interval. Such restriction will limit its accuracy when compared to directly reading the tag by RFID readers. It is however very cost effective to have the geo-fence as a virtual gate, since the only capital cost involved would be the GPS which is now a standard tool in most logistics operations.

4.3 QR Code

QR Code or 2D Barcode, was developed by Denso-Wave in 1994 (Walsh, 2009). The term "QR" is derived from "Quick Response". It is essentially a printed "image" containing data (Figure 7), which when photographed by a camera phone, can decode the information encoded in the image. It is developed for high demand, fast moving environment. QR codes has an error correction encoding, and can hold the data intact, even if 30 percent of the printed image is damaged or obscured.

QR Code can store a lot more data in a given area when compared to conventional barcode. Information such as product code, serial number and manufactured date can be encoded using a QR Code. The content of the encoded message can be stored as text format which can then be parsed using a delimiter to extract the field content. It is important to acknowledge that if data were to be stored in this manner, an agreed field format between the recipient and sender must be established. For this reason, it is a common practice to encode only the URL link, as pointer to a particular web portal, where detailed information can be retrieve externally. Since the information is stored externally, it can be updated in real-time environment.

QR code can be quickly decoded on virtually any computing device that connects to the Internet. Apart from normal desktop and laptop computers, most mobile phone have an in-build camera which can read the QR Code, decode it with a free software inside the phone and then either link them to a website, or store the entire business information somewhere in a server. Typical mobile phones such as Apple iPhone and most Nokia phones have been proved to work seamlessly with this code.

Fig. 7. QR Code encoded URL: http://www.rmit.edu.au

Since all SMEs will have this type of infrastructures (e.g. mobile phone with camera) the cost to implement within the business consortium is just the normal cost of printing, but the opportunity gained is substantial. The biggest potential benefit to the SMEs is the ability to link virtually everybody (carrying a mobile phone with access to the Internet) with digital data that can be immediately acted upon. This will eliminate any misunderstanding in verbal communications, legal and contractual fulfilment, errors in picture scanning, multimedia data, network breakdown and other issues related to reliability of business information. QR codes can be generated, re-generated, enhanced, added to, encrypted as necessary making it the most flexible and mobile data structure that can be used in business. The new data structure will also enable more variety of business models, especially in marketing, business networking and other day to day communication activities.

QR code area size is largely depended on the size of the encoded data. The accuracy of the QR Code is heavily depended on the quality of the input image. Thus, the bigger the data, the better the image need to be captures.

In the case that RFID infrastructure are not available, the shipment goods information can then be scanned manually with a digital camera from a mobile phone. QR code decoder could also be integrated as part of the security camera to decode the freight as it is being unloaded from the loading bay.

5. Transparent framework

Making use of the geo-fence and QR code technologies, the virtualised infrastructure requires integration of several key functions. Ubiquitous computing has been investigating since 1993 (Weiser, 1993). The major character of ubiquitous computing is to create a user centric and application orientated computing environment (Wang et al, 2008). The theory of ubiquitous computing is to integrate information from a large number of sources. By being everywhere and any time, huge amount of data can be collected and process via ubiquitous supply chain. Complex algorithm and computation can be utilised in real time, and only the relevant key information is then feed back to the physical systems and the stakeholders within their supply chain domain. However, the theory of ubiquitous computing is difficult to extend to real physical world unless there is a schema with defined scope.

We developed the system "Transparent" based on ubiquitous theory to manage the global logistics processes, with the application of RFID, mobile devices and virtualization technology. Figure 8 demonstrates the flow of information between various terminals, shipping lines and Australian Customs. Using the Transparent Gateway, all the public (vessel and voyage details) and private data (container status) for each carrier are synchronized. The data feedback from the private Transparent Gateway can also trigger an event which can update container status inside the each company's enterprise resources planning (ERP) system.

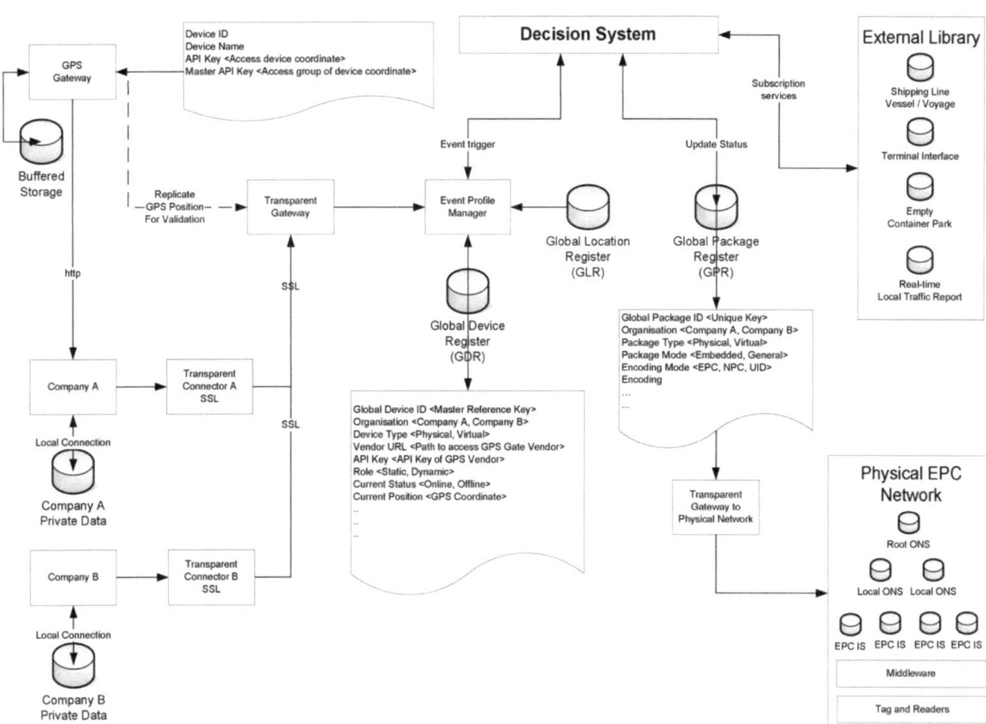

Fig. 8. Overall Design of the transparent virtual framework

The focus in Figure 8 is to keep track of each Stock Keeping Unit (SKU), and not on where it is coming from or going to. A clear record on the location of a SKU and trade unit number (TUN) in the supply chain must be kept. When an advance shipping notice (ASN) is received, the warehouse needs to accurately account for each item inside the package, and put them away at a set location inside the warehouse.

5.1 Distribution service

The supply chain process is reversed when an order is issued to the warehouse to deliver goods to another party. The SKU from a defined location is packed onto a pallet with a Serial Shipping Container Code (SSCC) attached. An ASN is then sent out to the recipients with a list of SSCCs and the items within. At this stage, a freight label is created and then applied to the SSCC. The freight label does not have any correlation to the SSCC but simply tells the truck driver how many packages he is picking up per consignment. It is within this transition that the link is broken, and in most cases two separate systems are used to capture this information. When the freight arrives at the recipient, only package details are captured, not the SSCC. The recipient will then need to re-identify the SSCC by visually checking or scanning the actual package.

The problem becomes more complex when barcodes are replaced by RFID tags. Since the warehouse and the transport company may not be the same entity, two different RFID systems may be used for the same purpose. This adds to the cost and defeats the integrity of the identity. There are security concerns that the package will need to be kept intact between dock locations. This issue has been investigated in some of the RFID pilot studies. A proven approach is to enhance the normal verification process known as "pick face" to ensure identification of items to the SSCC. A typical "pick face" process can be described in the following steps (Figure 9):

1. Bring up the interface in the browser and enter the purchase order number. When the order number is accepted, the system expects to scan a pallet.
2. The operator uses a mobile reader to scan the pallet tag. The system would check the EPC and only accept a GRAI tag. The operator then selects the product that would be packed.
3. The operator applies the tags to the cartons and uses the mobile reader to scan the tags until the required number is reached.
4. Products scanned are displayed on the screen.
5. The operator scans and applies a shipment tag to the package.
6. The system asks for the next pallet. The operator could either continue for another pallet or close the pick process. In the latter case, the system would return to Step 1.

All EPC information is captured immediately to the local server and subsequently uploaded to the global EPC-IS. Containment information is associated with the pallet which is registered prior to this pick face process.

Once the items are validated as one containment, the package is wrapped and sealed to prevent further altercations.

Using Transparent Gateway, the SSCC and the order fulfilment can be easily updated by querying the Transparent Framework. Since the SSCC as well as the physical RFID tags number are embedded inside the Transparent virtualized tags, the same set of information can be filtered across to the physical network, including the status and event of the physical system that can also be transposed to the virtual system.

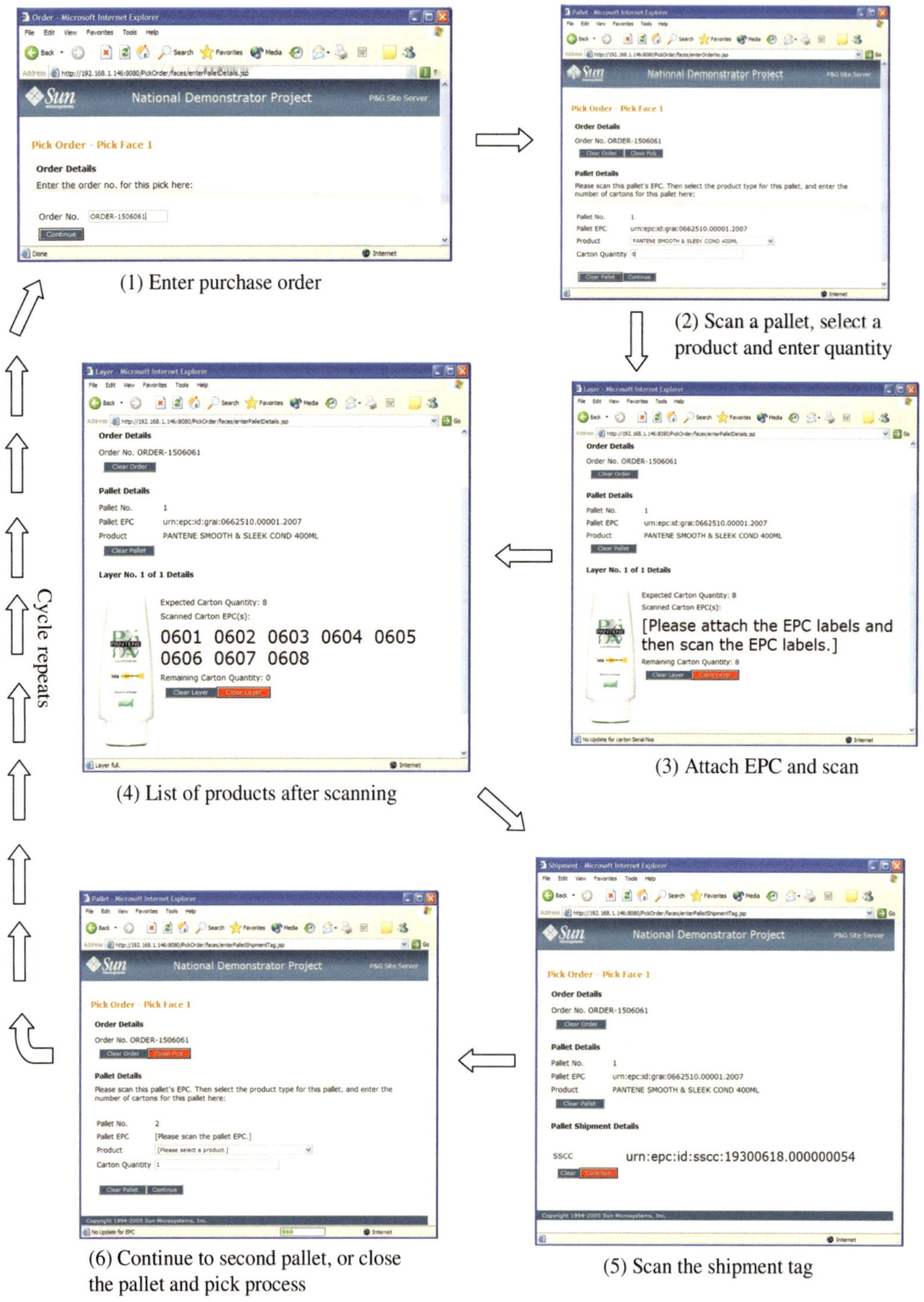

Fig. 9. Pick face process and screens

5.2 Data validation

Using Instamapper (http://www.instamapper.com) API (Application Program Interface), GPS data are synchronised to the local company's FMS or ERP database to trigger an event when a vehicle travel within the geo-fence at a defined moment in time. Since this GPS data are critical in terms of creating an event, such as changing ownership, etc, it is not desirable to depend on one source of data. Data that are stored locally can be tampered, which could then lead to data integrity and creditability issue. One solution is to allow the same set of data to also pass to the Transparent Gateway. When an event is triggered from a local database, it is verified by the data that are collected directly from the buffered storage within Instamapper. Since the data from Instamapper (or any other GPS Gateway companies) are independent and cannot be tampered with, if both set of data are identical, the event triggers are valid, otherwise, the relevant parties are notified and security investigations will be initiated.

5.3 Global registers

There are three primary registers that exist within the Transparent Framework. These registers are used to maintain instances of physical objects that would exist in various ERP, FMS and web application. Transparent then use this registers as access keys to connect to various systems.

5.3.1 Global Device Register (GDR)

When a local device is registered with Instamapper, a device ID is created within the GPS Gate local network. In a real-world environment, there could be multiple GPS gate providers. Therefore, the device ID cannot be assumed unique. Encapsulation of all device IDs in a Global Device Register (GDR) class is to manage ID information in order to ensure uniqueness in the system. Since the GDR can be referenced to a company, the GPS device can be assigned dynamically to trucks and other ID devices. Each GDR can have a defined event profile build in and managed by an Event Profile Manager. An event profile is a set of rules and parameters that controls the behaviour on how each GDR should behave when interacting with a Global Location Register (GLR) or another GDR at a specified condition.

5.3.2 Global Package Register (GPR)

The same principle applies to the Global Package Register (GPR). Since EPC network is not the only network in the global supply chain, the system will need to encapsulate each tags in a GPR class. For example, RFID information can flow to and from different networks, such as NPC, UID as well as within the virtual network. Data travel between physical networks is also managed by Transparent Gateway.

QR code may be a good medium to encode "Global Package Register Identification Number" (GPRID) from the proposed "Transparent" framework.

5.3.3 Global Location Register (GLR)

Unlike GDR and GPR, the global location register (GLR) operate almost in the reverse. Since a particular location, such as an office building or a consolidate warehouse could house multiple organisation unit. Therefore GLR is used to identify a location that resides within the domain of a particular organisation.

5.4 Decision system

The aim of the virtual network is to allow non-conforming systems to interact with conforming systems such as those of EPCglobal. However, the primary goals should not only be limited to traceability and visibility, but management of information so that decision paths could be established. Information gathered across the framework could then be synchronised with shipping line, terminal operators, customs agencies, empty container parks or even local traffic to further optimise the performance and the transparency of each process within the entire supply chain.

6. Traceability in a virtual infrastructure.

In order for both physical and virtual systems to be coordinated with each other, a communication path is required between a physical to virtual system (P2V) and vice versa. The physical scanners are not only given a unique location code within the EPC network, they are also given a global device register (GDR) number which is a unique attribute across the Transparent framework. Since there could be a single physical RFID scanner that acts as a host for multiple network such as NPC and UID, by obtaining a unique GDR, we can keep track of the physical device, regardless of which network it is operating in.

Figure 10 illustrates how a physical EPC infrastructure interacts with a virtual infrastructure environment in a typical distribution environment. A container is unpacked and the product encoded with EPC tags are then scanned (via GDR:P00001, which is type 'static') and stored as inventory. An order is then created and the product is issued out via a typical scan pack system. SSCC labels are generated and displayed as barcode to the outside packaging. At the same time, a global package register identification number (GPRID) is also created, which acts as a key between different networks, physical and virtual. The GPRID is then scanned to create a manifest of the vehicle, which is managed by an internal FMS. The GPS unit inside the truck contains a GDR number (GDR:V00001), which is type 'dynamic'. Within the FMS, the geo-fence of the destination has a global location register number (GLR:L00012). The virtual EPC network constantly checks for any device of type 'dynamic' that falls inside the geo-fence. Thus, if GDR:V00001 is inside location GLR:L00012, within the specified delivery time window in the FMS, a scan event from the emulated scanner GDR:P00003 is raised and the location of the GPRID is then updated in global package register (GPR) database. The GPR acts as an intermediate storage between the physical and virtual environments, since GDR:P00003 may not exist as a registered device in the physical EPC network.

The final order is issued from GLR:L00012 to a delivery location GLR:L00014. A new GPRID is created for the order based upon the original GPRID, once the SSCC is scanned from GDR:P00005 (physical, static device). The scan event is then raised within the physical EPC network and the GPR is updated.

It is important to note that we will only update the physical EPC network when a physical reader is utilised. We cannot update the physical EPC network directly from the virtual system, because the virtual reader may not exist as a register device in the physical EPC network. Thus, the only way to accurately query the physical and virtual data is via the global package register (GPR).

Since both physical and virtual environments can coexist in the supply chain, GPS data from GDR:V00001 can be used to forecast the delivery time accurately. This is particularly useful for those who operate a just-in-time (JIT) operation, where supply punctuality is critical. A virtual enterprise system not only improves the visibility of the supply chain, but also increases the performance of the overall supply chain with the utilisation of the real-time data to drive business decision in a time sensitive operation.

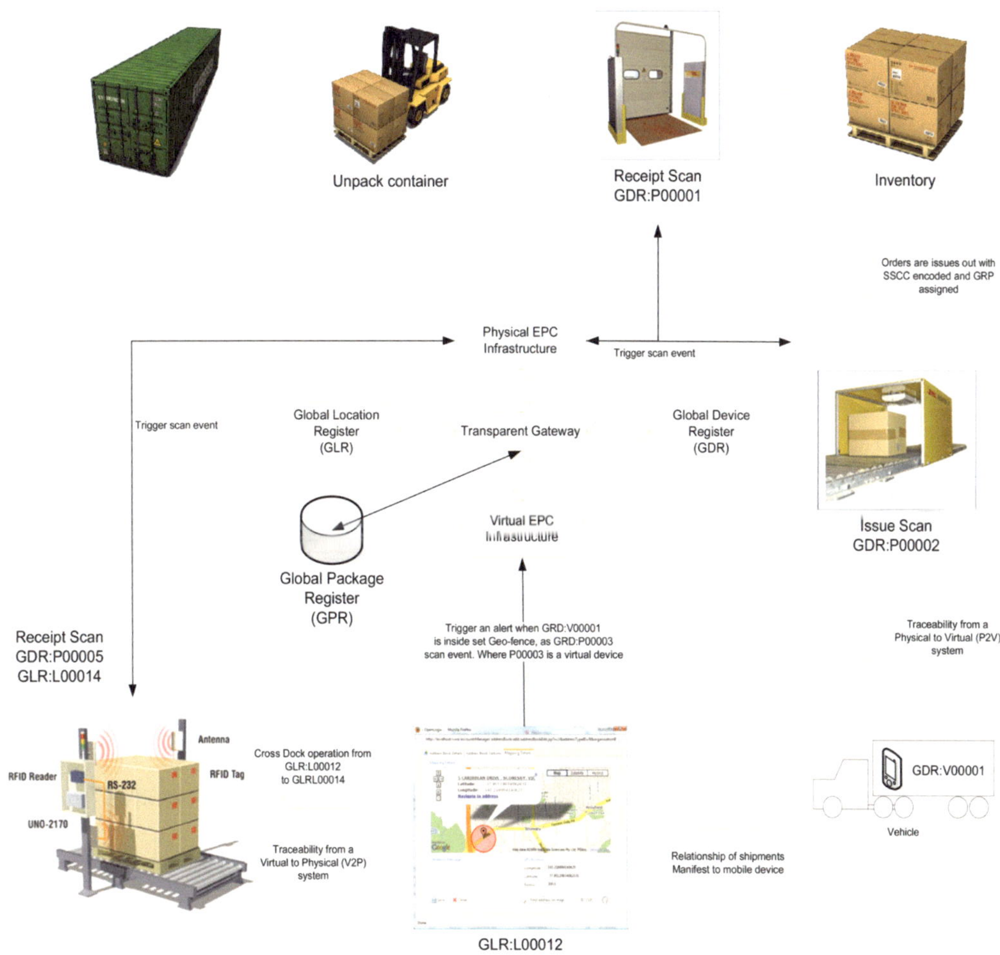

Fig. 10. Traceability of items in the virtualised supply chain

7. Conclusion

In this chapter, we have discussed the use of RFID system in two Australian national demonstrator projects (NDP and NDP Extension). The ability to scan without line of sight proved is the key advantage of RFID over conventional barcode scanning. However, the capital investment and maintenance cost is too much for some SME organisations.

To overcome these issues, this chapter demonstrates how GPS can be used to develop a geo-fence system that tracks consignments at locations where RFID systems are in accessible. This chapter further illustrates QR code, which supports multiple scanning via image recognitions. QR code can store much more information when compared to conventional barcode. With the utilisation of a modern mobile phone with a build in camera, QR code can be scanned, showing text, SMS, contact details or a link to a website. This enable QR code to act as a pointer to extract information stored externally. These information can be updated and manage in real-time, similar to those of RFID.

These technologies are then integrated with the model Transparent, which is based on ubiquitous computing. Transparent act was a router between physical and virtual infrastructure. It facilitates backend ERP and FMS systems using GPS technology and geo-fence to emulate event that are feed back to the physical RFID network. This ensures that data are captured along the supply chains, even if the receiving or sending parties may not be RFID ready. Transparent can also support multiple physical RFID system such as NPC, and UID.

In a supply chain that operates at high volume and low profit margin, very few businesses are willing to invest into new and evolving technologies such as those of RFID. Most of these businesses are SMEs that do not have the capital resource to purchase and maintain such technologies. Transparent offers a low cost virtualisation solution to such supply chains, by providing a communication path to those larger companies that have already invested heavily in such technologies.

8. References

Behzadan, A.H., Aziz, Z., Anumba, C.J., Kamat, V.R. (2008). Ubiquitous location tracking for context-specific information delivery on construction sites. *Automation in Construction*. August, 17(6):737-738

Chauhan, S.S., Proth, J.M. (2005). Analysis of a supply chain partnership with revenue sharing. *International Journal of Production Economics*. 97(2):44-51

Chiang, K.W. and Huang Y.W. (2008). An intelligent navigator for seamless INS/GPS integrated land vehicle navigation applications. *Applied Soft Computing*. 8(1):722-733

Demchenko, Y. (2004). Virtual organisations in computer grids and identity management. *Information Security Technical Report*. 9(1):59-76

EPCglobal Inc. (2006). EPCglobal Tag Data Standards, Version 1.3.1, March 8, 99 pages

Gajzer, S. (2007). Process Modelling and Development of the Key Performance Indicators for the National Electronic Product Code Network Demonstrator. *MEng Dissertation*. RMIT University

Hamilton, J.W. (1993). Wireless communication systems: a satellite-based communications approach for competitive advantage in logistic and transportation support services. *Computer in Industry.* 21(3):273-278.

Kelepouris, T., Pramatari, K., Doukidis, G. (2007). RFID-enabled traceability in the food supply chain. *Industrial Management and Data Systems.* 107(2):183-200

Kim, G. (2008). Seven steps to a secure virtual environment. *Network Security.* 2008(8):14-18

Liu, C.H., Sia, C.L., Wei, K.K. (2008). Adopting organizational virtualization in B2B firms: An empirical study in Singapore. *Information and Management,* Nov., 45(7):429–437

Mo, J.P.T. (2008a). Development of a National EPC Network for the tracking of fast moving consumer goods. *International Journal of Enterprise Network Management,* 2(1):25-46

Mo, J.P.T. (2008b). Modelling and Simulation of National Electronic Product Code Network Demonstrator Project. *International Conference on Information Systems (ICIS 2008).* 14-17 December, Paris, France

Mo, J.P.T., Gajzer, S., Fane, M., Wind, G., Snioch, T., Larnach, K., Seitam, D., Saito, H., Brown, S., Wilson, F., Lerias, G. (2009a). Process integration for paperless delivery using EPC compliance technology. *Journal of Manufacturing Technology Management,* 20(6):866-886

Mo, J.P.T., Sheng, Q.Z., Li, X., Zeadally, S. (2009b). RFID Infrastructure Design: A Case Study of Two Australian National RFID Projects. *IEEE Internet Computing.* 13(1):14-21.

Pedro, M., Reyes, P.J. (2007). Is RFID right for your organization or application? *Management Research News.* 30(8):570-580

Pennsylvania University RFID Study Group (2006). Challenges in RFID enabled supply chain management. *Emerald Management Reviews,* ISSN:0033-524X, Nov, 22-28

Quiroga, C.A., Bullock, D. (1998). Travel time studies with global positioning and geographic information systems: an integrated methodology. *Transportation Research Part C: Emerging Technologies.* 6(1-2):101-127

Soppera, A., Farr, J., Kasten, O., Illic, A., Zanetti, Z., Harrison, M. (2007). Supply Chain Integrity (D4.6.1), Building Radio frequency IDentification for the Global Environment (BRIDGE Project Report), Dec, WP04, European Commission contract No: IST-2005-033546, 22 pg.

Visser, H. G. (1991). Terminal area traffic management. *Progress in Aerospace Sciences.* 28(4):323-368

Walsh, A. (2009). Quick response codes and libraries. *Library Hi Tech News.* 26(5-6):7-9

Wang, H., Zhang, Y., Cao, J. (2008). Access control management for ubiquitous computing. *Future Generation Computer Systems,* 24(8):870-878.

Weiser, M. (1993). Hot topics-ubiquitous computing. *IEEE Computer.* 26(10):71-72.

Zarour N., Boufaida M., Seinturier L., Estraillier P. (2005). Supporting virtual enterprise systems using agent coordination. *Knowledge and Information Systems,* September, 8(3):330–349,

Zito, R., D'Este, G., Taylor, M.A.P. (1995). Global positioning systems in the time domain: How useful a tool for intelligent vehicle-highway systems? *Transportation Research Part C*, 3(4):193–209

2

A RFID Based Ubiquitous-Oriented 3rd Party Logistics System: Towards a Blue Ocean Market

Changsu Kim[1], Kyung Hoon Yang[2], and Jae Kyung Kim[3]
[1]*Yeungnam University,*
[2]*University of Wisconsin-La Crosse,*
[3]*SUNY College at Oneonta*
[1]*South Korea,*
[2,3]*USA*

1. Introduction

Companies have been competing based on how to gain the largest share of the market space, which causes intense competition with industry-wide over-supply, and even profit decrease in the case of shrinking market space. Blue Ocean strategy provides companies with guidelines on how to escape from intense competition over the same market space where there are limited customers with an increasing number of competitors by creating a new market space where there is less competition, if any (Kim & Mauborgne, 2005).

Systematic and efficient logistics service has become one of the core support services of e-businesses, and many innovative strategies utilizing globally expanding Internet technology and e-businesses have been proposed such as new business models with less distribution layers resulting in customer-based logistics, Internet-based logistics, logistics for small-batch production, zero-inventory logistics, and 3rd party logistics (3PL)' reverse logistics model and GRID services based marketplace model (Bhise et al., 2000; Bruckner & Kiss, 2004; Krumwiede & Sheu, 2002; Lee & Whang, 2001; Lee & Lau, 1999; Simchi-Levi et al., 2004). As a relatively new business model in logistics, 3PL companies provide outsourcing service of transportation, warehousing, freight consolidation, distribution, inventory management, and logistics information systems to companies who used to operate their own logistics network (Kimura, 1998; Rabinovich et al., 1999; Sink & Langley, 1997; Vaidyanathan, 2005).

CJ-Global Logistics Service (CJ-GLS) is a late comer in intensively competitive Korean logistics industry. Entered into the 3PL industry from the start, however, it has the largest client bases and ranked fourth in the market due to its strength in market analysis, customer requirement analysis, and constructing logistics information systems (LIS) including successful implementation of radio frequency identification (RFID)-based ubiquitous LIS. This chapter analyzes CJ-GLS' business model with Blue Ocean strategy to show how a company in the Red Ocean reinforces its competitive advantage to move toward a less competitive new market space by utilizing information technologies. For this case study, we

interviewed the CEO, CIO, CJ-GLS logistics strategy research manager, the team manager of the information strategy team, and several line workers both, in July and September 2005.
The structure of this paper is as follows. In the following section, we describe the case company CJ-GLS, including an explanation of the information systems and its evolving path to e-business and m-business supporting LIS. In section three, CJ-GLS' ubiquitous-oriented 3PL LIS with its fundamental RFID technology is explained. Furthermore, we project the forthcoming new business model based on the ubiquitous-oriented 3PL LIS. In the fourth section, we introduce the framework of analysis, Blue Ocean strategy, and analyze the case using this framework. Problems and success factors found throughout the implementation of the ubiquitous oriented 3PL LIS are stated. Finally, we summarize the CJ-GLS' strategic path and its capabilities in creating an uncontested Blue Ocean with its superb business processes and supporting technological capabilities.

2. The first mover: CJ-GLS

2.1 Introduction to CJ-GLS

Starting its business as a subsidiary of Samsung Corporation in 1988, CJ-GLS adopted a cross-docking system in South Korea for the first time in its industry in 1991, built LIS in 1994, and introduced coastal shipping logistics in 1996. In 1998, the logistics systems and technology subsidiary unit of Samsung Corp. renamed itself CJ-GLS and launched its separation with an initial public offering of stock issued in 1998.
The core business units of CJ-GLS are 3PL service and domestic and international small parcel services that are all based on corporate clients. 3PL service and the largest portion of its business executes the logistics support for client companies and includes freight consolidation, distribution, transportation, warehousing, product marking, labeling, packing, order management, etc.
Clients of CJ-GLS are twofold: the first group are companies who have their own logistic systems but utilize CJ-GLS' superb 3PL and international small parcel service capabilities, and the second group are startups without their own logistics systems and who utilize 3PL service for B2B logistics service and small parcel services for B2C logistics. Clients are composed of 220 companies from a variety of industries, including pharmaceutical, food, chemical, and e-marketplaces. Some of the international clients of CJ-GLS include Sony, National Panasonic, Lego, Nestle, and Osram Sylvania. Table 1 shows the growth of CJGLS' revenue. It increased from $64 million in 1998, to $751 million in 2008, which placed CJ-GLS in the top rank in 3PL service and third rank in Korean logistics industry. Fig. 1 and Table 2 shows the logistics market of Korea.

Service	1998	2004	2008	2009*
3PL	64	197	438	345
Parcel Delivery Service	-	139	212	295
Intenational Delivery Service	-	43	101	410
Total	64	379	751	1,050

*: Estimate

Table 1. Sales revenue of CJ-GLS (Unit: $ million)

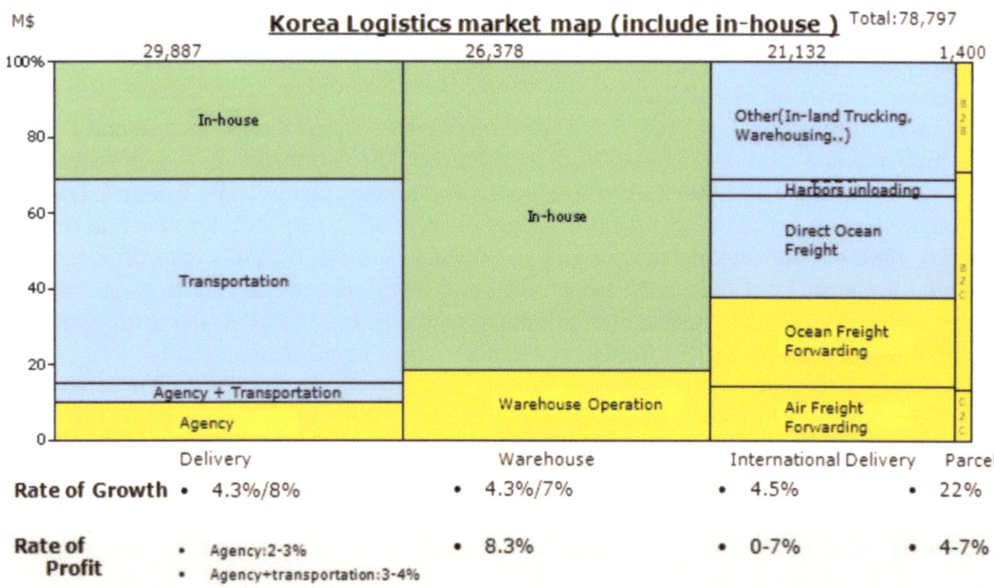

Fig. 1. Korean logistics market map

Company	Sales ($Mil)	Growth Rate (%)	Investment ($Mil)
DaeHan Tongwoon	1,266	8	80
Hanjin Haewoon	6,936	13	441
CJ-GLS	751	28	30
Hanjin	748	-2	60
Hyundai Taekbae	664	15	61
Hansol CSN	295	9	21
KCTC	173	15	15

Table 2. Sales revenue of the major Korean logistics service providers in 2008

2.2 Logistics information systems of CJ-GLS

Logistics information systems (LIS) in CJ-GLS have evolved from fundamental LIS, to e-business supporting LIS, and to the ubiquitous computing-oriented LIS. To avoid other incumbents' mistakes of adopting foreign solutions which failed to be customized to South Korea, CJ-GLS followed an in-house development approach through its sister company, CJ Systems. In-house development reduced the erroneous design causing conflicts between the actual business processes and the information flow.

Fundamental Logistics Information Systems: OMS, TPS, and WMS - From its early stage of market entrance, CJ-GLS concentrated on building business operation systems targeting to 3PL processes, composed of Order Management Systems (OMS), Warehouse Management Systems (WMS), and Transportation Management Systems (TMS). OMS is constructed based on the business process regarding order receipt, order tracking, performance analysis, and generating ad hoc reports. WMS is inventory management through tracking the movements, the storing of materials within a warehouse, sharing accurate inventory information with the clients, and directing the movement of goods based on urgency for space, equipment, inventory, and personnel. TMS is connected to and shares information with WMS and OMS for accurate delivery of ordered materials.

E-Business Supporting Logistics Information Systems - Upon these fundamental LIS, CJ-GLS implemented new e-business support features. The storefront is a web-based user interface for clients and other IS users to access the system through the Internet. Delivery routing systems powered by artificial intelligence were added to TMS for safe and on-time delivery. OMS is connected to finance and accounting modules of SAP's enterprise resource planning systems. Data generated from WMS and TMS are analyzed with a data mining technology and applied to customer relationship management (CRM) systems for managing uncertainties such as preorder forecasts and delivery routing adjustments.

CJ-GLS added mobility to its e-business-based LIS, specifically to TMS. Core strategy of TMS mobilization is threefold: 1) to enable clients and CJ-GLS to track the shipments in and out of warehouses or distribution centers 2) to enable clients and CJ-GLS to predict the accurate arrival time of the shipments and 3) to reflect the arrival information of delivery scheduling in order to increase the efficiency of the system. TMS provides schedules locally and nationwide, and it sends the scheduling information to the carriers' PDA. Carriers download the shipping schedules from TMS, and the shipping status is administered back into TMS through a Global Positioning System (GPS). This mobile technology-based logistics management enables not only clients to access shipping status and scheduling information through the storefront using their web browsers, but also the provider to manage issues in shipping.

3. Toward deep blue ocean: ubiquitous-oriented 3PL system

Although e-business support and mobilization of OMS and TMS improved the efficiency of order-and freight-related business processes, productivity gains of the entire 3PL business process (Fig.2) is not maximized because of inefficiency of the bar code-based WMS business processes where reading errors cause inaccurate information and extra operating time and operating cost. To improve the warehouse business process and maximize efficiency and synergy from the integration of three LIS subsystems, RFID based WMS was developed.

As Carr (2003) argued, ICT-driven competitive advantage is not sustainable unless the ICT strengthens the existing competitive advantage. The companies such as Hanjin and Hyundai, who entered into the 3PL market in 2000, also integrated their 3PL system with the clients' and their IS (e.g., Hanjin's e-HaNex and Hyundai's HiTexII). The strategy CJ-GLS is presently using to gain a competitive advantage over incumbents in the industry involves developing and applying a new ICT-driven LIS. This innovative approach is always accompanied by a great risk of failure, and if one of two aspects is not successful, whether it is the ICT or the application, it would result in a disastrous outcome for CJ-GLS.

To continue along a strategic path, there was a pressing demand for an ubiquitous-oriented 3PL system and new technological innovation. The core technology necessary to accomplish promoting the strategy for an ubiquitous computing 3PL system is RFID technology, which remotely recognizes large volumes of inventory and manages inventory information. A collaborative business model was developed to harmonize and coordinate all the business processes. The RFID-based ubiquitous 3PL system was promoted to apply a collaborative business model into the field operation of warehouses and distribution centers.

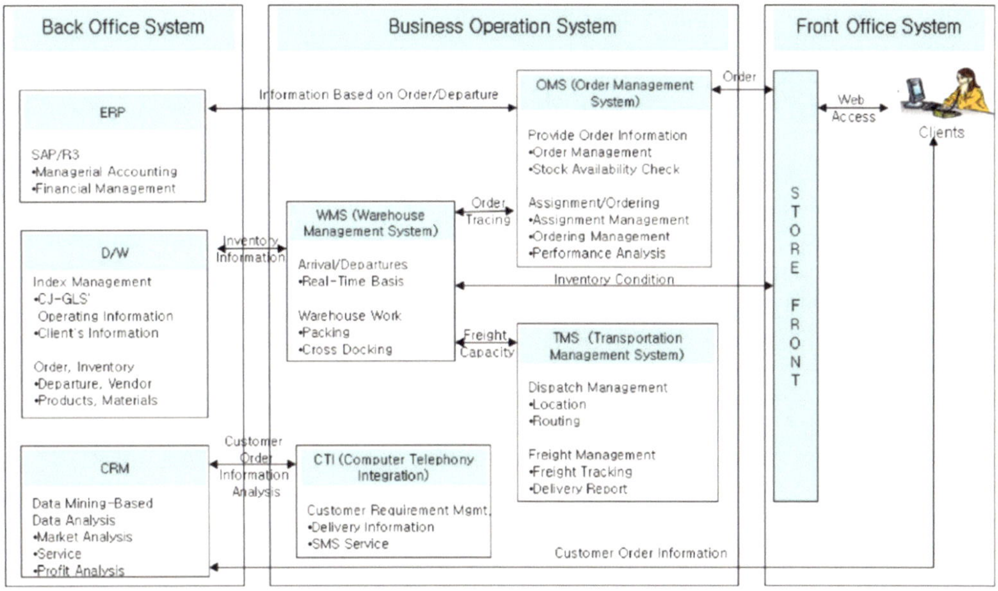

Fig. 2. Information flow in 3PL LIS of CJ-GLS

3.1 RFID technology

RFID uses radio frequency to transmit stored information to a remote reader. Information about the material from the beginning of its production up to its distribution is stored in a RFID tag that is traceable through certain wireless frequency. Its advantage over the bar code is noncontact based. In the bar code system, operators need to scan every single bar code with bar code readers in order to recover such information. The RFID system enables operators to retrieve the information on inventory at once regardless of its quantity. The storage capacity of RFID tags is about 6,000 times larger than that of bar codes, which enables us to assign different IDs to the same product and other variable information. While the bar code can only be read, the RFID allows rewriting updated information over previously recorded information. These advances make RFID the next exciting innovation in logistics industry.

3.2 RFID and WMS

In CJ-GLS, the starting point has been the replacement of the bar code system in warehouses and distribution centers with RFID tags. The information flow in warehouse is composed of seven business processes.

Receiving Inspection - materials are inspected before they get unloaded, and each bar code is scanned in order to be matched with the information in WMS.

Stocking - unloaded materials are placed into storage space designated by WMS.

In-Storage Handling - material information, such as volume, weight, and storage requirements, is handled, and each time transfer occurs, the scanned bar code information should be transferred to WMS.

Selection - bar codes of ordered materials are received from WMS and matched by scanning each material.

Picking – selected materials are staged for shipping.
Shipping Inspection – materials are inspected before they get loaded, and each bar code is scanned to be matched with information in WMS.
Shipping – loaded equipment departs to the next destination when all requirements are met.
For each of these seven processes as shown in Fig. 3, warehouse operators are required to scan bar codes of the materials continuously to check the inventory, both previous and additional, and to calculate what is missing. This frequent bar code reading causes many problems from scanning omissions resulting in an inaccurate inventory and unclear boundary of liability between processes. These problems cause delays in all the business processes of LIS, including OMS and TMS.

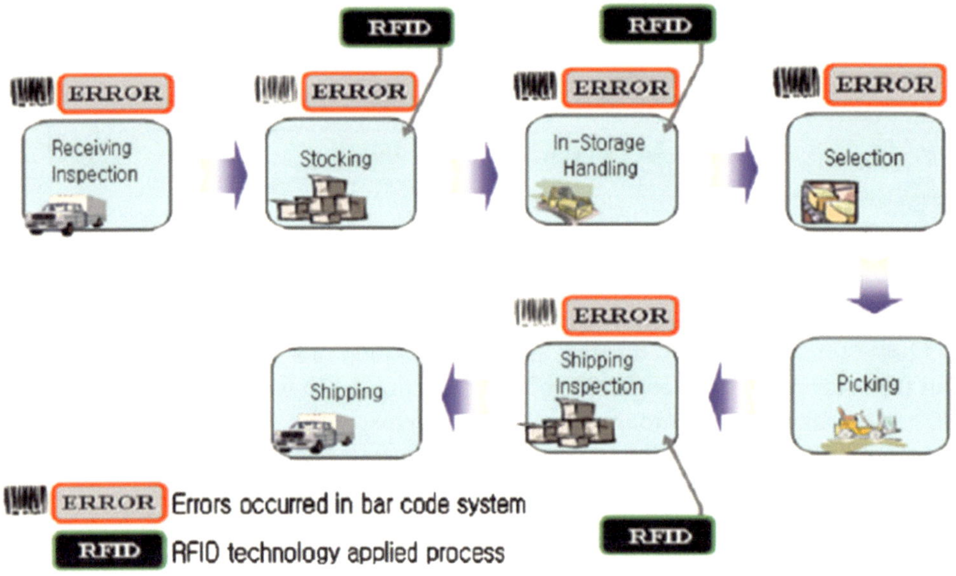

Fig. 3. Business processes in WMS prior to RFID implementation

RFID-based WMS reduced the seven-step warehouse processes into four steps (Fig. 4). It enabled managing inventories on real-time basis tracking, as well as updating inventory information, which resulted in accurate inventory information. Along with WMS, accurate information minimized the additional costs from returned materials. Overall, management innovation, work time reduction, optimized inventory, zero errors, and labor cost reduction were achieved.

Fig. 4. Business processes in WMS after RFID implementation

The RFID-based WMS increased the accuracy of inventory management and cut down the work time to one-third of the previous system. For example, receiving inspection time has been reduced from 10 to 3 seconds, shipping inspection time has been reduced from 600 to 3 seconds, and in-storage handling time has been reduced from 10 to 3 seconds. As the amount of materials increase, inspection time in the bar code-based WMS increased in multiples of the amount not including the added time due to errors and omissions. However, the inspection time in the RFID-based WMS was not influenced by the changes in the amount ordered and was free from added time caused by errors and omissions. These effects become clearer as the amount of inventory increases. Qualitative benefits include the transcendental operation of value-added service from the connection of existing information systems and RFID-based ICT innovation and its application into the collaborative business model, systematic knowledge acquisition through collaboration and knowledge-sharing among participants (universities and research institutions), and guiding experience through the first application of RFID in logistics operations in South Korea. Fig. 5 shows the details of the benefits CJ-GLS acquired from RFID-based WMS.

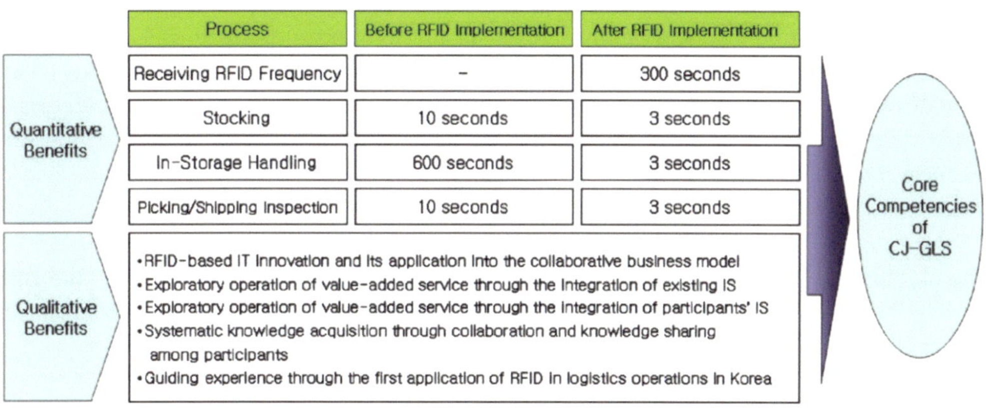

Fig. 5. Benefit of RFID-based WMS

3.3 New business model

CJ-GLS' strategic movement includes the improvement of its internal IS with RFID technology, which strengthens its competitive position in the existing market (Red Ocean) and in the development of the new market called RFID-based electronic logistics business. CJ-GLS plans to develop an application service provider (ASP)-based, new business model through utilizing its connection between the manufacturing industry and distribution industry.

Current major distribution companies like Wal-Mart, Tesco, and Costco require major manufacturers to attach RFID tags on the products they want to distribute through these distribution companies. Major manufacturers may be capable of doing this. However, small and medium enterprises (SMEs) may face technological difficulties and a considerable cost increase for doing this. In the evolving electronic logistics service market, CJ-GLS will provide solutions covering the entire logistics service from tag installation, to ONS registration of the clients' products, to product history management, eventually becoming a 4th party logistics (4PL) service provider. 4PL increases the SMEs' opportunity for marketing their products to major distribution companies, provides efficiency in the logistics process, and allows them to concentrate on their core competencies. An ubiquitous electronic logistics business model will allows CJ-GLS to create an uncontested new market space where new demand and high growth is possible.

4. Analysis of case

4.1 Framework of analysis: blue ocean strategy

There are two opposing market space concepts: Red Ocean, which is the current existing industry, where boundaries and competitive rules are well known to competitors who try to achieve a greater share of existing demand and Blue Ocean, which is any new industry where "demand is created rather than fought over" (Kim & Mauborgne, 2004). Blue Ocean strategy is the strategy which can avoid severe competition by creating an uncontested market space. Two approaches for creating Blue Ocean have been proposed. The first approach is to create completely new industries (e.g., online auction service in auction industry). The other approach is to create a new niche market within the existing industry by altering the existing competitive rules (e.g., ubiquitous-oriented 3PL in logistics industry). Kim and Mauborgne (2004) suggested the Four Actions Framework, which provides guidelines to create a new strategy profile and which defines the characteristics of Blue Ocean strategy. Using this framework, we have identified CJ-GLS' electronic logistics business model as a case of Blue Ocean strategy.

4.2 Four actions framework for CJ-GLS's RFID service

The ubiquitous electronic logistics business model of CJ-GLS is analyzed based on the Four Actions Framework of "Eliminate," "Reduce," "Raise," and "Create" (ERRC) proposed by Kim & Mauborgne (2005). The results are summarized in Table 3.

Eliminate	Raise
- Errors caused by manual handling of warehousing activities - Errors in recognizing bar codes resulting from repetitive work - Work delays involving deliveries to or collections from warehouses	- Automated handling rates of logistics activities - Information service optimized for the company's activities - Speed of process
Reduce	**Create**
- Labor costs and fixed costs for warehouse management - Inefficient work activities - Steps and errors	- Blue Ocean market of electronic logistics business - New cyber distribution/logistics - New customer satisfaction and loyalty

Table 3. ERRC chart of CJ-GLS

First, the introduction of RFID-based electronic logistics business has resulted in the "elimination" of certain elements that had been taken for granted in the industry, such as the occurrence of errors caused by manual handling of warehousing activities, errors in recognizing bar codes resulting from repetitive work, and work delays involving deliveries to or collections from warehouses. Secondly, the concerning elements that should be reduced below standards of the industry (namely, labor costs for warehouse management and fixed costs) have been reduced thanks to automation and optimized work handling, respectively. Such "elimination" and "reduction" of inefficient work activities have been effective in sharply cutting both time and costs. The third aspect concerns elements that have been "raised" higher than industry standards compared to rival companies, and they include increases in automated handling rates of logistics activities and in information service optimized for the company's activities. More specifically, large volumes of loaded cargo are recognized in warehouses within a short time, and information on cargo taken in or out of warehouses is delivered to the operation system in real time, thus enhancing efficiency in the entire logistics business and improving information service. Finally, disproving the existing generally accepted notion that logistics companies only provide logistics-related services, CJ-GLS has successfully realized an ubiquitous-oriented Third Logistics System based on RFID, which has yet to be offered by other companies, thereby "creating" a Blue Ocean market of electronic logistics business that is a new cyber distribution/logistics area where competition does not exist. Hence, CJ-GLS created new customer satisfaction and loyalty by achieving accurate prediction and information on shipments due to the location information of carriers on a real-time basis, which optimized delivery schedule and dispatch.

In summary, as can be seen from Table 3, we apply the Four Actions Framework to analyze the CJ-GLS ubiquitous oriented-electronic logistics model. CJ-GLS creates the Blue Ocean in terms of an electronic logistics business model, which is within their core businesses. CJ-GLS focuses on noncustomers, uncontested market space, and new demand for the electronic logistics business. The case analysis of CJ-GLS provides analytic frameworks and useful insights for capturing the Blue Ocean opportunity.

4.3 Comparison with red ocean strategy

In this section, CJ-GLS' strategy is compared to that of the competitor's in view of Blue Ocean strategy.

Market Space - Other logistics service providers tried to compete in existing market space, Red Ocean, but CJ-GLS explored uncontested market space using the RFID tag agency service. The new, uncontested market space which CJ-GLS' new business model will create expands from the existing 3PL industry, with new RFID-enabled services. In addition to the fulfillment of existing 3PL service, these services include the assignment of unique ECP codes to physical RFID tags and the attachment of them to the manufacturers' products.

Competition - Other Korean logistics service providers tried to attract more market share from the existing competition based on accurate and timely delivery of products at the lowest freight cost. CJ-GLS made the competition irrelevant. While the incumbents have deployed more physical distributions, with the new RFID outsourcing service, CJ-GLS adds more value to the clients, along with providing more advanced service in the physical delivery of the product. This additional value to the clients makes the competition based on physical delivery irrelevant.

Demand - Red Ocean strategy exploits existing demand, while Blue Ocean strategy creates and captures new demand. Other competitors try to raise demand in the current existing market, but CJ-GLS creates new demand using RFID. Even though RFID was new technology that other companies had not tried, CJ-GLS took a risk creating a new market.
Value-Cost Trade-Off - Red Ocean strategy emphasizes the current model of trade-off between value and cost, while Blue Ocean strategy looks at value and cost in a new way. Incumbents try to add value by increasing the additional cost of physical resources, such as storage and trucks. But Blue Ocean strategy tries to remove human errors. Removing human errors in inventory management using WMS, as well as OMS and TMS, brings additional value to clients, in stead of incurring the additional cost of physical resources.
Alignment of the System - Red Ocean strategy aligns the whole system of a firm's activities with its strategic choice of differentiation and low cost, while Blue Ocean strategy aligns the whole system of a firm's activities in pursuit of differentiation and low cost. CJ-GLS' RFID service, in addition to the current 3PL service, will bring exclusive values to clients. To achieve this, CJ-GLS analyzed and redesigned its internal business process and educated its employees. Integrating the two services will bring synergy to clients, such as convenience from one-stop service and efficiency from concentrating on clients' core business.

4.4 Findings: problems and success factors

During implementation of the new technologies into existing IS and business processes, several organizational and technological problems are bound to occur. To solve such problems, CJ-GLS involved all levels of its employees in searching out solutions and effective alternatives. After surveying possible solutions and/or alternatives, the information strategy team conducted several strategic seminars to establish an efficient development direction to solve problems. In this section, we list the bottlenecks and success factors that we faced in building the Blue Ocean strategy.
Bottlenecks - Bottlenecks, whether they are social-oriented or technology-oriented, are common when new technology is introduced into an existing business process and organizational structure.

a. *From the Organizational Perspective* - CJ-GLS experienced two problems from the mobilization of TMS. The carrier's equipment operators complained about the delivery schedule downloaded to their PDAs from TMS, considering it an invasion of privacy. The operators also argued that the impracticality of the schedule created from freight information, distance between customers, and the expected delivery time is a result of the lack of actual work experience. The vehicle location tracking system, enabled by GPS or GIS, which provides real-time tracking for clients, was also considered to infringe on the operators' privacy.

   Another difficulty arouse when the RFID technology was implemented to WMS. Warehouse operators felt that RFID would eventually replace their jobs. This was the source of their uncooperative attitudes. For the scheduling problem, CJ-GLS upgraded their routing engine to include additional fixed and variable unloading time. GPS and GIS replaced the geolocation service to improve the accuracy of vehicle locations. Continuous education and learning resolved the issues of privacy invasion and the possibility of jobs lost from adopting the technology. Workers were assured that the new system was implemented to increase customer value and accurate delivery, and was for the common benefit of all.

b. *From the Technical Perspective* - The biggest concern at CJ-GLS was having no prior experience in RFID technology and the fact that there were no practical cases of its successful application. The absence of a technology standard, a successful application case, and a shortage of insight on the forthcoming business process changes were the main shortfalls. All of these implied a high possibility of failure when using a RFID-based 3PL system.

The major technical problems of RFID utilization in a warehouse setting are multiple readings and omission errors. As freight with RFID tags is being processed at one of the steps of the warehouse operation, the designated RFID antenna for the step should automatically receive the information of the freight being processed and this information should be transferred to WMS to show where and at which step of the operation the freights are. However, multiple receiving by non-designated antennas or antennas designated for other steps of warehouse operation misleads WMS to recognize the freight to be at multiple steps at the same time. Multiple receiving occurs due to the interference of noise and obstacles in the warehouse and multidirectional frequencies sent by RFID tags.

Multiple receiving errors were corrected by reprogramming the middleware that processes information received by RFID antennas. The middleware allowed WMS to recognize information sent by the RFID antenna designated for the second step only when the information from the antenna designated for the first step had been entered and so forth with the rest of the steps.

Another technical bottleneck occurs when the RFID antenna fails to receive freight information sent by RFID tags. Such failures are caused when the frequencies from the RFID tags are weakened or blocked by the stocking of freight or metallic packaging materials. After several discussions and benchmarking of other similar cases, CJ-GLS installed a conveyor belt system, with RFID antennas attached directly above the belt. Freight is placed on the belt individually and transported into the warehouse after being removed from their pallets and packaging materials, which block RFID frequencies. This business process reengineering enabled the RFID antennas to receive the information sent by RFID tags without any failures.

Success Factors - The success factors of CJ-GLS' ubiquitous-oriented RFID 3PL system showed how the company supports complementary assets (Brynjolfsson & Hitt, 2000; Marchand, 2004; Teece, 2009), such as cultivating the organizational learning culture and managerial advancement, in addition to new technological innovations to achieve a competitive advantage against leading competitors.

a. *Adventure to the Blue Ocean: Highly Motivated Information Strategy Team Culture* - CJ-GLS realized IS as one of its core competencies early on. They developed and operated their core IS in-house, mobilized TPS, and recognized that the potential impact of ubiquitous technology and an information strategy team is the center for developing a successful IS. Highly motivated team members and a progressive culture within the information strategic team is the driving force. Following is the quote from CIO of CJ-GLS "*We all agreed that RFID technology would be a solution to many logistics problems. However, we could not find a single successful case of RFID application in the logistics industry, and the RFID system vendors were not able to provide any useful application insights. We found that Wal-Mart and Tesco, just like us, are in the process of applying RFID technology to their logistics system and testing it. Instead of searching for and adopting prior applications, we (the information strategy team at CJ-GLS) chose to develop an RFID logistics system of our own with collaboration from outsourced research institutions. All team members of CJ-GLS and the researchers searched, shared, and learned RFID references, which later turned into a learning and proactive culture. Even though we have no experience in this field, we are highly motivated to move in a new direction. The team members' attitude and CJ-GLS' culture is the driving force for individual and organizational growth."*

b. *Advantages of the First Mover and Top Management Leadership* - In a global e-business environment, leadership is the driving force behind the revolution of the entire organization. While accomplishing the RFID-based, ubiquitous-oriented 3PL system project, top management at CJ-GLS manages participants consistently to prevent them from losing focus with the objectives of the project, which is highly possible when multiple participants are involved. Following is the quote from strategy research manager of CJ-GLS "*This RFID project was our first large scale consortium. Therefore, we have less experience in collaborating with outside participants. Since all participants have their own interests in regards to the project, it was not an easy job to reconcile them to focus on the objectives. Our top management played an important role as a neutral consultant for moving the project in the right direction."*

c. *Open Mindset for New ICT* - CJ-GLS was aware of the importance of the RFID-based, ubiquitous-oriented 3PL system and started research on technology by establishing a RFID center. The research initiative is located in the same physical space occupied by the information strategy team for easy communication among researchers and employees. Following is the quote from the interview with CIO of CJ-GLS "*Our core competencies are not from the number of vehicles or warehouses but from a Korean-style, 3PL system development capability, top management's open mind to new technological innovations, and the system development capabilities of our human resources."*

5. Conclusion

This case shows how a 3PL provider develops an ubiquitous, electronic logistics business model which pioneers into an uncontested market by not only defining customers from an existing logistics market but also by recognizing future customers.

Unlike numerous companies that have tried to mingle foreign solutions into their IS and business processes, CJ-GLS developed their own systems, managed the organizational changes, and utilized ICT's impact toward strengthening its 3PL business model. CJ-GLS evolved from e-business supporting LIS to m-business supporting ubiquitous-oriented LIS, which achieved efficiency in OMS and TPS. Whenever they faced bottlenecks and problems, they overcome those problems through organizational and technological change. When inefficiencies occurred from the existing business process, new technology was deployed. Once a new business process is enabled by technology, employees are educated and organizational changes follow.

CJ-GLS has built its competitive advantage through expanding capabilities, such as human resources and new technology innovations, rather than an expansion of tangible assets, such as warehouses and freight fleets, and has focused on its core competency, as well as strengthening its competitive advantage through utilizing technological innovation.

Because expansion of global e-business has brought about infinite global competition, companies must devote themselves to finding a new growth engine. Under these circumstances, Blue Ocean strategy provides a guideline for how companies can survive by creating new uncontested market space instead of competing in the existing market.

6. References

Bhise, H. ; Farrell, D. ; Miller, H.; Vanier, A. & Zainulbhai, A. (2000). The duel for the doorstep. *The McKinsey Quarterly*, Vol.2, 33-41, 0047-5394.

Bruckner, L.K. & Kiss, T. (2004). Grid solution for e-marketplaces integrated with logistics, *Proc. of the DAPSYS 2004 Conference*,pp. 155-163, 0387230947, September, 2004, Budapest, Hungary.

Brynjolfsson, E. & Hitt, L.M. (2000). Beyond computation: information technology, organizational transformation, and business performance. *Journal of Economic Perspectives*, Vol.14, No.4., 23-48, 0895-3309.

Carr, N. (2003). IT doesn't matter, *Harvard Business Review*, Vol.81, No.5, 41-49, 00178012. Kim, W.C. & Mauborgne, R. (2005) Blue ocean strategy: from theory to practice. *California Management Review*, Vol. 47, No.3, 105-121,0008-1256.

Kim, W.C. & Mauborgne, R. (2004) *Blue ocean strategy: how to create uncontested market space and make competition irrelevant*, Harvard Business School Press, 1591396190, Boston, MA.

Kimura, T. (1998) The emergence of third party logistics, Vol. 120. Tokyo, Japan: NLI Research Institute.

Krumwiede, D.W. & Sheu, C. (2002). A model for reverse logistics entry by third party provides. *Omega*, Vol. 30, No.5, 325-333, 0305-0483.

Lee, H.L. & Whang, S. (2001). Winning the last mile of e-commerce. *MIT Sloan Management Review*, Vol.42, No.4, 54-63, 1532-9194.

Lee, W.B. & Lau, H.C.W. (1999). Factory on demand: the shaping of an agile production network, International. *International Journal of Agile Management Systems*, Vol.1, No.2, 83-87, 1465-4652.

Marchand, D.A. (2004). Extracting the business value of IT: it is usage, not just deployment that counts. *Capco Institute Journal of Financial Transformation*, Vol.11, 125-131.

Rabinovich, E.; Windle, R.; Dresner, M. & Corci, T. (1999). Outsourcing of integrated logistics functions. *International Journal of Physical Distribution and Logistics Management*, Vol.29, No.6, 353-373, 0960-0035.

Simchi-Levi, D.; Bramel, J. & Chen, X. (2004). *The logic of logistics: theory, algorithms, and applications for logistics and supply chain management*, Springer, 0387949216, New York.

Sink, H.L. & Langley, C.J. (1997). A managerial framework for the acquisition of third-party logistics services. *Journal of Business Logistics*, Vol.19, No.1, 121-136, 07353766.

Sjöstedt, L.M. (2002). Managing sustainable mobility: a conceptual framework. In: *Information systems in logistics and transportation*, Tilanus B, (Ed.), 19-31, Elsevier, 0080430546, Kidlington, Oxfordshire.

Teece, D. (2009). *Economic Dynamic Capabilities and Strategic Management: Organizing for Innovation and Growth*, Oxford University Press, 019954512X , USA.

Vaidyanathan, G. (2005) A framework for evaluating third-party logistics. *Communications of the ACM*, Vol.48, No.1, 89-94, 0001-0782.

3

Monitoring Cold Chain Logistics by Means of RFID

Luis Ruiz-Garcia[1] and Loredana Lunadei[2]
[1]*Physical Properties and Advanced Technology in Agrofood. Universidad Politécnica de Madrid,*
[2]*Dipartimento di Ingegneria Agraria. Università degli Studi di Milano*
[1]*Spain*
[2]*Italy*

1. Introduction

Every day, millions of tons of temperature sensitive goods are produced, transported, stored or distributed worldwide. For all these products the control of temperature is essential. The term "cold chain" describes the series of interdependent equipment and processes employed to ensure the temperature preservation of perishables and other temperature-controlled products from the production to the consumption end in a safe, wholesome, and good quality state (Zhang, 2007). In other words, it is a supply chain of temperature sensitive products. So temperature-control is the key point in cold chain operation and the most important factor when prolonging the practical shelf life of produce. Thus, the major challenge is to ensure a continuous 'cold chain' from producer to consumer in order to guaranty prime condition of goods (Ruiz-Garcia et al., 2007).These products can be perishable items like fruit, vegetables, flowers, fish, meat and dairy products or medical products like drugs, blood, vaccines, organs, plasma and tissues. All of them can have their properties affected by temperature changes. Also some chemicals and electronic components like microchips are temperature sensitive.

1.1 Problematic

The quality of these products might can change rapidly, when inadequate temperature and relative humidity conditions during transport and storage. Temperature variations can occur in warehousing, handling and transportation. Studying and analyzing temperature gradients inside refrigeration rooms, containers and trucks is a primary concern of the industry. Any temperature disturbance can undermine the efforts of the whole chain (Meenke, 2006). In refrigerated trucks or containers, temperatures rise very quickly if a reefer unit fails. A recent study shows temperature-controlled shipment rise above the specified temperature in 30% of trips from the supplier to the distribution centre, and in 15% of trips from the distribution centre to the store. Lower-than required temperatures occur in 19% of trips from supplier to distribution centre and in 36% of trips from the distribution centre to the store (White, 2007).

The supply chain management for temperature sensitive goods requires fast decisions; goods are forwarded within hours. Appropriate planning calls for more information than that which could be provided by standard RFID tracking and tracing. It is essential to ensure that temperatures are adequate. Reports from literature indicate gradients of 5°C or more, when deviations of only a few degrees can lead to spoiled goods and thousands of Euros in damages (Tanner & Amos, 2003; Nunes et al., 2006; Rodríguez-Bermejo et al., 2007). Loss and damage of perishable goods during storage and transportation is a substantial global issue. Inadequate temperature is second on the list of factors causing foodborne illness, surpassed only by the initial microflora present in foods (Sánchez López & Daeyoung, 2008). It is estimated that 300 million tons of produce are wasted annually through deficient refrigeration worldwide (IIR/UNEP, 2002). The Cool Chain Association (CCA) estimates that 30% of temperature-sensitive products are lost during transport (Hoffman, 2006).

International refrigerated trade continues growing at 6% per annum (Coates, 2003). Quality control and monitoring of goods during the cold chain is an increasing concern for producers, suppliers, transport decision makers and consumers. Commercial systems are presently available for monitoring containers, refrigerated chambers and trucks, but they do not give complete information about the cargo, because they typically measure only a single or very limited number of points (Ruiz-Garcia et al., 2007). Also, there is an increasing demand of traceability in the cold chain, statutory requirements are growing stricter and there is increasing pressure to develop standardized traceability systems. Each event in the chain, like production of transportation, packing, distribution or processing results in a different product which can have its own information associated within the tracing system. From the raw material to the sale of goods, more and more information like temperature, humidity or tamper needs to be gathered and made available. Supplementary information may also be collected at any step, in order to provide data for analysis and optimization of production practices (Thompson et al., 2005).

Parties involved need better quality assurance methods to satisfy customer demands and to create a competitive point of difference. Successful cold chain logistics calls for automated and efficient monitoring and control of all operations. The monitoring should allow the establishment of better knowledge of the cold chain, the detection of weakness and the optimization of the whole process, all things that potentially would have a significant impact on the supply chain (Vervest et al., 2005).

The cold chain is involved in many industries as it was mentioned previously. However, existing applications on the cold chain topic are associated mainly with food and pharmaceuticals. Therefore, the scope of this chapter will focus on these products.

1.2 Perishable food products

The internal biological and chemical processes of fresh produce continue after harvesting. Produce is a living, breathing commodity, which emits heat and carbon dioxide. The risk of a failure in the cold chain could cause excessive ripening, weight loss, softening, color and texture changes, physical degradation and bruising, and attack by rot and molds. These factors affect freshness, desirability, and marketability. Strict temperature control throughout the supply chain can minimize the risk of food-borne illnesses because low temperatures drastically reduces the growth rate of most human pathogens (Ukuku & Sapers, 2007).

Depending on the temperature requirements we can identified four groups of products (Fernie & Sparks, 2004):

- Frozen is -25°C for ice cream, -18°C for other foods and food ingredients.
- Cold chill is 0°C to 1°C for fresh meat and poultry, most dairy and meat-based provisions, most vegetables and some fruit.
- Medium chill is 5°C for some pastry-based products, butters, fats and cheeses.
- Exotic chill is 10-15°C for potatoes, eggs, exotic fruit and bananas.

EN 12380 (1999) legislation demands class one temperature measurement for refrigerated food transport. Measurement has to be feasible in the range between -25 °C up to +15 °C with an accuracy ± 1 °C and resolution ≤ 0.5 °C (CEN, 1999).

It is not easy to maintain appropriate conditions over the whole chain, negligence or mishandling in the logistic of perishable food products is very common, including goods poorly or excessive cooled. Roy et al. (2006) analyzed the supply of fresh tomato in Japan and quantified product losses of 5% during transportation and distribution (Roy et al., 2006). Tanner and Amos (2003a and 2003b) studied thermal variations during long distance transportation of fruits from New Zealand to Europe. The results showed that the products were out of the set-point more than 30% of the time, with a significant variability both spatially across the width of the container as well as temporally along the trip. In those experiments monitoring was achieved by means of the installation of hundreds of wired sensors in a single container, which makes this system architecture commercially unfeasible (Tanner & Amos, 2003).

Global food safety policies concerning the cold chain have been stipulated by Governmental authorities with the aim of protecting the safety and quality of food. In the European Union Manufacturers, distributors and other supply chain partners must follow the ATP (Agreement on the international carriage of perishable foodstuff and on the special equipment to be used for such carriage) (UNECE, 2003). In the United States of America the HACCP (Hazard Analysis and Critical Control Point) (HACCP, 1997), Bioterrorism Act (FDA, 2002) and COOL (Mandatory Country of Origin Labeling) (USDA, 2009).

1.3 Medicinal cold chain

As it was mentioned previously, medical products like vaccines, blood, some pharmaceuticals and others are temperature sensitive. Most of these products require temperatures in the range of 2-8°C. So, they need a cold chain logistic in order to prevent damage caused by heat exposure. Thus, temperature control and monitoring is one of the most critical factors in the pharmaceutical supply chain and is gaining more attention from the industry. For example, in the case of vaccines, keeping too cold can be just as harmful as keeping them too warm, since many vaccines may be damaged by freezing. For safety and quality, temperature needs to be carefully and continuously monitored and controlled in each stage of the supply chain. The effort expended in reaching children with immunizations services will be lost if vaccines are improperly handled so that they are damaged by incorrect temperature maintenance (Bishara, 2006).

The cold chain of pharmaceutical products is complex. This complexity is engendered by the globalization of pharmaceutical companies. Various packagers, primary and secondary distributors, retail pharmacies, hospitals and others may have handled the drug before getting the final consumer. In this framework counterfeit is a main problem. For example, of the top 34 drugs that are of most concern for counterfeiting according to Florida's Bureau of Statewide Pharmaceutical Services and Drug Wholesaler Advisory Council, more than 61% have strict cold-chain requirements. Cold chain drugs are most commonly counterfeited due to their higher added value (Ames, 2006; Floridashealth, 2006).

All the members of the pharmaceutical supply chain have various global regulatory requirements to meet while handling, storing, and distributing environmentally sensitive products. The World Health Organization specifies: "where special storage conditions (e.g. temperature and relative humidity) are required during transit, these should be provided, checked, monitored and recorded." And also states: "temperature mapping of vehicles (where applicable) should support uniformity of the temperature across the vehicle. Recorded temperature monitoring data should be available for review." Other regulations like the Canadian Guidelines for Temperature Control of Drug Products during Storage and Transportation remark that "temperatures should be controlled and monitored using calibrated monitoring devices and records of temperature and alarms, were applicable, should be maintained"(Health Canada, 2006). Similar statements are in the "Guidelines on Good Distribution Practice of Medicinal Products for Human Use" of European Union (European Union, 2003) or in the recommendations for "Good Storage and Shipping Practices" of the United States Pharmacopeia (USP, 2005).

2. The role of RFID in cold chain

RFID is entering in a new phase. RFID technologies are said to improve the performance of the cold chain. Recent advances offer vast opportunities for research, development and innovation in the cold chain. This is the consequence of lowering costs of ownership, engineering increasingly smaller sensing devices and the achievements in radio frequency technology and digital circuits.

RFID was originally developed for short-range product identification, typically covering the 2 mm - 2 m read range and has been successfully applied to food logistics and supply chain management processes. However, recent developments in RFID hardware outfitted with sensors extend its range of application. Adding sensors to the same tags used to track items moving through the supply chain may also alert if they are not stored at the right temperature and predict the remaining shelf life. There are active and semi-passive tags that can measure temperature (Amador et al., 2008; Jedermann et al., 2009), humidity (Chang et al., 2007; Abad et al., 2009), shock/vibration (Todd et al., 2009) or light (Cho et al., 2005; Abad et al., 2009). Moreover, the last generation of Class 4 RFID tags can be configured in a mesh network. In this type of network, the tags can communicate each other to get to a reader circumventing environmental obstacles and extend the size of the system (Sarma & Engels, 2003).

2.1 Advantages of RFID monitoring

Current temperature monitoring systems like strip chart recorders or temperature dataloggers are usually expensive and not automated, thus requiring manual inspection. RFID devices are more accurate and can be read without opening the container or package (Abad et al., 2009). Specialized RFID monitoring devices promise to revolutionize the shipping and handling of a wide range of perishable products. They can be placed in transport vehicles in order to monitor the on-the-go environment and can be the basis for distributed systems, enabling environment sensing together with data processing (Ruiz-Garcia et al., 2007); giving suppliers and distributors continuous and accurate readings throughout the chain. Precise, frequent and automated readings, interpreted by software and coordinated with existing and planned product inventories, should translate into more intelligent goods management and fewer rejected batches.

The lowering cost of RFID will provide the opportunity to track and trace not only large and expensive products, but small and cheap ones, creating a new generation of intelligence products (Meyer et al., 2009). From the raw material to the sale of goods, more and more information can be gathered and made available. Moreover, this information can be linked with a traceability system in each step of the life of the product, *"from farm to fork"*, tracking and tracing products from the field to industry in a new exhaustive way. The concept of "cold traceability" has been introduced to trace groups of temperature-sensitive products are transported in different atmosphere requirements (Ruiz-Garcia et al., 2009).

2.2 FIFO vs. FEFO

If a direct access to the means of transport is not possible, online notifications offer new opportunities for improve transport planning. If fixed delivery commitments require ordering of a replacement, the time of information is very crucial. Improved cool chain management methods such as the Quality Oriented Tracking and Tracing Systems (QTT) offer new features (Scheer, 2006). An example of this approach is the Safety Monitoring and Assurance System (SMAS) that was developed to reduce customers' risk of consuming microbiologically contaminated meat (Koutsoumanis et al., 2005). The growth rate of pathogens was estimated based on temperature history. At a control point the package was either sent to the local or the export market. A case study of cooked ham was carried out based on previous surveys of distribution chain conditions. Following the SMAS approach, the number of products with zero shelf life could be reduced from 12 % to 4% in the export store compared to normal FIFO (First In, First Out) handling. A retailer that knew which of the cases had the shorter shelf life could put it out before the one with the longer shelf life. This is known as FEFO (First Expire, First Out) (Emond & Nicometo, 2006).

3. Applications of RFID in cold chain tracking and monitoring

Several applications for monitoring cold chain logistics by means of RFID have been reported. The majority are oriented to perishable food products. Here are the most representative to our knowledge.

The use of microbial growth models combine with information from active RFID has been faced. These models allow the prediction of microbiological safety and quality of foods, by monitoring the environment without recourse to further microbiological analysis. Thus immediate decisions on the quality and/or safety of fresh produce can be made based on the temperature profile of the supply chain. Three different cases were studied: frozen dairy product, meat carcass chilling and fermented meat processing (McMeekin et al., 2006).

An important step in cold chain management is recording the temperature throughout the supply chain. Implementation of HACCP requires measurements to ensure that the prescribed control limits are not being violated. Ogasawara and Yamasaki (2006) reported a cold chain solution that uses RFID tags with embedded temperature sensors. It also introduced a temperature-managed traceability starter kit that contributes to effective risk management by easily enabling consistent temperature management throughout transportation processes (Ogasawara & Yamasaki, 2006).

The integrity of the cold chain must be maintained from the very beginning of production or processing, through each link (loading, unloading, transport, handling, storage) to the consumer end. Gras (2006) monitored a cold chain of frozen goods using semi-passive and active RFID instrumented with temperature sensors. The experimental work covers four

steps of the cold chain: production, transportation, storage and delivery. Data was linked with computerized cold chain management system (Gras, 2006).
Environmental temperature can differs from each other depending in the location of the logger, packing material or heat dissipation of the product (Moureh et al., 2002; Raab et al., 2008). RFID tags can be also used to measure, not just the walls of the vehicle, but also inside the boxes. Amador et al. (2008) showed the use of RFID for temperature tracking in an international shipment of pineapples from a packing house, in Costa Rica, to a wholesale storage, in USA. They studied the use of RFID in temperature monitoring by comparing the performance of RFID temperature tags versus conventional temperature tracking methods, as well as RFID temperature tags with probe versus RFID temperature tags without probes and their utilization along the supply chain. The temperature mapping of a shipping trial comprising pallets of crownless pineapples instrumented using different RFID temperature dataloggers and traditional temperature dataloggers and packed in two kinds of packages (corrugated boxes and reusable plastic containers) inside a container was performed. The results showed that RFID temperature tags are analogous with regards to accuracy to the conventional methods, but have a superior performance because they allow quick instrumentation and data recovery and the possibility of accessing the sensor program and data at any point of the supply chain without line of sight (Amador et al., 2009).
The fresh fish logistic chain has been also monitored using RFID. Abad et al. (2009) validated a RFID smart tag instrumented with light, temperature and humidity sensors. The system provides real-time traceability information of the product to the different fish distribution chain links. The RFID tag was placed on a corner of the fish box to ensure a correct real-time reading of the temperature and relative humidity measurements (maximum reading distance of about 10 cm) from the outside without the need of opening the fish box (Abad et al., 2009).
The accuracy of data loggers is a critical issue in cold chain management. This accuracy becomes even more important if the objective is early detection of temperature changes and gradients. Standards for food distribution allow deviations of ± 0.5°C from the set point (CEN, 1999). Jedermann et al. (2009) compared three different RFID loggers in order to find the most appropriated one for monitoring cold chain logistics. Several tests were carried out in a climatic chamber. In separate experiments with 20 to 40 samples per logger type, the temperature in the chamber was increased stepwise. The sampling rate of the loggers was set to 5 minutes. Constant temperatures of -10 °C, 0 °C, 15 °C and 30 °C were maintained over a time span of at least of 30 minutes, giving a minimum of 7 valid samples per logger and temperature step. For the test temperatures, the average and standard deviation δ (root mean square deviation) were calculated separately for the three logger types in Table 1. Part of the difference between the average temperature and the set point might have resulted from tolerances of the climatic chamber, which were specified by the manufacturer to be 0.1 °C. The percentage of measurements with a difference to the average value less than the deviation $\pm\delta$ was between 66% and 73%, which was very close to the expected value for a Gaussian distribution of 2/3. The iButtons produced the best results with a deviation of ±0.09 °C, followed by the TurboTags with ±0.19 °C, and the KSW tags with ±0.36 °C.
Jedermann et al. (2009) also studied the temperature distribution inside a packed and sealed pallet. For evaluation of the penetration depth of temperature changes, two test pallets were equipped with 50 or 70 KSW Variosens data loggers. The tags were placed in pre-cooled boxes of dairy products. 180 boxes were loaded in 5 layers onto a pallet of 80 cm by 120 cm to a height of 95 cm. The tagged boxes were placed in the bottom, middle and top layers.

	Experiment 1: KSW (23 units)	Experiment 2: Turbo-Tag (36 units)	Experiment 3: IButton (20 units)
Temperature (°C)	Mean and Standard deviation (°C)	Mean and Standard deviation (°C)	Mean and Standard deviation (°C)
-10	-10,00±0,41	-9,81±0,25	-9,90±0,15
0	0,13±0,32	0,16±0,18	0,12±0,06
15	15,17±0,29	15,05±0,17	15,02±0,06
30	30,23 ±0,42	30,05± 0,18	29,99±0,09

Table 1. Test of different logger type in climatic chamber (Jedermann et al., 2009)

After one day at the optimal temperature of 6.5 °C the pallet was moved to a non-refrigerated storage space with a temperature of about 20 °C. Because of the lack of air ventilation, the surface of the pallet warmed up to only 16.5 °C. After 3 days the pallet was moved back to the refrigerated room. The positions and temperatures of different measurement points inside the densely packed pallet of the first setting are displayed in Figure 1. Due to the large thermal mass of the goods, the effect of changes in the ambient temperature on the temperature of the pallet core was delayed for several days. After 60 hours the change in core temperature had only reached 50% of the change in ambient temperature.

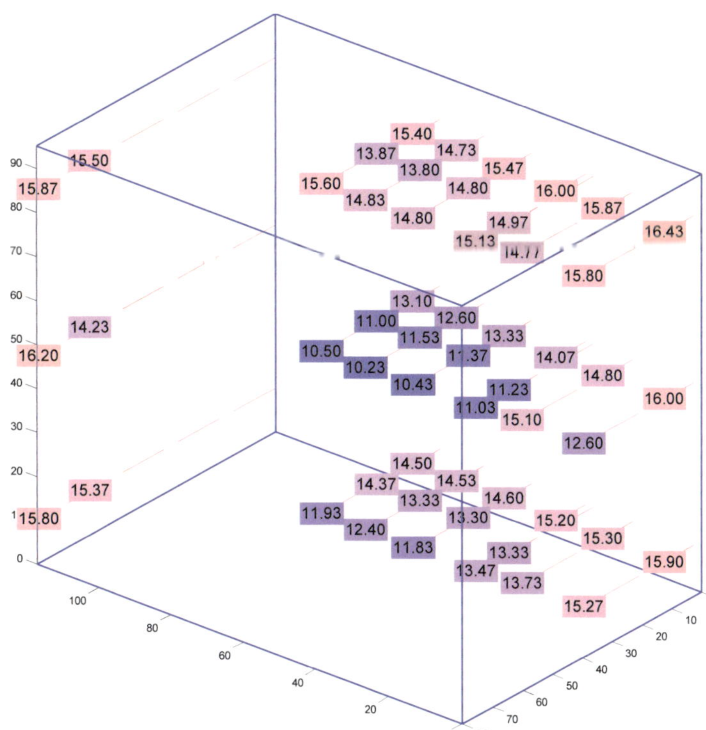

Fig. 1. Temperature rise inside a sealed palette after 60 hours without refrigeration (Jedermann et al., 2009).

Demands for mixed loads of products require different storage temperatures and the trend of refrigerated transport is to use multi-compartmental vehicles. Jedermann et al. (2009) monitored 16 multi-compartmental trucks using semi-passive RFID instrumented with temperature sensors (Turbo Tag) detecting temperature gradients. The authors concluded that semi-passive tags can be used to monitor environmental variables such as the temperature of chilled food refrigerated goods, to identify problem areas and to raise alarms. RFID loggers are good tools are cost-effective for the characterization of refrigerated transport units such as trucks or containers (Jedermann et al., 2009).
Average temperature often lies. The average information does not provide information about gradients, maximum of minimum temperatures during a shipment. Moreover, if two shipments have the same average temperature does not means that both have were of the same performance. Jedermann et al. (2009) make use of a normalized temperature difference (*ΔTnij*) developed by Rodriguez-Bermejo et al. (2007) , which is computed with respect to the set point and to the outside temperatures (Equation 1). This value gives a normalized measure with respect to the varying ambient conditions of the experiments in order to judge the difference between them. Values over zero indicate that the temperature inside the transport was higher than the set point, which was the case of majority of the shipments (Rodríguez-Bermejo et al., 2007).

$$\Delta Tn_{ij} = \frac{\overline{T_{ij}} - Ts_i}{\overline{T_{ai}} - Ts_i} \ (\text{with} \ -1 < \Delta Tn_{ij} < 1) \qquad (1)$$

ΔTnij =Normalized temperature difference for experiment i and logger j
Tij = Temperature for experiment i and logger j (°C)
Tsi = Set point (°C)
Tai= Ambient temperature (°C)
One statistical parameter that provided interesting information is the variance of *ΔTnij*, known as *"Mean Normalized Variance"*. If this value is more than one that means temperature has fluctuated more inside the transport than outside. For the experiment of the 16 trucks all values were under 1, which indicates that the fluctuations inside were less than outside (Jedermann et al., 2009)
In the intermodal transportation, the performance of radio waves inside metal enclosed areas was studied. Furh and Lau (2005) tested a radio frequency device in a metal cargo container and demonstrated that it is possible to communicate with the outside world (Fuhr & Lau, 2005). Jedermann et al. (2006) presented a system for intelligent containers combining wireless sensor networks and RFID (Jedermann et al., 2006). Laniel et al. (2008) focuses on the 3-D mapping of RFID signal strength inside a 12m (40') refrigerated marine container. Three different types of radio frequency configurations were tested: 2.4GHz, 915MHz and 433MHz.Tests were performed with an empty container and the main goal was to find a frequency and configuration that would allow real time reading of the temperature in a shipment of perishable products using RFID. Results obtained in this study showed that wave propagation inside a closed marine container is significantly higher at 433MHz than at 915MHz or 2.4GHz, with attenuation averages of 19.57 and 18.20 versus 36.49 and 35.91 and 29.91 and 29.78 dBm respectively (Laniel et al., 2008). At 433 MHz the wavelength is approximately a meter, enabling signals to diffract around obstructions. The level of diffraction depends on the size of the object versus the signal wavelength. At

2.4GHz the diffraction is very limited and therefore not recommended for most cold chain applications which are in crowded environments (Technologies, 2008).

3.1 Pharmaceutical applications

Most of the top pharmaceutical companies are running RFID cold-chain pilot projects but results and details are not published. The typical application is RFID loggers or active tags are placed inside test packages to record temperature and sometimes also humidity at specified intervals and wirelessly download the data to a web-accessible database when they pass a reader. For example, CSL, Australia's biggest pharmaceutical company, is using an RFID-based temperature sensing system that features a read-write, credit-card-size, 13.56 MHz active tag, which can be attached to or inserted in shipments (Forcinio & Wright, 2005). Also some specific RFID devices for monitoring some products like for example blood were developed. The system helps track the blood in space and time from extraction to transfusion. In this way, the blood is monitored according to certain parameters: the time elapsed and the temperature that could determine its state before transfusion (Abarca et al., 2009).
Another interesting application in high value products is to integrate the RFDI tag together with the sensors in paperboard packaging. Smart packages with temperature and tamper sensing capabilities for the pharmaceutical industry have been reported (Figure 2) (CYPAK, 2009).

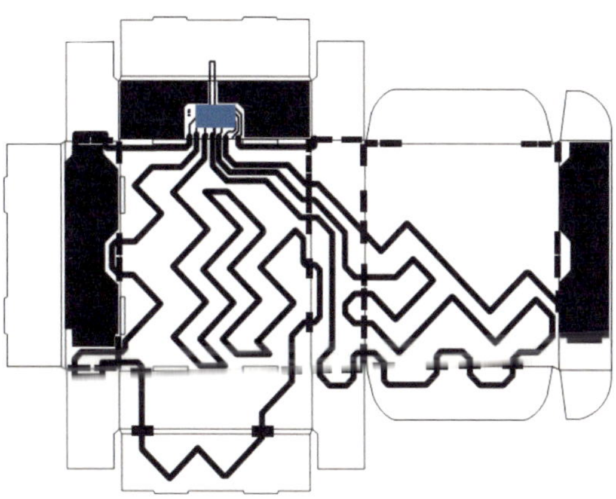

Fig. 2. Smart package (Source: SecurePak)

Any commercial application of RFID in the pharmaceutical cold chain must include anti-counterfeit mechanisms such as ePedigree, providing means of authenticating legitimate product together with temperature information (Ames, 2006).

4. Challenges and limitations

RFID has limitations and its implementation involve multitude of challenges. A significant proportion of RFID deployments remain exploratory. There is a need to know the long-term behavior of the systems. Most of the applications reported have short experimental periods. Longer testing and experimentation is necessary for validate some of applications presented. One limitation might be that these monitoring systems create huge volumes of data that are difficult to manage, causing a huge increase in the daily volume of data in a corporate information technology system. Database administrators need to be able to deal with the potential stresses on the databases, both in terms of speed and volume. Even so, data volume can be overwhelming to the network. If a product have 1,000 bytes of data associated with it, the RFID monitoring system would generate 10 terabytes of data per year. If the data of five years is stored, that means a database of 50 terabyte. Thus, the solution lies in implementing a decentralized data management system. Data can be pre-processed and duplicate information eliminated close to their point of origin by intelligent systems, which could be sited at the level of the tag or reader (Roberti, 2003; Ruiz-Garcia, 2008).

An important research topic that must be faced is fault detection and isolation. In a remote sensing application is essential to detect the erroneous measurements. False reads can be done as a result of radio waves being distorted, deflected, absorbed, and interfered with. Wrong information provided by the monitoring system should be identified and skipped. Also the implementation of artificial intelligence in the core of the system can block the transmission of erroneous data (Angeles, 2005).

RFID data loggers are available in high quantities, but they require manual handling because of their low reading range. Another disadvantage is that temperature loggers are only available for the 13.56 MHz HF-Range. The major drawback of this band is the limited reading range of about 20 cm. If a gate reader scans items automatically upon arrival at the warehouse, the reading range has to cover several meters. Also these tags take around five seconds to transfer recorded temperature values over the RFID interface (Jedermann et al., 2009). A higher data rate is required according to a normal flow of goods in a warehouse.

Another important issue is to deal with the physical limitations of RFID. Metals and liquids inhibit the propagation of electromagnetic waves. This is particularly true for UHF and microwave frequencies (2.4GHz). Some temperature sensitive products such as fruits, vegetables or juices have high water content, sometimes more than 90%. As a result, performance can be affected by the item on which the tag is attached (Angeles, 2005; Ruiz-Garcia, 2008). However, reflections and product dimensions is important for liquid and dielectric materials as well since much of the power loss occurs at the interfaces between air and the medium (Fletcher et al., 2005).

The lack of uniformity in global standards makes the RFID implementations more difficult. Managing multiple readers and related hardware can be a challenge, especially across multiple facilities. There is practically no part of the spectrum available worldwide because governments have assigned different uses for the various parts of this spectrum, with the exception of the ISM (industrial, scientific and medical) bands. In order to solve this problem, the Auto-ID Center has explored the concept of "agile" readers that will allow the network to operate at different frequencies in a wide variety of geographical locations (Haller & Hodges, 2002).

The level of granularity is a limitation in most of the applications. Normally three levels of granularity are considered: pallet, case or item-level. The primary advantage of case or item-level tagging over pallet-level tagging is more detailed and accurate information, since each pallet in a load and each carton on a pallet can experience temperature variations. Instead of reject an entire shipment goods can be considered on a pallet-by-pallet or case-by-case basis. But high granularity also means much more tags to handle with, higher costs and huge data to be processed. Pharmaceutical cold chain is the leading candidate for RFID tagging at item level. The high relative value of pharmaceutical products relative to the price associated with RFID-equipped make it possible (Angeles, 2005).

5. Conclusions

The use of RFID in cold chain monitoring provides new features that have the potential to be an economically viable. RFID and sensor technology is evolving rapidly and enabling greater capabilities at lower costs. The value of technology can be best realized when integrated with decision support systems. Also improving operations by providing early warning of equipment failure and a predictive maintenance tool, improving energy management, providing automatic record-keeping for regulatory compliance, eliminating personnel training costs or reducing insurance costs. Growers, shippers, processors, distributors and retailers that can establish the continuity of the cold chain have a powerful advantage in a highly competitive marketplace. The collaboration and synergy of sensing, processing, communication and actuation is the next step to exploit the potential of these technologies.

Another important benefit of the systems is the visibility that it can give along the cold chain. This kind of systems can be used in a warehouse, container or vehicle for remotely monitoring and tracking environmental characteristics. Measurements obtained are consistent and provide valuable information on the conditions encountered during the life cycle of the products. It is possible to address, at regular time increments, what is happening with the product, whether it is temperature, humidity, acceleration, etc. Another advantage is providing effective support in legal situations as well as safety inspections.

The suitability of this technology for monitoring refrigerated cold chain is clear. RFID has the potential to be a revolution although the adoption of RFID in the industry is at an early stage. However, this technology, as any other, has limitations that have been explained in this chapter. A lot of work remains to be done on overall approaches involving harmonization of standards, long term experimentation and hardware and software improvements.

Sharing information is crucial for an effective cold chain but also privacy. To protect data of unjustified access there are methods in information technologies but in addition an agreement has to be found, that opens the possibility of a fast and detailed rebuild of the trace of a product along the chain, without giving sensitive information to the public or competitors

6. References

Abad, E., Palacio, F., Nuin, M., González de Zárate, A., Juarros, A., Gómez, J. M. & Marco, S. (2009). RFID smart tag for traceability and cold chain monitoring of foods: Demonstration in an intercontinental fresh fish logistic chain. *Journal of Food Engineering*, 93,4.

Abarca, A., de la Fuente, M., Abril, J. M., García, A. & Pérez-Ocón, F. (2009). Intelligent sensor for tracking and monitoring of blood temperature and hemoderivatives used for transfusions. *Sensors and Actuators A: Physical*, 152,2, 241-247.

Amador, C., Emond, J. P. & Nunes, M. C. (2008). Application of RFID Technologies in the Temperature Mapping of the Pineapple Supply Chain. In Food Processing Automation Conference. Providence, Rhode Island (USA).

Amador, C., Emond, J. P. & Nunes, M. C. N. (2009). Application of RFID technologies in the temperature mapping of the pineapple supply chain. *Sensing and Instrumentation for Food Quality and Safety*, 2009,3, 26-33.

Ames, H. (2006). Authentication From a Cold Chain Perspective. *Pharmaceutical Commerce*, 2006,July/August.

Angeles, R. (2005). RFID technologies: Supply-chain applications and implementation issues. *Information Systems Management*, 22,1, 51-65.

CEN (1999). 12830. Temperature recorders for the transport, storage and distribution of chilled, frozen, deep-frozen/quick-frozen food and ice cream - Tests, performance, suitability. European Committee for Standardization.

Coates, J. D. (2003). Cold Chain Challenge: Profit Potential in a "Chilly" New World. *World Trade*, 16,4, 44.

CYPAK (2009). SecurePak. www.cypak.com (last accessed 15-September-2009).

Chang, K., Kim, Y. H., Kim, Y. & Yoon, Y. J. (2007). Functional antenna integrated with relative humidity sensor using synthesised polyimide for passive RFID sensing. *Electronics Letters*, 43,5, 259-260.

Cho, N., Song, S. J., Kim, S. & Yoo, H. J. (2005). A 5.1-mu W UHF RFID tag chip integrated with sensors for wireless environmental monitoring. *Esscirc 2005: Proceedings of the 31st European Solid-State Circuits Conference*, 279-282.

Emond, J. P. & Nicometo, M. (2006). Shelf-life prediction and FEFO inventory management with RFID. In Cool Chain Association Workshop. Temperature measurements-when, where and how? Knivsta, Sweden.

European Union (2003). Guidelines on Good Distribution Practice of Medicinal Products for Human Use. European Union.

FDA (2002). Bioterrorism Act. ed. U. S. F. a. D. Administration.

Fernie, J. & Sparks, L. (2004). Logistics and retail management: insights into current practice and trends from leading experts.

Fletcher, R., Marti, U. P. & Redemske, R. (2005). Study of UHF RFID signal propagation through complex media. In Antennas and Propagation Society International Symposium, 2005 IEEE, 747-750.

Floridashealth (2006) Pedigree Paper Requierements. *Florida Deparment of Health, Statewide Pharmaceutical Services.* http://www.doh.state.fl.us/mqa/DDC/Pedigree_Paper_Requirements.htm (last accessed 30-September-2009).

Forcinio, H. & Wright, C. (2005). Cold Chain Concerns. *Pharmaceutical Technology*, 29,4, 44-50.

Fuhr, P. & Lau, R. (2005). Mesh radio network performance in cargo containers. In Sensors Magazine Online.

Gras, D. (2006). RFID based monitoring of the cold chain. In 2nd international Workshop Cold Chain Management,, ed. J. P. Kreyenschmidt, B., 81-82. Bonn, Germany.

HACCP (1997). Hazard Analysis and Critical Control Point (HACCP) System and Guidelines for its Application.

Haller, S. & Hodges, S. (2002) The need for a universal smart sensor network. *Auto-ID Center* www.autoidcenter.org (last accessed 7-September-2009).

Health Canada (2006). Guidelines for Temperature Control of Drug Products during Storage and Transportation. In GUIDE-0069, ed. H. Canada.

Hoffman, W. (2006). Hot Market, Cool Freight. *Journal of Commerce*, 2006,1, 1.

IIR/UNEP (2002). Industry as a Partner for Sustainable Development. Refrigeration.

Jedermann, R., Behrens, C., Westphal, D. & Lang, W. (2006). Applying autonomous sensor systems in logistics - Combining sensor networks, RFIDs and software agents. *Sensors and Actuators A: Physical*, 132,1, 370-375.

Jedermann, R., Ruiz-Garcia, L. & Lang, W. (2009). Spatial temperature profiling by semi-passive RFID loggers for perishable food transportation. *Computers and Electronics in Agriculture*, 65,2, 145-154.

Koutsoumanis, K., Taoukis, P. S. & Nychas, G. J. E. (2005). Development of a safety monitoring and assurance system for chilled food products. *International Journal of Food microbiology*, 100,1-3, 253-260.

Laniel, M., Emond, J. P. & Altunbas, A. E. (2008). RFID Behavior Study in Enclosed Trailer/Container for Real Time Temperature Tracking. In Food Processing Automation Conference. Providence, Rhode Island (USA).

McMeekin, T., Smale, N., Jenson, I. & Tanner, D. (2006). Microbial growth models and temperature monitoring technologies. In 2nd international Workshop Cold Chain Management, ed. J. P. Kreyenschmidt, B., 71-78. Bonn, Germany.

Meenke, C. P. (2006) Always Keep Cool. Germanischer Lloyd Certification (GLC). http://coolchain.org/uploads/certified!2006-01ECCQI.pdf (last accessed 11-September-2009).

Meyer, G. G., Främling, K. & Holmström, J. (2009). Intelligent Products: A survey. *Computers in Industry* 60,2009, 137-148.

Moureh, J., Laguerre, O., Flick, D. & Commere, B. (2002). Analysis of use of insulating pallet covers for shipping heat-sensitive foodstuffs in ambient conditions. *Computers and Electronics in Agriculture*, 34,1-3, PII S0168-1699(01)00181-8.

Nunes, M. C. N., Emond, J. P. & Brecht, J. K. (2006). Brief deviations from set point temperatures during normal airport handling operations negatively affect the quality of papaya (Carica papaya) fruit. *Postharvest Biology and Technology*, 41,3, 328-340.

Ogasawara, A. & Yamasaki, K. (2006). A temperature-managed traceability system using RFID tags with embedded temperature sensors. *NEC Technical Journal* 1,2, 82-86.

Raab, V., Bruckner, S., Beierle, E., Kampmann, Y., Petersen, B. & Kreyenschmidt, J. (2008). Generic model of shelf life dynamics in support of cold chain management in pork and poultry supply chains. *Journal on Chain and Network Science*, 8,2, 59-73.

Roberti, M. (2003). Seven Reasons to Act Now. In RFID Journal.

Rodríguez-Bermejo, J., Barreiro, P., Robla, J. I. & Ruiz-Garcia, L. (2007). Thermal study of a transport container. *Journal of Food Engineering*, 80,2, 517-527.

Roy, P., Nei, D., Okadome, H., Nakamura, N. & Shiina, T. (2006). Effects of cultivation, transportation and distribution methods on the life cycle inventory (LCI) of fresh tomato. . In ASABE/CSBE North Central Intersectional Conference. , ed. ASABE. Saskatoon, Saskatchewan, Canada.

Ruiz-Garcia, L. (2008). Development of Monitoring Applications for Refrigerated Perishable Goods Transportation. In Ingenieria Rural. Madrid: Universidad Politécnica de Madrid.

Ruiz-Garcia, L., Barreiro, P., Rodríguez-Bermejo, J. & Robla, J. I. (2007). Monitoring intermodal refrigerated fruit transport using sensor networks: a review. *Spanish Journal of Agricultural Research*, 5,2.

Ruiz-Garcia, L., Steinberger, G. & Rothmund, M. (2009). A Model and Prototype Implementation for Tracking and Tracing Agricultural Batch Products along the Food Chain. *Food Control*, 21,2010, 112-121

Sánchez López, T. & Daeyoung, K. (2008). Wireless Sensor Networks and RFID integration for Context Aware Services. In White Paper series 2008 ed. A.-I. L. W. Paper.

Sarma, S. & Engels, D. W. (2003) On the Future of RFID Tags and Protocols. http://www.autoidlabs.org/uploads/media/MIT-AUTOID-TR018.pdf (last accessed 23-September-2009).

Scheer, F. P. (2006). Optimising supply chains using traceability systems. In *Improving traceability in food processing and distribution*, ed. W. p. limited, 52-64. Cambridge, England.

Tanner, D. J. & Amos, N. D. (2003). Heat and Mass Transfer - Temperature Variability during Shipment of Fresh Produce. *Acta Horticulturae*, 599193-204.

Technologies, S. (2008) Selecting the Right Active Frequency. http://www.autoid.org/2002_Documents/sc31_wg4/docs_501-520/520_18000-7_WhitePaper.pdf (last accessed 5-September-2009).

Thompson, M., Sylvia, G. & Morrissy, M. T. (2005). Seafood Traceability in the United States: Current Trends, System Design, and Potential Applications. *Comprehensive Reviews in Food Science and Food Safety* 1,2005, 1-7.

Todd, B., Phillips, M., Schultz, S. M., Hawkins, A. R. & Jensen, B. D. (2009). Low-Cost RFID Threshold Shock Sensors. *IEEE Sensors Journal*, 9,4, 464-469.

Ukuku, D. O. & Sapers, G. M. (2007). Effect of time before storage and storage temperature on survival of Salmonella inoculated on fresh-cut melons. *Food Microbiology*, 24,3, 288-295.

UNECE (2003). Agreement on the international carriage of perishable foodstuffs and on the special equipment to be used for such carriage (ATP).

USDA (2009). COOL. Country of Origin Labeling. ed. U. S. D. o. Agriculture.

USP (2005). Genearl Chapter <1079> Good Storage and Shipping Practices. ed. U. S. Pharmacopeia.

Vervest, P., Van Heck, E., Preiss, K. & Pau, L.-F. (2005). Smart Business Networks.

White, J. (2007). How Cold Was It? Know the Whole Story. *Frozen Food Age*, 56,3, 38-40.

Zhang, L. (2007). Cold Chain Management. In Cranfield Centre for Logistics & Supply Chain Management. Cranfield University.

4

RFID-enabled Supply Chain Traceability: Existing Methods, Applications and Challenges

Dongmyung Lee and Jinwoo Park
Department of Industrial Engineering, Seoul National University
Republic of Korea

1. Introduction

Radio Frequency Identification (RFID) technology promises to revolutionize various areas in supply chain. Recently, many researchers have investigated on how to improve the ability to track and trace a specific product along the supply chain in terms of both effectiveness and efficiency with the help of this technology. To enable traceability over the entire lifecycle a robust and seamless traceability system has to be constructed. This requires for the following three elements: (1) data model and storage scheme that allows unique identification and scalable database, (2) system framework which enables to share the traceability data between trading partners while maintaining a sovereignty over what is shared and with whom, and (3) a tracing mechanism in order to achieve end-to-end traceability and to provide the history information of products in question.

Along with the studies addressing the requirements and design of traceability system architecture, applications in the real environment have also been reported. Due to the strong regulation in EU which states that food business operators shall be able to identify any person who supplied them and any business which takes food from them, RFID-enabled traceability systems are well implemented in the food supply chain. However, there exist other industries adopting RFID to enhance traceability such as pharmaceutical and aviation and even in the continuous process industry like iron ore refining.

Despite the promising nature of RFID in tracking, there are several challenges to be addressed. Since an RFID tag does not require line-of-sight, multiple tags can be read simultaneously but also tag collisions may occur. Therefore there is no guarantee that a tag will be continuously detected on consecutive scans. Moreover, the use of RFID tags can be of serious threat to the privacy of information. This may facilitate the espionage of unauthorized personnel.

In this chapter, we analyze the main issues of RFID-enabled traceability along the supply chain mentioned above: existing methods, applications and future challenges. Section 2 starts with pointing out the characteristics of RFID data and the requirements for RFID-enabled traceability. Subsequently, we introduce data types, storage schemes and system frameworks proposed in the existing literatures. Then, we discuss tracing methods based on the traceability system architecture. Section 3 presents current applications in real settings of both discrete and continuous production. We also discuss challenges that are preventing companies from adopting RFID for their traceability solutions in section 4. Finally, we conclude our study in section 5.

2. System architecture

Traceability, according to ISO standard (ISO, 1995), is the ability to trace the history, application or location of an entity, by means of recorded identifications. It also may be defined in general as the "ability to trace and follow any product through all stages of production, processing and distribution (EC, 2002)." Traceability itself can be divided into three types (Perez-Aloe *et al.*, 2007): (1) Back traceability (supplier traceability) (2) Internal traceability (process traceability) (3) Forward traceability (client traceability). Having the end-to-end traceability encompasses all three types of traceability and since traceability is defined over every stages of a supply chain, several researchers pointed out various elements that should be taken into account. Bechini *et al.* (2005) posited that traceability systems are constituted by three elements: (1) univocal identification of units/batches (or lots) of every product components, (2) information collection about time and location for every batch transfer/transformation, and (3) a method to relate this kind of data. Similarly, Jansen-Vullers *et al.* (2003) suggest the four elements of traceability in order to manage information on manufacture: (1) physical lot integrity, which determines the traceability resolution, (2) collection of tracing and process data, (3) product identification and process linking, and (4) reporting/system data retrieval.

In practice, traceability systems should be able to store information and show the path of a particular item/part of interest along the whole supply chain from the supplier to the retailer and eventually to the customer. Throughout this process, secure, reliable and automatic product identification is crucial to provide effective and efficient traceability. Barcode technology, in the past, has been used for the identification of products. However, in order to meet the traceability requirements pronounced by the governments, a new technology that allows automated recording of information was needed.

Today, in this spirit, new opportunities for traceability come from the RFID technology. Unlike barcode technology, RFID has several advantages such as multiple reading at a rate of 1000 tags per second that makes it not only a feasible and cost-effective candidate for object identification but also a significant tool to provide visibility along different stages of the supply chain. A typical RFID system is made up of: a reader, which forms an electromagnetic field, and some passive tags without a power supply. They can be read only if they are in the vicinity of a reader which supplies the power required (Bernardi *et al.*, 2007). Every RFID tag has a unique identification number and, depending on the type, it can have a rewritable memory (Gandino *et al.*, 2007).

2.1 Requirements for RFID-enabled traceability

Given the increasing importance of RFID as a means to monitor and manage products for traceability, it has been extensively used for a diversity of applications. For RFID to be used in traceability systems the characteristics of RFID data should be considered in advance. Lee & Park (2008) pointed out two peculiar characteristics of RFID data which are:

- *Dynamic*: Not only the word dynamic means that data changes over time but it also imposes the concept of the item being merged with other subparts and being divided into several components through the production process.
- *Procedural*: Items will leave a trail of data as they pass through several different locations. As an object is equipped with a tag, when it passes by the interrogation zone of a reader it will send information messages to the interrogator. This process will make

the identification of a tag for the reader at every location creating a consistent path of the object.

Along with this particular elements RFID data possesses there are a number of requirements that ought to be respected for the construction of RFID traceability system. Identifying the requirements is important for the data model and storage scheme on which the system is based as it reveals the information needs that the model should meet. Although the requirements might slightly differ based on which application the RFID technology is used, the common requirements can be inferred from the ones that can be applied to the food supply chain (Kelepouris *et al.*, 2007) and are as follows:

- *Item identification*: Many researchers agree that all traceable items in the supply chain should be uniquely identified (Jansen-Vuller *et al.*, 2003; Moe, 1998; van Dorp, 2003). The item identification should be consistent for all partners in the chain. In a different case, data synchronization should take place, which results in significant cost increase and poor data quality.
- *Bill of lots – batch distribution*: For the production process that is concerned with a sequence of activities transforming a listing of raw materials, parts, intermediaries and subassemblies into a particular end item, a complete BOL that contributed to the composition of a product batch must be registered in order to support full traceability. If a supply chain is consisted of stages in which the product is not subject to any transformations, then the BOL need not be registered. However, the registration of the distribution of the batches in items, cases or pallets is mandatory.
- *Operations and capacity units*: Traceability requires the recording of both the variables and their values under which the operation took place.
- *Item Observation*: Apart from the information regarding product composition, information representing product location throughout the chain should also be recorded.

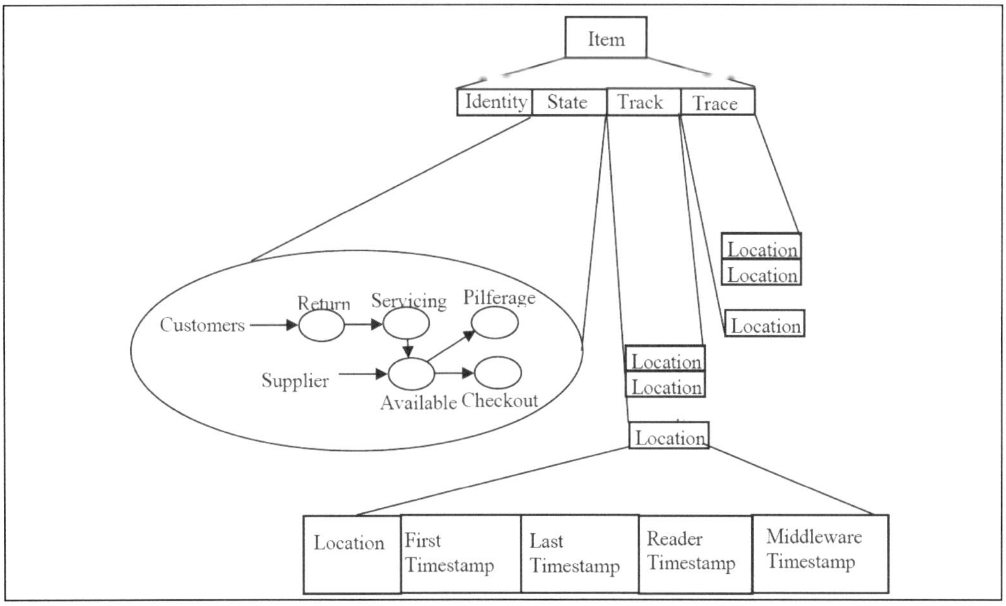

Fig. 1. A schematic view of item record (Gan *et al.*, 2006)

2.2 Data model and storage scheme

Taking these characteristics and requirements into account, data model as well as storage scheme can be built. A data model is essential in that it efficiently models different collection of identifiers that typically occur in item-tracking applications. Numerous researchers have proposed various data models for tracking items. A simple approach is to capture the identity code by RFID readers at both source and destination locations as item moves from one interrogation zone to another. A reader normally records the time of identifying the code based on its internal clock. Fig. 1 illustrates a schematic view of an item record in the database (Gan *et al.*, 2006). Each item record includes an identity code field, a state field, a track record pointer and a trace record pointer. The state field includes several values (e.g., Available, Checkout, Pilferage, Checkin, Return and Servicing) that commonly occur along the supply chain. The track record updates the timestamp fields of the specific location record with the most recent timestamps but if an item visits the same location several times it is not added. On the other hand, the trace record stores every location visited by each item up to the present location in order to trace the history locations.

Another approach came from the key observation that although individual items need to be tracked, they can be tracked more efficiently by tracking the groups to which they belong (Hu *et al.*, 2005). For example, a group can have several items in the same proximity (e.g., items of an identical product). One alternative to tracking a group of identifiers is to maintain an item count for the group, where the count can be viewed as a transformation obtained from an identifier collection (Hu *et al.*, 2005). However, in combining this approach with RFID, there are issues to be addressed.

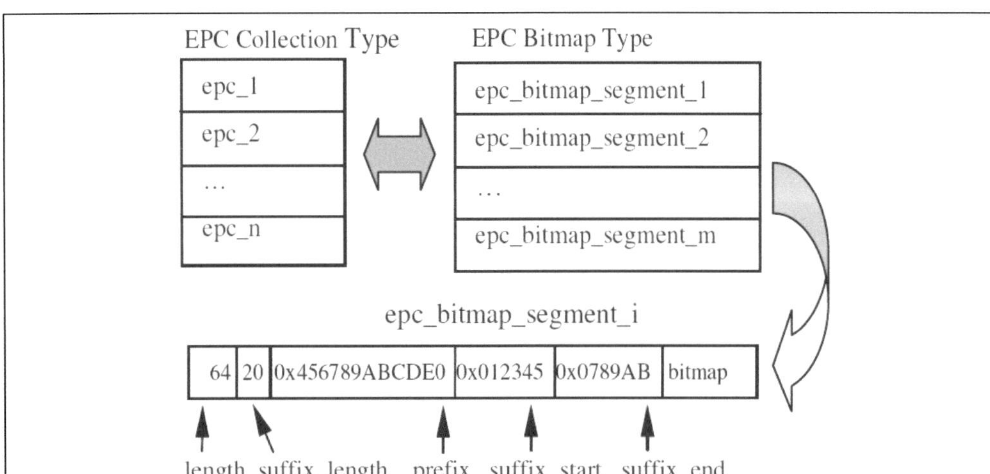

Fig. 2. EPC collection datatype and bitmap datatype (Hu *et al.*, 2005)

In this sense, Hu *et al.* (2005) proposed a bitmap datatype that compactly represents a collection of RFIDs, that is, their EPCs. The Electronic Product Code (EPC) is a simple naming scheme that uniquely identifies objects (items, cases, pallets, locations, etc.). Like many numbering schemes used in practice, the EPC can be divided into digits that identify the manufacturer and product type as well as the unique item within a group by using extra digits or a serial number (Engels, 2003). An EPC code generally contains (Brock & Cummins, 2003):

- *Header*, which identifies the length, type, structure, version and generation of EPC.
- *Manager number*, which identifies an organizational entity.
- *Object Class*, which identifies a "class", or type of thing.
- *Serial number*, the specific instance of the Object Class being tagged.

Hu *et al.* (2005) defined a new *epc_bitmap_segment* type to represent a collection of EPCs, which share a common *epc_prefix*. This type can be used to represent different classes of EPCs (i.e., 64, 96, and 256 bit) since the individual attributes within the epc_bitmap_segment type capture EPC type specification information (EPC length, EPC prefix length, EPC suffix length, etc.). Fig. 2 shows that the epc_bitmap_segment type is used to model a collection of EPCs that generates closely together. Additionally, a multiset type called epc_bitmap is defined that can hold collections of bitmap segments in order to model a collection of EPCs that can be arbitrarily dispersed.

The above scheme is suitable for applications that maintain summary information at periodic intervals such as retail store management, supply chain management (SCM), and asset management (Hu *et al.*, 2005). However, for the method to be exploited in other areas, it should be modified or used with other supplementary schemes.

In addition to providing the means to identify information about lots (or products), a data model should also be able to represent activities or processes (Kim *et al.*, 1995). Fig. 3 designates that the lot behaviour can be modelled by the following six activity patterns using UML (Unified Modelling Language) diagram (Ministry of Agriculture, Forestry and Fisheries of Japan, 2003):

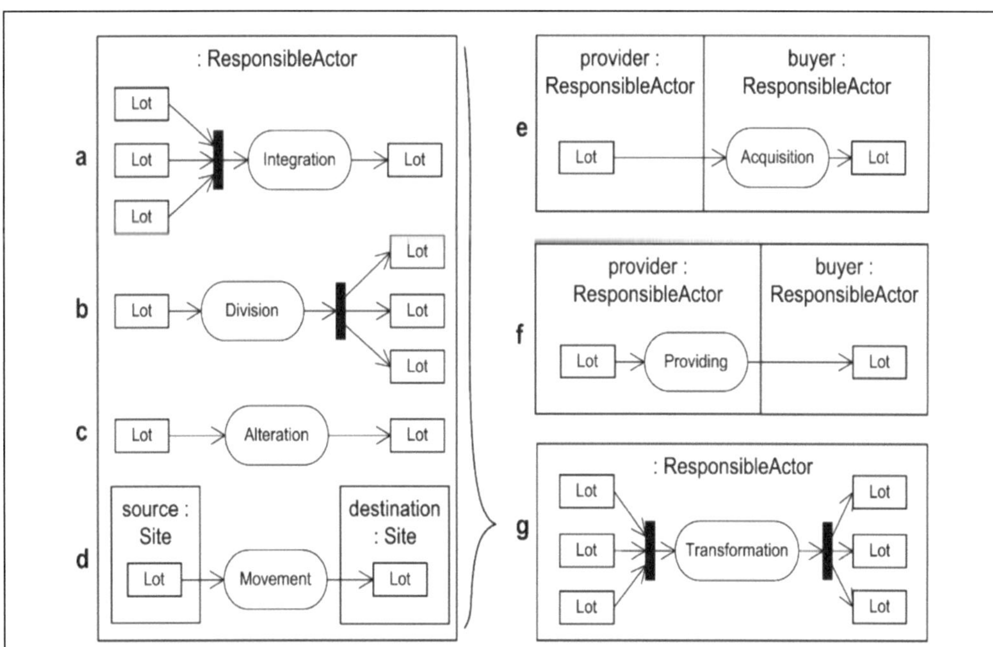

Fig. 3. Basic behavioural patterns of a lot (Bechini *et al.*, 2008)

1. *Lot integration* (e.g., mixing and packing): A number of lots are integrated into a single lot. The responsible actor of the lot creates an association between the pre-integration lots and the post-integration lot.

2. *Lot division* (e.g., cutting and splitting): A lot is divided into a number of lots. The responsible actor of the lot creates an association between the pre-division lot and the post-division lots and vice versa.
3. *Lot alteration* (e.g., heating and drilling): A new lot is generated from a lot by an alteration activity. The responsible actor of the lot creates an association between the pre-alteration lot and the post-alteration lot, and vice versa.
4. *Lot movement*: A lot is moved from one location to another under the same responsible actor. Since a lot can be associated with a unique site, the responsible actor has to create a new lot with a new identifier.
5. *Lot acquisition*: An actor of the supply chain acquires a lot from another actor. This association allows implementing the tracing process and therefore determining origin and characteristics of a particular product.
6. *Lot providing*: An actor provides another actor with a lot. This association allows implementing the tracking process and hence following the downstream path of a product along the supply chain.

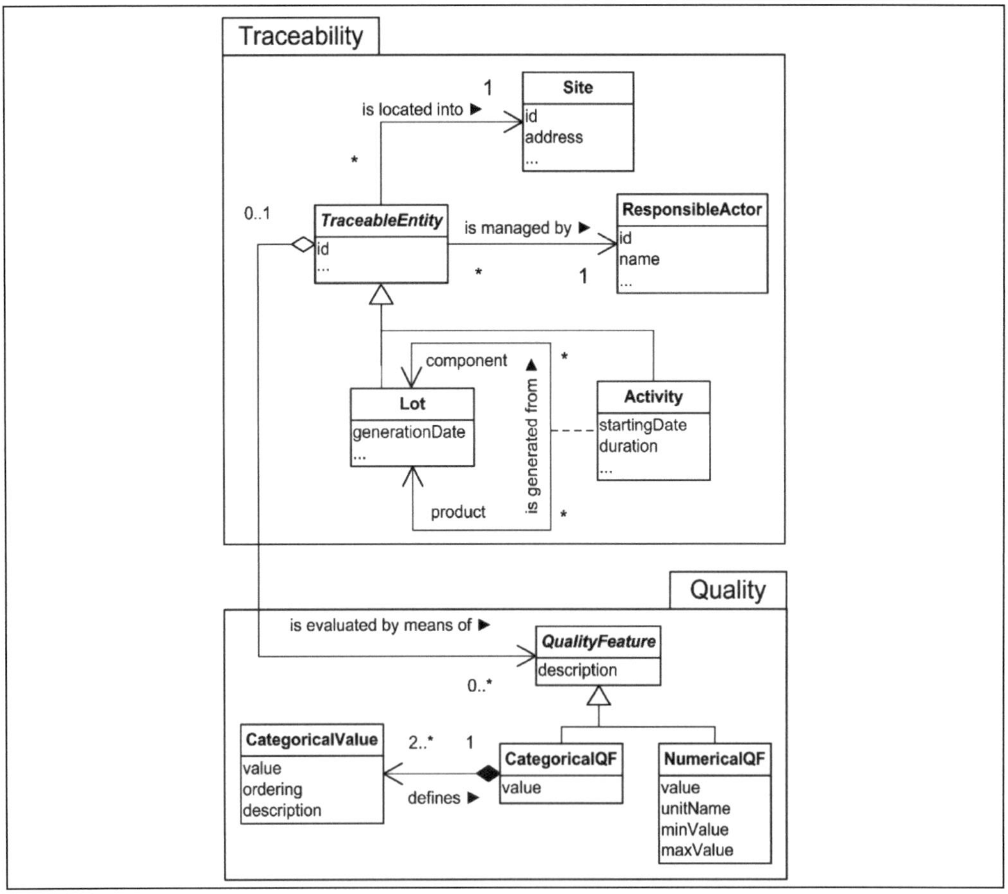

Fig. 4. UML class diagram of the traceability model (Bechini *et al.*, 2008)

Quality is also recommendable to be taken explicitly into account in the data model. The data model shown in Fig. 4 uses UML to reflect these requirements (Bechini *et al.*, 2008).

Here, classes are grouped into two packages: *Traceability* and *Quality*. The entities that allow tracing and tracking the product path is contained in the former while, in the latter, the information related to lot quality is included. The Quality package contains the abstract class *Quality Feature* (QF), which includes a description of the feature itself and a collection of methods to set and retrieve its values of either categorical (*CategoricalQF*) or numerical (*NumericalQF*).

2.3 System framework

Using the above models, the overall system framework should be constructed. The system architecture may vary based on which data model and, especially, standard it uses. Among the organizations that provide standardization of the RFID network, EPCglobal is the leading consortium which defines standards to enable data sharing of electronic product code related information within and between enterprises.

The EPCglobal Network (Fig. 5) promises a scalable, secure and easy to use architecture in which mechanisms to query traceable data are included. It describes components and interfaces for the RFID data interchange between servers called EPCIS (EPC Information Service) that contain information related to items identified by EPC numbers (EPCglobal, 2006). Each EPCIS contains the read time (at the reader location), capture time (at the EPCIS), reader location, action, and several optional attributes of all the EPCs of interest which can be specifically defined by companies (e.g., manufacturers). The standardization of the interface encourages the partners to provide vendor-specific implementations, which increase the scope of the traceability along the supply chain.

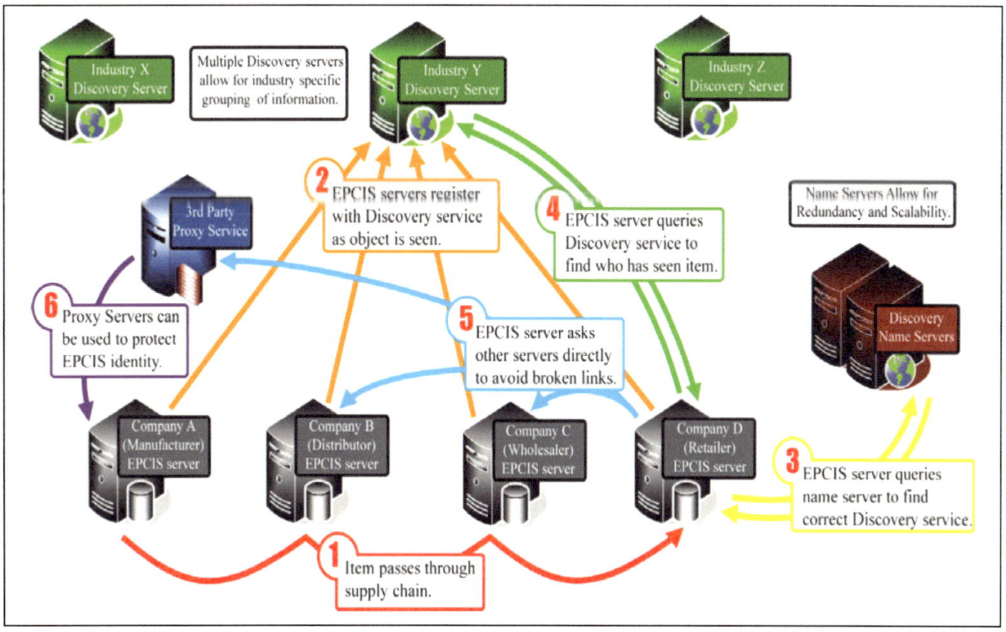

Fig. 5. EPCglobal network architecture (Beier *et al.*, 2006)

In the EPCglobal Network, Discovery Services are the core components to provide traceability by simplifying the data exchange process and by offering trading partners the ability to find all parties who had the possession of a given product and to share RFID events about that product (Beier *et al.*, 2006). As an instantiation of Discovery Services, EPCglobal proposed an Object Name Service which is based on the idea of what Domain Name Service is to the Internet (EPCglobal, 2005). However, unlike the Internet where domain addresses are freely available to anyone, EPC-related information need to have a privacy protection and so be shared selectively by the partners' agreement.

Discovery Services are consisted of a database and a set of web services interfaces (EPCglobal, 2006). The web service interfaces will allow an authorized firm to register EPCs and EPCIS URL links when the company has manufactured or been supplied with a new product. Then, authorized companies may be allowed to retrieve links to all EPCIS servers that contain events for a specific EPC (Beier *et al.*, 2006). The record attributes that Discovery Services store are as follows (Beier *et al.*, 2006):

- *EPC number* of the item
- *URL* of the EPCIS that submitted this record to indicate that it had a custody of the item
- *certificate* of the company whose EPCIS submitted the record
- *visibility*, a flag indicating if this record can be shared with anybody or merely with parties who submitted records about the same EPC
- *timestamp* when this record was inserted

2.4 Tracing methods

Several researchers proposed detailed mechanisms for traceability that take into account more specific characteristics occurring in every nodes of the supply chain. For their practicality, these approaches have the advantage over a general one and can be a promising candidate for being included in future standards.

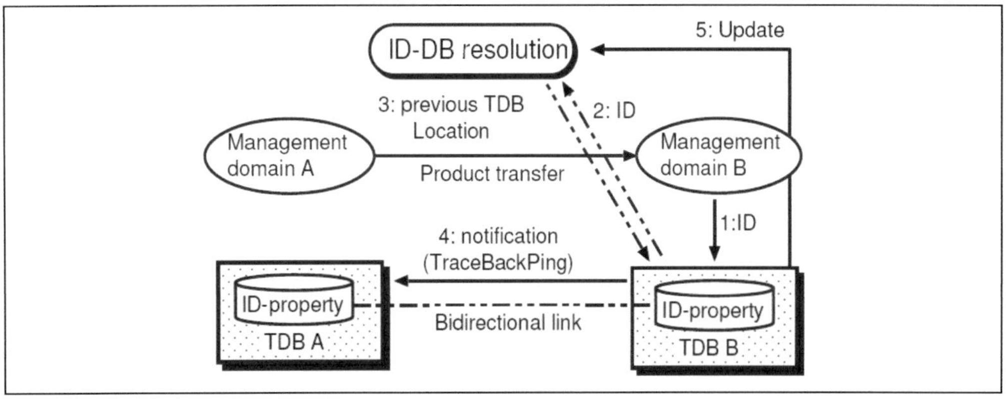

Fig. 6. TraceBack mechanism for traceability systems (Shirou *et al.*, 2007)

To be used in real applications, tracing methods should consider the peculiar characteristics of items as they traverse down the supply chain. Specifically, they must be able to track and trace not only the end items but also the components that compose them. Shirou *et al.* (2007) proposed a mechanism called "*TraceBack*" for traceability systems which uses a TDB (Traceability DataBase) and provides a tracing mechanism that are capable to reflect the product 'merge and division' mentioned above. TraceBack mechanism (Fig. 6) can be illustrated as the following flow of messages (Shirou *et al.*, 2007):

1. When the Management domain B receives the product, the TDB B in the Management domain B receives the ID of that product using an RFID reader.

2. The TDB B sends a query to the ID-DB (A mechanism which is used in the models proposed by standardization groups such as EPCglobal to find the database matching a specific ID) resolution mechanism to get the previous TDB location.
3. The ID-DB resolution mechanism sends back the last TDB location (TDB A).
4. The TDB B sends notification to the TDB A and creates a link to the TDB A. The TDB A receives the notification and creates a link to the TDB B.
5. The TDB B updates the location of the product in the ID-DB resolution mechanism.

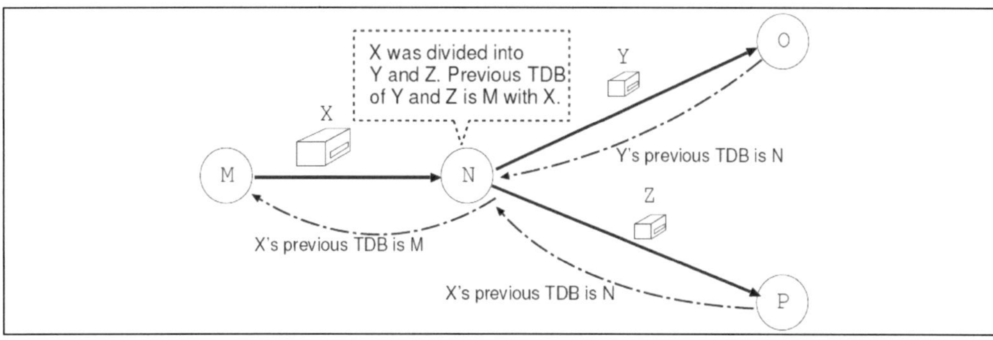

Fig. 7. Tracing mechanism for product division (Shirou *et al.*, 2007)

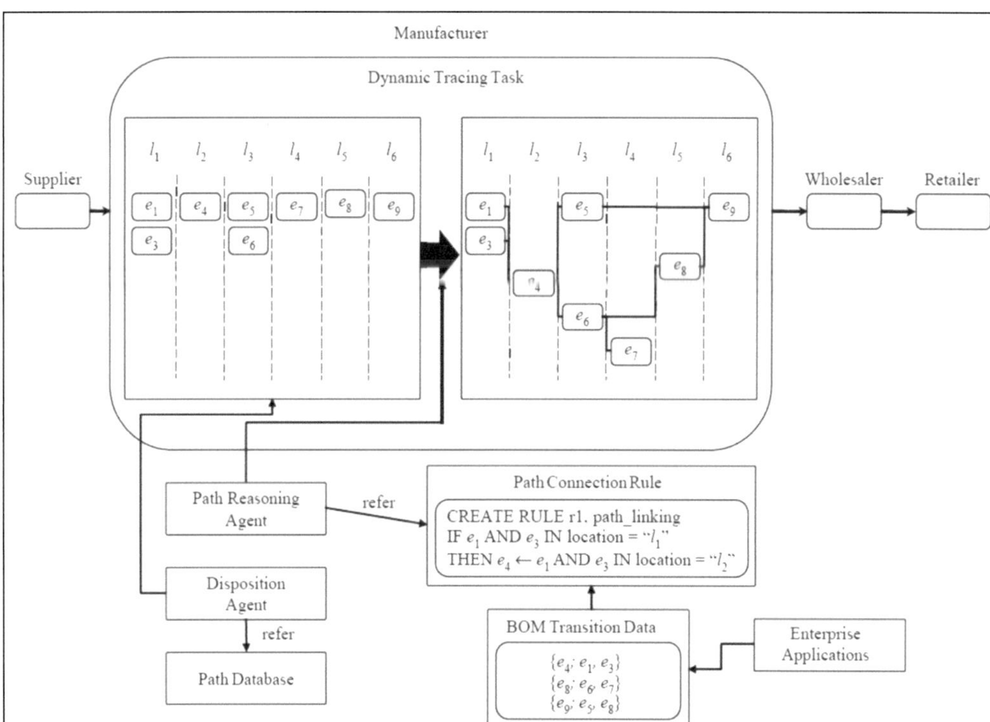

Fig. 8. An example of DTTM (Lee & Park, 2008)

Based on this model they suggested a mechanism regarding "divide and construct" of a product. Fig. 7 shows how this works for a product division (Shirou *et al.*, 2007). In this figure, product X is divided into product Y and Z. In this special case, TDB_N knows that product X was divided into two parts and transferred to TDB_O and TDB_P. After TDB_N receives a link message on Y and Z, TDB_N makes a response with link information on X. From this, the message sender can track product Y and Z correctly.

Lee & Park (2008) proposed a different model, DTTM (Dynamic Tracing Task Model) to get a full end-to-end traceability. The motive for their model and the previous one is almost the same in that they both did not disregard the critical feature of product "merge and split" process occurring in the supply chain. Fig. 8 shows an example of DTTM procedure. The distinctive advantage of this approach is that it uses the existing BOM (Bill of Material) for the tracing mechanism which can be adopted in an efficient way because of the fact that firms already have this information in their enterprise applications (ERP, SCM etc.). This will not only save the implementing cost of an RFID traceability system but also the effort and burden of changing the legacy system to adapt to it.

With the data model, system framework and tracing methods, managers who are in charge of tracking and tracing now have the ability to specify queries across the entire network. To illustrate traceability query execution in a network, a relational model is needed. Agrawal *et al.* (2006) called their relational model as the *global schema* (Fig. 9) and assumed that every participant in the supply chain has access to this information. This model allows organizations to make a query without knowing where and how data has been stored in the past. Physically, the global relations *Observed, ObsPropertySet, Assembled, AsmPropertySet, Disassembled, DsmPropertySet* are partitioned horizontally such that each partition belongs exclusively to each organization, and the attributes *parent* and *child* are object identifiers (*oids*) and the attribute *ts* is a timestamp (Agrawal *et al.*, 2006). As an example of traceability queries, a recall query has the following general form expressed in relational algebra (Agrawal *et al.*, 2006):

$$Qr: \pi_{lid}(max_{ts}(\sigma_{oid=o}(Observed)))$$

Recall queries are used to detect the current location of an object. They need to be executed at each node where there is an entry with *oid=o* in the respective partition of the relation *Observed* (Agrawal *et al.*, 2006).

Property(name)
Organization(gid)
OrgPropertySet(gid, propertyName, value)
Location(lid, parentLid, gid)
LocPropertySet(lid, propertyName, value)
Object(oid)
ObjPropertySet(oid, propertyName, value)
Observed(oid, lid, ts)
ObsPropertySet(oid, lid, ts, propertyName, value)
Assembled(parent, child, ts)
AsmPropertySet(parent, child, ts, propertyName, value)
Disassembled(parent, child, ts)
DsmPropertySet(parent, child, ts, propertyName, value)

Fig. 9. Relational schema (Agrawal *et al.*, 2006)

3. Applications

3.1 Food supply chain

In many countries traceability systems are mandatory for enterprises in the food chain. Based on the EU requirement, food business operators shall register the origin and the destination of the alimentary commodities they manage, and they shall label food to facilitate its tracking process (EC, 2002).
In the Italian cheese (Parmigiano Reggiano) supply chain several actors are involved including milk farm, dairy processors, and warehouses. The first step in this Parmigiano Reggiano traceability process is the identification of the characteristics of product in its different aspects along the entire supply chain (Regattieri *et al.*, 2007). The relevant information (such as cooking boiler temperature, humidity, and fodder batch code) is collected automatically using sensors, PLCs and barcodes but a sensor-augemented-RFID tag can also be used (Fig. 10).

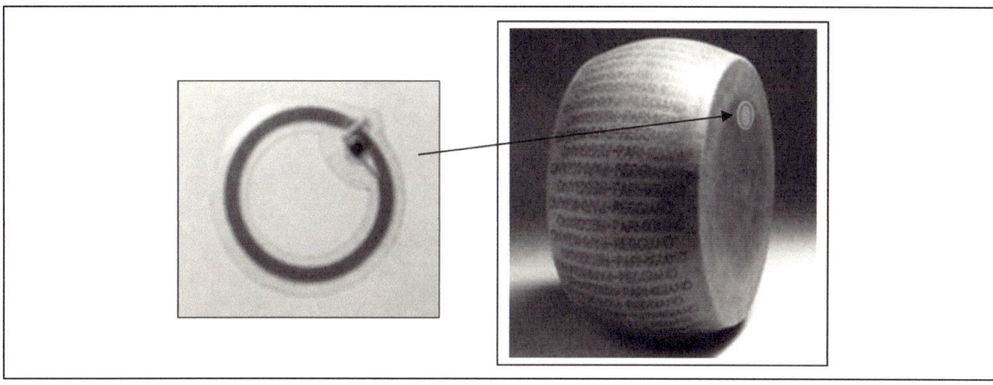

Fig. 10. A tag applied to cheese (Regattieri *et al.*, 2007)

Fig. 11. MTs at warehouses and Exposition in a supermarket (Martinez-Sala *et al.*, 2009)

Martinez-Sala *et al.* (2009), in their paper, showed the work developed in collaboration with a Spanish company ECOMOVISTAND in order to turn the MT (an innovative and ecological packaging and transport unit) into an intelligent product platform by embedding active RFID tags. MT is mainly made of last generation plastics with high mechanical endurance and an expected lifespan of more than 10 years. In their solution, an active RFID is the core element to achieve the business needs. MTs can be tracked by attaching an RFID tag. The RFID constraints posed by the ECOMOVISTAND business model is as follows (Martinez-Sala *et al.*, 2009):

- *Long range readings.* MTs are carried by forklifts and moved through large, open warehouses and dock gates (Fig. 11). Besides this capability is key to future service extensions, such as smart shelves (Fig. 11)
- *Monitoring of perishable goods.*
- *Full read/write capabilities.*
- *Ensuring reader a seamless integration into the customers facilities with minimum impact to their information systems.*

Depending on the type of RFID tags used (e.g., passive or active) the capabilities it can provide in the food supply chain may differ. Thus, companies in the network should collaborate in order to make the RFID system properly implemented.

3.2 Air baggage handling

RFID technology can also be used to enhance the ability for baggage tracking, dispatching and conveyance inside the airport. Due to the management efficiency and user satisfaction, RFID was deployed in the Beijing Capital International Airport (BCIA), the largest and most important airport in China.

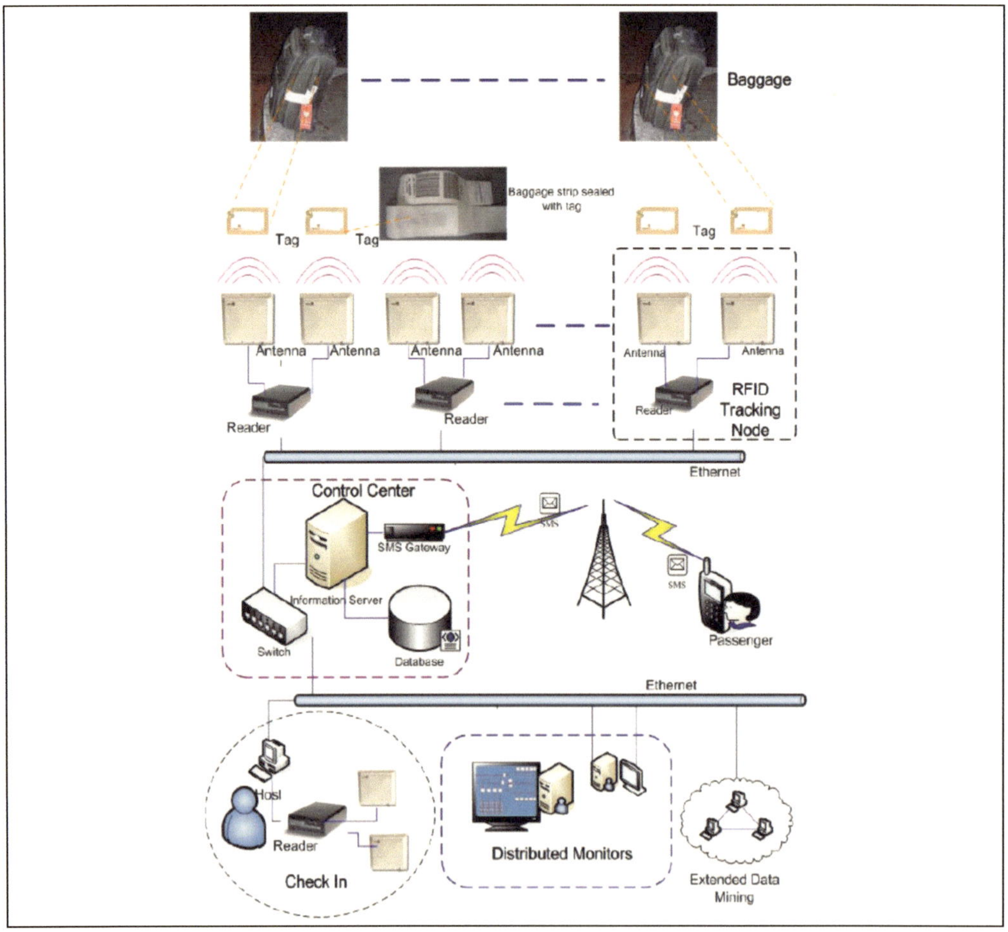

Fig. 12. RFID-based traceable air baggage handling system topology (Zhang *et al.*, 2008)

The system (Fig. 12) was developed by Zhang *et al.* (2008), which was expected to resolve the problems including delayed, lost, stolen, and not-located baggage. Their approach exploits not only RFID but also the existing barcode-based automated handling systems with less efforts and investments. The functionality of each component can be summarized as follows (Zhang *et al.*, 2008):

- *RFID tracking node* is consisted of one TCP/IP supported reader and a couple of fixed antennae transmitting RF signal
- *Control center* takes charge for the coordination of large-scale and heterogeneous readers
- *Information server* supplies the service interfaces and APIs to human interaction related applications by means of Web Services
- *Distributed Monitors* display the query results in the form of GIS map

The application was implemented inside the Airport Terminal 2. The LED screen was installed above each conveyor exit and it displays the amount of the sorted and sorting baggage of the flight. According to the statistics from Air China between March 19, 2005 and April 18, 2006, RFID saved costs related to lost baggage, baggage damage, and temporary costs. Also it has been expected to save the industry US $760 million annually upon full implementation. Moreover, the processing cost has been decreased due to the boosted productivity with a more automatic workflow provided by RFID (Zhang *et al.*, 2008).

3.3 Iron ore distribution chain

Traceability is common in part production and often easy to achieve by using RFID. However, obtaining traceability information in continuous processes implies many challenges. In a continuous process, the products are gradually refined through a series of operations and there are no natural product lots (Fransoo & Rutten, 1994; Dennis & Meredith, 2000). Moreover, the physical characteristics of the material are often changed during the refinement process, which also makes it difficult to place the definition of a unit of measure (Fransoo & Rutten, 1994). Due to these characteristics, RFID alone cannot give the ability to trace in the continuous process. Instead, various tracer methods (e.g., off-line methods or on-line methods) are used to support the whole distribution chain of a flow process.

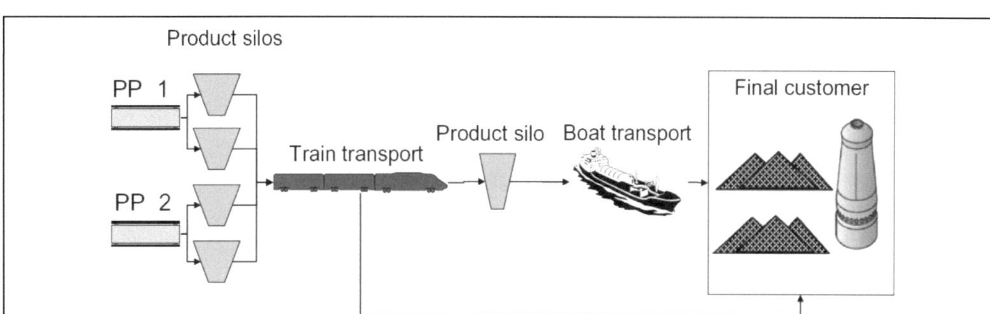

Fig. 13. Flow chart of the iron ore pellets distribution chain (Kvarnstrom & Oghazi, 2008)

Nevertheless, RFID may be very promising in one part of the whole supply chain, the distribution chain. Fig. 13 shows a flow chart of the distribution chain for iron ore pellets produced in Malmberget, Sweden. Together with traceable unit method which creates virtual batches by dropping some type of marker in the material flow with regular intervals,

RFID offers the possibility to measure precise residence time in the process. The RFID tags would be used as the start and the end points of each batch. And process data can then be linked to a virtual production batch (Kvarnstrom & Oghazi, 2008). From the test results in the iron ore distribution chain, in order to achieve a sufficient read rate (more than 50%), it is necessary to use RFID tags that are larger that the pellets (Kvarnstrom & Oghazi, 2008). However, they proved that their technique may be used to create traceability in the continuous process chain.

4. Challenges

There are several challenges in deploying RFID for traceability. One of the main challenges comes from the inherent technical nature of RFID. RFID tags may collide with each other when multiple tags try to respond to the reader simultaneously. Thus, there is no guarantee that a tag will be continuously detected on consecutive scans (Floerkemeier & Lampe, 2005). Given the fact that it does not provide 100% read rate, most of the data models and tracing mechanisms to provide an efficient way for traceability would almost be useless. However, lots of efforts poured by engineers have increased the level to almost 99% and in the near future the technology will be matured to the level of the existing barcode system which has no problem of reading items.

RFID tags are subject to price and size constraints. Although the tag price is continuously decreasing companies are hesitant to use RFID because of the price disadvantage over barcode. Since tags derive energy from readers (in the case of passive tags) their communication bandwidth is limited (Zarolostas *et al.*, 2007). In addition, due to their small size requirements, tags have limited memory which indicates that the stored information on tags is also limited (Sarma *et al.*, 2001).

Another important facet of RFID-enabled traceability is that the amount of information generated by the infrastructures as items pass through the supply chain is vast. This leads to the need of heavy filtering and aggregation so as to generate meaningful events or information that could impact the operational procedures and business strategy (Zarolostas *et al.*, 2007). Middleware which is responsible for this filtering should be carefully designed so that the trade-offs between communication overhead and processing capabilities have the desirable value to the firms.

As each node in the network would implement the system produced by different RFID vendors the information flow is likely to be heterogeneous. One solution to this problem might be the establishment of standards. However, the standardization of RFID in all aspects is still in progress. Therefore researchers are trying to propose methods that can seamlessly make various types of information interoperable. Also two groups, EPCglobal and ISO are the leading organizations that are expected to establish the de facto standards for RFID, so most of the vendors make their products in accordance to the standards they provide.

Since traceability can only be obtained through all the participants' cooperation along the supply chain, privacy issues have been reported by several practitioners and governments. In US, many organizations, such as Consumer Privacy and Civil Liberties Organizations, are requesting attention to privacy threats (Privacy Rights Clearinghouse, 2003). In Canada, the Annual Report to Parliament 2005 of the Privacy Commissioner underlines the importance to make sure that RFID does not harass informational privacy rights (Privacy Commissioner of Canada, 2005). In European Union, the national authorities established guidelines needed for a safe use of RFID in compliance with the Working Document adopted by 2005 by the European Data Protection Working Party (Data Protection Working Party, 2005). Due to the privacy threats made by RFID (e.g., personal belongings monitoring and industrial espionage) many solutions have been analyzed (Bernardi et al., 2007):

- *Killing the tag*: a command that kills the tag at the POS (point-of-sale)
- *Using passwords or encryption*: To avoid unauthorized readings of tags (Weis *et al.*, 2003)
- *Changing tag ID*: Using different IDs makes difficult to recognize a tag (Juels, 2004)
- *Blocking the anti-collision system of the reader*: A special tag stops the correct functioning of the reader (Juels *et al.*, 2003)

Recently, researchers are considering RFID as a device to develop "intelligent" products, and when combined with other sensors, it is expected to provide capabilities far beyond mere identification function of today's RFID (Zuehlke, 2008). This means that they record all pertinent product and operating data and exchange this information with all participants as well as other products and environments to enable the so-called "Internet of things." However, to make this feasible, the memory capacity of an RFID chip should be large enough and this will incur more cost. Thus, in designing tags for the future scenario, an efficient storage scheme should not be overlooked.

5. Conclusion

RFID technology gives several advantages for enterprises to effectively operate their daily operations (Lee & Park, 2008). One of the benefits is the product traceability. If a product can be tracked and/or traced, product recall, product pedigree analysis and consumer visibility can be improved. Design and implementation of traceability systems need preliminary investigations to point out not just requirements but also problems and solutions at different abstraction levels (Bechini *et al.*, 2008). In order to achieve coordination along the supply network, appropriate and widely accepted data models and system framework should be constructed before the actual tracing query is made.

In spite of the promising nature of RFID, numerous applications in the actual supply chain has not been reported. Only few pilot studies as well as experimental tests have been proved that RFID would be a successful tool to enable supply chain traceability. The reasons why companies are yet reluctant to have confidence in adopting the technology to gain their product visibility may be attributed to the several challenges such as lack of standards, immaturity of RFID, and privacy issues.

Researchers as well as practitioners are proposing methodologies that address those current problems. Also more and more opportunities from RFID are being investigated. Thus in spite of the obstacles facing RFID, the fact that the technology promises to revolutionize the way we track items in the supply chain will not likely to change.

6. References

Agrawal R.; Cheung, A.; Kailing, K. & Schonauer, S. (2006). Towards traceability across sovereign, distributed RFID databases, *Database Engineering and Applications Symposium, 10th International*, pp. 174-184, 0-7695-2577-6, Dec. 2006

Bechini, A.; Cimino, M.G.C.A. & Tomasi, A. (2005). *IFIP International Federation for Information Processing*, Springer Boston, 978-0-387-28753-9

Bechini, A.; Cimino, M.G.C.A.; Marcelloni, F. & Tomasi, A. (2008). Patterns and technologies for enabling supply chain traceability through collaborative e-business. *Information and Software Technology*, Vol. 50, pp. 342-359, 0950-5849

Beir, S.; Grandison, T.; Kailing, K. & Rantzau, R. (2006). Discovery Services - Enabling RFID traceability in EPCglobal networks. *International Conference on Management of Data*, Delhi, India, December 2006

Bernardi, P.; Demartini, C.; Gandino, F.; Montrucchio, B.; Rebaudengo, M. & Sanchez, E.R. (2007). Agri-food traceability management using a RFID system with privacy protection. *International Conference on Advanced Networking and Applications*, 0-7695-2848-5/07, 2007

Brock, D. & Cummins, C. (2003). EPC™ Tag Data Specification. http://www.autoidlabs.com/whitepapers/MIT-AUTOID-WH025.pdf, June 2003

Data Protection Working Party (2005). Working document on data protection issues related to RFID technology, Article 29, *Working Party on the Protection on Individuals with regard to the Processing of Personal Data*, Jan. 2005

Dennis, D. & Meredith, J. (2000). An empirical analysis of process industry transformation systems. *Management Science*, Vol. 46, No. 8, pp. 1085-1099

EC (2002). Regulation (EC) No 178/2002 of the European Parliament and of the Council of 28 Article 18. *Official Journal of the European Union*

Engels, D. (2003). The use of the Electronic Product Code™. http://www.autoidlabs.com/whitepapers/mit-autoid-tr009.pdf, February 2003

EPCglobal (2005). Object Naming Service (ONS) Version 1.0, ratified version. October 4, 2005

EPCglobal (2006). EPC Information Services (EPCIS) Version 1.0 Specification, working draft. June 8, 2006

Floerkemeier C. & Lampe, M. (2005). RFID middleware design addressing application requirements and RFID constraints, *ACM International Conference Proceedings of the 2005 joint conference on Smart objects and ambient intelligence*, Vol. 121, pp. 219-224, 2005

Fransoo, J.C. & Rutten, W.G.M.M. (1994). A typology of production control situations in process industries. *International Journal of Operational and Production Management*, Vol. 14, No. 12, pp. 47-57

Gan, O.P.; Zhao, Y.Z.; Luo, M.; Zhang J.B. & Zhou, J.H. (2006). System architecture and data design for RFID-based item level track and trace, *Proceedings of the 1st World Congress on Engineering Asset Management (WCEAM)*, pp. 1098-1103, 978-1-84628-814-2, July 2006, Springer London

Gandino, F.; Montrucchio, M.; Rebaudengo, M. & Sanchez, E.R. (2007). Analysis of an RFID-based information system for tracking and tracing in agri-food chain. *RFID Eurasia*, pp. 1-6, 978-975-01566-0-1, Sept. 2007

Hu, Y.; Sundara, S.; Chorma T. & Srinivasan, J. (2005). Supporting RFID-based item tracking application in oracle DBMS using a bitmap datatype, *Proceedings of the 31st VLDB Conference*, pp. 1140-1151, 1-59593-154-6, Trondheim, Norway, 2005

Jansen-Vullers, M.H.; van Dorp, C.A. & Beulens, A.J.M. (2003). Managing traceability information in manufacture. *International Journal of Information Management*, Vol. 23, pp. 395-413

Juels A. (2004). Minimalist cryptography for RFID tags, *4th Conf. Security in Comm. Networks (SCN)*, Springer-Verlag, pp. 149-164, 2004

Juels A.; Rivest, R. & Szydlo, M. (2003). The blocker tag: Selective blocking of RFID tags for consumer privacy, *8th ACM Conf. Computer and Comm. Security*, ACM Press, pp. 103-111, 2003

Kelepouris, T.; Pramatari, K. & Doukidis, G. (2007). RFID-enabled traceability in the food supply chain. *Industrial Management & Data Systems*, Vol. 107, No. 2, pp. 183-200, 0263-5577

Kim, H.M.; Fox, M.S. & Gruninger, M. (1995). Ontology of quality for enterprise modelling, *Proceedings of IEEE International Workshop on Enabling Technologies: Infrastructures for Collaborative Enterprises*, pp. 105-116, Los Alamitos, CA, USA 1995

Kvarnstrom, B. & Oghazi, P. (2008). Methods for traceability in continuous processes - Experience from an iron ore refinement process. *Minerals Engineering*, Vol. 21, pp. 720-730, 0892-6875

ISO, E.S. (1995). EN ISO 8492.1995. *European Committee for Standardization*, Point 3.16

Lee, D. & Park, J. (2008). RFID-based traceability in the supply chain. *Industrial Management & Data Systems*, Vol. 108, No. 6, pp. 713-725, 0263-5577

Martinez-Sala, A.S.; Egea-Lopez, E.; Garcia-Sanchez, F. & Garcia-Haro, J. (2009). Tracking of returnable packaging and transport units with active RFID in the grocery supply chain. *Computers in Industry*, Vol. 60, pp. 161-171, 0166-3615

Moe, T. (1998). Perspectives on traceability in food manufacture. *Trends in Food Science and Technology*, Vol. 9, pp. 211-214

Ministry of Agriculture, Forestry and Fisheries of Japan (2003). Guidelines for Introduction of Food Traceability Systems. http://www.maff.go.jp/trace/guide_en.pdf, 2003

Perez-Aloe, R.; Valverde, J.M.; Lara, A.; Carrillo, J.M.; Roa, I. & Gonzalez, J. (2007). Application of RFID tags for the overall traceability of products in cheese industries. *RFID Eurasia*, pp. 1-5, 978-975-01566-0-1, Sept. 2007

Privacy Commissioner of Canada (2005). Annual report to parliament 2005 - Report on the personal information protection and electronic documents act, *Privacy Commissioner of Canada*, pp. 39-42

Privacy Rights Clearinghouse (2003). RFID position statement of consumer privacy and civil liberties organizations, *Privacy Rights Clearinghouse*, Nov. 2003

Regattieri, A.; Gamberi, M. & Manzini, R. (2007). Traceability of food products: General framework and experimental evidence. *Journal of Food Engineering*, Vol. 81, pp. 347-356, 0260-8774

Sarma, S.; Brock, D. & Engels, D. (2001). Radio frequency identification and the electronic product code. *IEEE Micro*, Vol. 21, No. 6, pp. 50-54

Shirou, W.; Yusuke, D.; Satoshi, O. & Atsushi, I. (2007). Cost-effective product traceability system based on widely distributed databases. *Journal of Communications*, Vol. 2, No. 2, pp. 45-52

van Dorp, C.A. (2003). A traceability application based on gozinto graphs, *EFITA 2003*, Debrecen, 2003

Weis S.A..; Sarma, E.S.; Rivest R.L. & Engels, D.W. (2003). Security and privacy aspects of low-cost radio frequency identification systems, *Security in Pervasive Computing*, 2003

Zarokostas N.; Dimitropoulos, P.D. & Soldatos, J. (2007). RFID middleware design for enhancing traceability in the supply chain management, *18th Annual IEEE International Symposium on Personal, Indoor and Mobile Radio Communications*, 1-4244-1144-0/07, 2007

Zhang, T.; Ouyang, Y. & He, Y. (2008). Traceable air baggage handling system based on RFID tags in the airport. *Journal of Theoretical and Applied Electronic Commerce Research*, Vol. 3, No. 1, pp. 106-115, 0718-1876

Zuehlke, D. (2008). SmartFactory - from vision to reality in factory technologies. *Proceedings of the 17th World Congress, The International Federation of Automatic Control*, Seoul, Korea, July, 2008.

5

Applications of RFID to Improve Traceability in Continuous Processes

Björn Kvarnström[1] and Johan Oja[2]
[1]*Division of Quality and Environmental Management, Luleå University of Technology,*
[2]*SP Trätek, SP Technical Research Institute of Sweden, Sweden and Division of Wood Technology, Luleå University of Technology, Sweden*

1. Introduction

To achieve traceability in discontinuous (discrete or batch) processes can be rather straightforward since identification markers, such as bar codes or serial numbers, can be attached to the product/batch or its package and be scanned at predetermined locations in the supply chain. However, continuous processes differ from discontinuous processes in a number of ways and these differences results in a number of difficulties concerning, for example, traceability. Traceability is important in continuous processes and the importance of traceability has been highlighted in, for example, the food and dairy industries due to food crises, such as, the BSE (mad-cow disease), foot-and-mouth disease, and the scandal of infant milk tainted with Melamine in China.

This chapter presents a number of traceability methods that can be used to improve traceability in continuous processes with a special focus on RFID (Radio Frequency IDentification). We then exemplify how RFID may be applied and combined with other methods to improve traceability in continuous processes with two examples from continuous processes. The first example demonstrates how RFID can be used to improve traceability in a continuous refinement process of iron ore by adding transponders with similar physical characteristics as the product to the product flow. The second example is taken from the forestry industry and gives an example of how RFID can be used to improve traceability through the wood refinement process by tagging individual logs with a transponder at the felling site.

In addition to the two examples additional RFID applications to improve traceability in continuous processes are described. Various benefits of using RFID to improve traceability in continuous processes are identified and discussed from the described RFID applications. Also, special challenges using RFID in continuous processes are identified and how to solve or avoid these challenges are discussed. Finally, we describe some ideas for future research projects and ongoing research projects.

2. Traceability in continuous processes

All production processes occasionally fail to produce products within specifications. Some defects will not be detected before the product is sold and a product recall may be necessary. A product recall results from an incident or an attempt to prevent an incident and the cost for a product recall can be astronomic and escalates dramatically, see, for example, Jacobs (1996). Traceability is an important tool to minimize the extent of product recalls, see, for example, Jacobs and Mundel (1975) and Fisk and Chandran (1975). Also, traceability is vital for continuous improvements (Mahoney and Thor, 1994), important for identification of root-causes and prevent their reoccurrence (Duffin, 1995), essential if effective methods of process control are to be applied (Oakland, 1995), et cetera. Hence, traceability is an important ability for all manufacturers (Fisk & Chandran, 1975).

In this chapter traceability is seen as the ability to *"preserve and access the identity and attributes of a physical supply chain's object"* (Töyrylä, 1999, p. 38). According to Moe (1998), traceability needs to be managed by traceability systems. A traceability system is the system that enables traceability in a process by combining process information with data covering the product flow throughout the supply chain. The product flow data can be continuously recorded or modelled with different traceability methods.

The research literature concerning traceability is dominated by applications in discontinuous processes, while descriptions of applications in continuous processes are scarce. Continuous processes can be described as production processes which primarily schedule long production runs where setups are fixed or seldom changed and the production systems are organized and sequenced according to a fixed production flow. Continuous processes are commonly found in process industries, and characterized by that the products are refined gradually and with minimal interruptions through a series of operations (Fransoo & Rutten, 1993; Dennis & Meredith, 2000). There are a number of characteristics in continuous processes that makes traceability complicated. For example, traceability systems normally rely on the definition of a batch, and the continuous flow in a continuous process implies that no batches exists (Skoglund & Dejmek, 2007). Also, the products in continuous processes are often in a non-discrete state and can undergo multiple changes of states prior to the delivery (Fransoo & Rutten, 1993; Dennis & Meredith, 2000). Another traceability concern is that the product flow in the sub-processes can be continuous as well as batch-wise (Lundqvist & Kubulnieks, 1995). Finally, the flow in continuous processes is not necessarily sequential, as the processes are typically designed to separate, mix, form or react (Hild et al., 2000). Yet, traceability is important in continuous processes and especially in the food and pharmaceutical industries, see Flapper et al. (2002). The special characteristics described imply that other types of traceability methods are needed for creating traceability in continuous processes compared to discontinuous processes.

3. Traceability methods for continuous processes

Kvarnström and Oghazi (2008) have identified five traceability methods that can be used in continuous processes to improve traceability as well as advantages and disadvantages with each method, see Table. 1. We will here give a brief description of each method with a focus on RFID, for further descriptions see Kvarnström and Oghazi (2008). However, it should be noted that the same traceability method is rarely suitable to use throughout the whole process, since, for example, the material often change state during the process. Hence, the suitability of a traceability method needs to be considered for each specific process stage.

Traceability method	Advantages	Disadvantages
Chemical tracer	- Flexible - Easy to use - Low-cost	- Dilutes - Needs sampling - Based on historical data
Radioactive tracer	- Flexible - No sampling needed - Interior flows can be measured	- Health hazards - Permits required - Based on historical data
Process data	- Easy to use - Low-cost - Based on real time data	- Hard to find - Low precision - Initial sampling needed
Material signature	- Flexible - Informative - High analysis precision	- Large amount of data handling - Time demanding - Costly
Traceable marker	- High precision - No sampling needed - Could be used in process sections with both batch and continuous flows	- Lower flexibility - Can not be used for fluids - Can only be used at shorter distance - Costly to implement

Table 1. Various traceability methods that can be used to improve traceability in continuous processes. Adapted from Kvarnström and Oghazi (2008, p. 729).

3.1 Off-line tracer methods

Off-line tracer methods imply that mathematical models are created for estimating the residence time of the products in a process section based on data from, for example, chemical or radioactive tracer experiment. Such experiments are usually performed the following way, a temporary modification is made to the input of a process section or stage and the effect of the modification are then studied in the output. For example, a chemical tracer can be added momentarily to the input section and the concentration of the tracer can then be measured at various times in the output to estimate the residence time of the tracer. Experiments to estimate the residence time distribution are often performed at various process settings to determine how the residence time distribution depend on different factors. A mathematical model can then be fitted to the estimated residence time distribution, assuming that the process conditions are known and can be quantified. The tracer method to estimate residence time distribution in a process section or stage has, for example, been used by Yianatos et al. (2001), Lelinski et al. (2002), Choi et al. (2004), and De Andrade Lima and Hodouin (2005). Also, data from product changes may be used to estimate the residence time distribution based on the transition time (the time it takes for a process to react on a change) in a process. Vanhatalo et al. (2009) describes and exemplifies a method were principal component analysis and time series analysis are combined to estimate the transition time in a continuous process based on data from the input and output of a process.

3.2 Process data

For some continuous processes there may exist known and repeatable changes in the raw material properties. These changes can often be seen at different positions in the process, and the residence time distribution may be estimated based on the propagation of the changes and then mathematically modelled or continuously monitored. Lundqvist and Kubulnieks (1995) have created traceability in a paper and pulp production industry by comparing the appearance time and forms of deviations in kappa number (a measure indicating the bleach ability of wood pulp) and brightness (a measure of how much light is reflected), two process parameters normally continuously measured at various positions in the process. Kvarnström and Bergquist (2009) describe a similar method using simulations based on ideal flows and existing process knowledge to improve traceability in a continuous process.

3.3 Material signature (fingerprint)

Instead of using knowledge of the process or process data, knowledge of the material may be used. Like all humans have unique DNA and fingerprints, similar unique signatures may be found in the structure of other materials, such as meat, grain, wood, and ore. For example, in a pork chop the exact amount and combination of chemical elements may depend on the origin, nourishment, soil, birth date and other variables. Oghazi et al. (2009) has suggested and exemplified how particle texture analysis and multivariate data analysis can be combined to identify and follow changes in particle morphology in the concentration process of iron ore. A similar approach for the sawmilling industry has been proposed by Flodin et al. (2008) using 3-D data, x-ray data, and discriminant analysis to match planks to logs based on, for example, knots.

3.4 Traceable marker (or unit)

Different types of markers are commonly used in discontinuous processes to mark either individual units or batches. The same approach may be used in continuous processes. Marking technique such as paint label, stamped codes, paper or plastic label, magnetic stripe card or smart card, RFID, microtaggant paint, and chemical tracer may for example be used, for an extensive list of marking techniques with descriptions of each technique see, for example, Dykstra et al. (2002). To mark all units may be too expensive or even impossible in continuous processes, and a batch approach may be a more appropriate solution. If the marking are to be based on a batch system, the flow needs to be divided into subgroups and, hence, the continuous process flow is problematic. A solution is to use the markers to divide the flow into batches, and the markers would then work as start- and endpoints of each batch. To do this the markers need to behave as the product in the process flow and to be identifiable within the product flow.

RFID transponders offer a possibility to mark a flow with markers that are automatically readable and the markers can be designed to have similar flow behaviour as the product. Also, it is possible to see exactly where each transponder is since the transponders has a unique identification number. Hence, RFID transponders can be used both to mark individual units and to create batches in continuous flows. Henceforth, the chapter will focus on the use of RFID to improve traceability in continuous processes. A number of examples describing the use of RFID in continuous processes are presented with a special focus on two examples, one in which RFID has been used to create batches and one in which RFID has been used to mark each unit.

4. Example 1 –Tracing iron ore pellets

The first example is taken from the refinement process of iron ore to iron ore pellets (henceforth pellets) and the distribution of the pellets from producer to customer. First, the studied process will be described as well as why it is important to achieve traceability in the process. Secondly, the design of the RFID application used to improve traceability in the process will be described and tested. Thirdly, results and experiences of various experiments investigating the possibility to use transponders to trace the movement of pellets in the distribution chain are described. Finally, on-going and future development of RFID for the pelletising process is described.

4.1 The production process of iron ore pellets

Luossavaara-Kiirunavaara AB (LKAB) is a Swedish iron ore mining company that extracts and refines iron ore from deposits in northern Sweden (Kiruna, Malmberget and Svappavaara). LKABs main product is various types of pellets used for iron making in blast furnaces and direct reduction furnaces. A pellet is a spherical particle created from iron ore concentrate with diameters ranging from 8-18 millimetres. The production process of pellets at LKAB is performed in four consecutive steps. First, crude iron ore is extracted from underground mines through blasting and drilling and hoisted to surface level. Secondly, the iron ore concentrate is separated from the tailing (gangue minerals) in the crude ore by crushing and magnetic separation. Thirdly, the iron ore concentrate is further refined by grinding and wet magnetic separation. Fourthly, the refined iron ore concentrate is transformed to pellets through agglomeration and thermal treatment. The pellets produced at LKAB are then shipped to the customers from two different harbours one in Luleå (pellets produced in Malmberget) and one in Narvik (pellets produced in Kiruna and Svappavaara). The distribution chain for the two harbours differs on a detailed level, but both distribution chains consists of a buffer system at the pelletizing plant where the pellets are stored before transportation, a train transport to the harbour, a buffer system at the harbour where the pellets are stored before shipment, and a shipment to the customers, see Figure 1. The RFID technique has been suggested to be used to improve traceability in the distribution chain, while other techniques has been suggested for the other parts in the production process, see Kvarnström and Oghazi (2008).

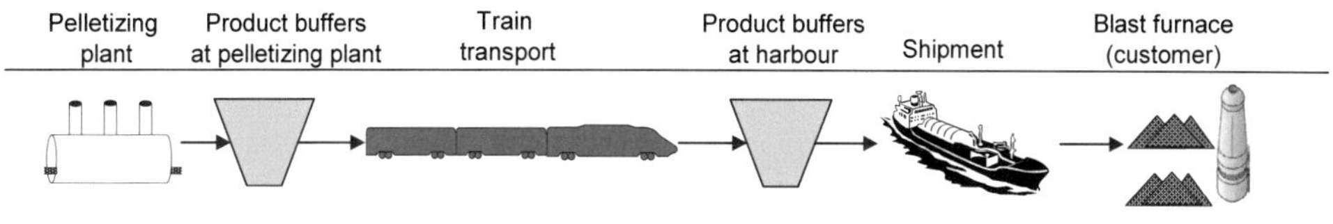

Fig. 1. The distribution chain of iron ore pellets at LKAB.

Improving the traceability in the distribution chain at LKAB would result in a number of benefits. For example, the knowledge of the flow mechanisms in the distribution chain would be improved and it would be possible to combine product analysis from different stages in the distribution chain. Also, traceability would make it easier to manufacture customer specific pellets. Perhaps most importantly, an improved traceability would make it easier to track customer complaints or praises back to the production process as well as trace pellets with quality deviations for disposal or downgrading.

4.2 Design of RFID application

In the pelletizing process the RFID application was intended to improve traceability in the distribution chain of iron ore pellets. To improve traceability, RFID transponders were to be added between the pelletizing plant and the product buffers at the pelletizing plant. The transponders were to be added directly into the product flow to create batches and indicate start and end point of each batch. The batch approach was selected because the individual pellet value is to low to motivate that each pellet was marked with a transponder. A transponder type with similar size as a pellet was necessary since the transponders need to behave as the pellet in the product flow. Also, passive transponders were to be used since no data except identification number needs to be transferred between the transponder and the reader. Furthermore, the transponders can not be removed from the product flow and, therefore, need to be harmless for later processes and approved by external customers.

The readers were decided to be placed at conveyors, used for transporting pellets, at different positions in the process, since the conveyors are the process section were the smallest read zones can be applied. The conveyors used in the process have a width between 1.2-2 meters and a height of approximately 0.5 meters. When lying on the conveyors the transponders moves with a speed of 3 meters/second and the orientation of the transponders and the distance between transponders are completely random on the conveyors. There was no water and metal within the read zone, but there was some metal in the proximity. Also, the reader needed to be designed to be insensitive to dust and large temperature shifts (between ±50 degrees of Celsius).

The prerequisites for the design of the RFID application imply that not all transponders were expected to be read. If a transponder is not read, data are lost and information about the change of batch is not received. To secure that information about the start and end point of each batch will be received several transponders are to be dropped simultaneously. Also, multiple RFID readers can be used at a single process stage to further reduce the risk of missing information about the passage of a batch. The readers would then be mounted with different orientation against the conveyor and with a distance large enough to avoid interference between readers but close enough to avoid that the transponder shift orientation. While moving through the distribution chain the pellets are extensively mixed during, for example, buffering and transports, and the batch boundaries will blur. Hence, in case of a disturbance within a batch resulting in that an action must be performed the action needs to be performed to the actual batch and the adjacent mixing zones, see Figure 2 for illustration. Furthermore, if the disturbance that needs an action occurs within a mixing

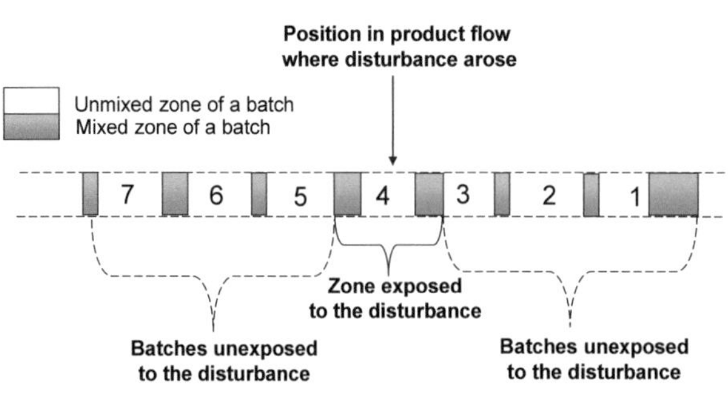

Fig. 2. Illustration of how a disturbance in the product flow within a batch affects other batches.

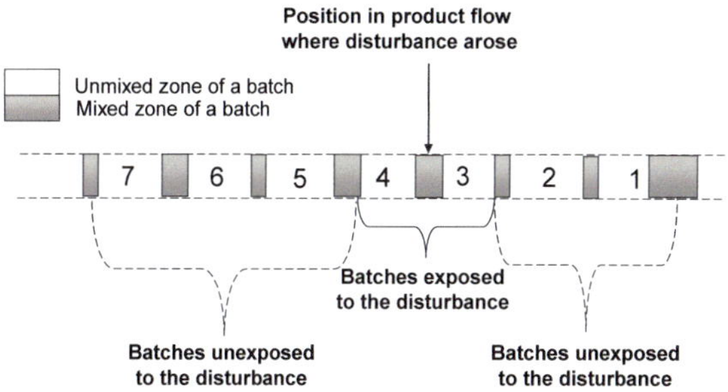

Fig. 3. Illustration of how a disturbance in the product flow within the mixed zone of a batch affects other batches.

zone the preceding and the following batch may be exposed to the same disturbance, see Figure 3 for illustration. When deciding the batch size for a traceability system the mixing zone must be known, since the selected batch size must exceed the size of the mixing zone. The mixing zone can be estimated by studying the residence time distribution for transponders added simultaneously.

4.3 Results from tests with the RFID application

To identify suitable antenna designs and transponder types for a traceability system in the distribution chain of pellets, laboratory tests were performed. The laboratory tests showed that RFID may be possible to use and that a low frequency (125 kilo Hertz) 12 millimetres long glass transponder with a 2.12 millimetres in diameter, see Figure 4, could be used, but the read rate was low (around 50 %) with the established read zone. Based on the results of the laboratory tests it was decided that initial tests in the distribution chain should be performed. For the tests in the distribution chain a longer glass transponders (22 millimetres long and 4 millimetres in diameter), with higher read rates anticipated, were decided to be tested in addition to the 12 millimetres transponders. Furthermore, the temporary reader antenna was to be mounted wrapped around the conveyor. The tests in the distribution chain indicated a decrease in read rate compared to the laboratory tests. Also, the tests showed that the 22 millimetres transponders had a better read rate than the 12 millimetres transponders and that it could not be rejected that the 22 millimetres transponders behaved as the 12 millimetres transponders in the product flow. For a closer description of these tests see Kvarnström and Vanhatalo (2010).

Fig. 4. The smallest glass transponder, 12 millimetres long and 2.12 millimetres in diameter, used for the tests.

After the initial tests a second set of tests were performed in the distribution chain with two readers, see Figure 5, and the same two types of transponders. The read rate for the tests is summarized in Figure 6. For the 12 millimetres transponders Reader 2 had a higher read rate, but for the 22 millimetres transponders no clear difference could be seen between the two readers. The read rate for the 22 millimetres transponders were higher than the read rate for the 12 millimetres transponders, which was expected. By using two readers with different antenna orientation against the conveyor the overall read rate was improved from 60 % to 80 % for the 22 millimetres transponders.

Fig. 5. The position of the two RFID readers used for the second set of tests. Reader ones antenna is mounted under the conveyor but bend slightly as a saddle so the distance between the conveyor and the antenna was minimized. Reader twos antenna is mounted around the conveyor.

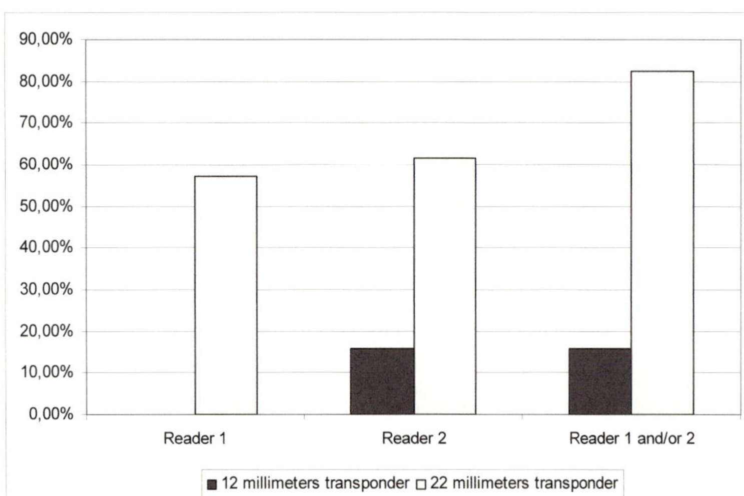

Fig. 6. The observed read rate for the two types of transponders tested at reader 1, reader 2 and reader 1 and/or reader 2.

4.4 On-going and future development of RFID technology for the pelletising process

The tests made within the distribution chain of iron ore pellets have shown promising results. However, there are still a number of questions that needs to be answered and solved before a final judgment about the suitability to use RFID to improve traceability can be made. A research project, within the Mining Research Programme at VINNOVA, aiming to further investigate the possibility to use RFID to trace granular products has been granted. The research project will, for example, address the possibility to develop RFID transponders that behave as a specific granular product.

5. Example 2

The second example is taken from the sawmill industry, following the chain from felling of trees in the forest until the sawn wood (boards) arrives at the secondary manufacturer (Figure 7). The structure of the chapter follows the same structure as Example 1. First a description of the sawmill process and the potential of using traceability technology, secondly a description of the RFID-application followed by some results, experiences and descriptions of future work and development.

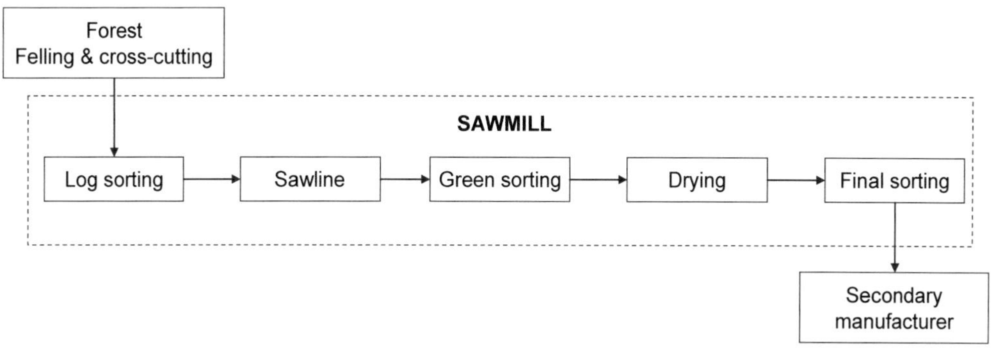

Fig. 7. A schematic description of the production of sawn wood.

5.1 The sawmill process

Wood is a biological material. Every stem, log and board has its unique individual properties. One effect of this is that while most industrial processes have a convergent flow, the sawmill industry has a divergent flow.

The production of sawn wood starts in the forest (Figure 7). In the Nordic forest industry an important decision is made already in the forest when the harvester cuts the stem to logs with different length. The cross-cutting of a stem results in a number of logs, each with individual properties and consequently best suited for different products. The individual logs are transported to different sawmills, and after arriving at the sawmill the logs are sorted and allocated to different groups of logs. The sorting of the logs improve the efficiency of the production since it allows sawing of log batches with similar properties such as dimension and knotstructure. In the sawline, each log is sawn into several boards, each with different dimension and properties. In the green sorting the boards are sorted based on customer demands. In this way the drying can be optimized for a specific dimension and product. The pre-drying sorting is important, for instance because that different products are dried to different moisture content and with varying demands on variation in moisture content. After drying, the boards are graded and sorted in the final sorting and then shipped to the customers, i.e. the secondary manufacturers.

The fact that the sawmill industry is based on a continuous flow of individual logs and boards, and that it is a divergent flow in several steps, makes it very interesting to apply traceability technology to the production of sawn wood (Chiorescu, 2003; Uusijärvi, 2003). There are three main benefits with tracing logs and boards through the process:

1. *Follow-up of decisions in the production process*, i.e. to create a system where it is possible to get feedback from stages later in the chain. Without traceability it is not possible to know for instance what happened when a specific log was sawn. With a traceability

system it would be possible to verify whether a log was sorted to the correct batch of logs, or if it was better suited for other products. In this way it becomes possible to improve and fine-tune the production process based on the feedback from previous production.
2. *To certify the origin of the products.* Traceability would make it feasible to prove, for instance, that a product is based on wood from a specific area that has been managed in a sustainable way.
3. *To base process control decisions on information from previous production steps.* With a real-time traceability system it would for instance be possible to base the decision in the log sorting also on information from the harvester.

To achieve benefit 2 and 3 requires traceability for every individual log and board. Hence, if benefits 2 and 3 are to be achieved the traceability system has to be based on a technology that provides a low cost for marking each log or board. The first benefit can on the other hand, be based either on a system where all logs and boards are traced, or a system where only a random sub-sample is traced.

For a sawmill, the raw material, i.e. the logs, represents more than 70 % of the total production cost. Because of this, there is a great need to optimize the utilization of the raw material. An accurate and efficient optimization of the production requires feedback from the process. Consequently, the first benefit has the highest economical potential and therefore also the highest priority.

For boards, there are solutions that make it possible to trace individual boards by printing a matrix code on the board and then read it later in the process. For logs, it is possible to print a code on the end-surface of the log, but snow and dirt make the identification unreliable. RFID has therefore been suggested as a suitable technology for tracing logs from forest to sawline (Uusijärvi, 2003).

5.2 RFID application

In 2006, Träcentrum Norr[1] initiated a project where the aim was to show how a traceability system can make it possible to get feedback from the production and use this information for optimization of production. Based on the suggestion by Uusijärvi (2003) it was decided to use RFID for tracing logs from log sorting (or forest) to sawline. All installations, tests and verifications of how to utilize the system were done at a large sawmill in northern Sweden. The sawmill has an annual production of 390 000 m^3 sawn wood and has high technology level with X-ray and 3D-scanning of logs, automatic grading of boards in both green and final sorting and a printer/reader system for tracing individual boards from green to final sorting.

When deciding on how to design a RFID system for the sawmill industry, there are a number of factors to take into account:

1. *The cost for the transponders.* The sawmill where the system was installed consumes approximately 5 million logs per year. Therefore, the application is very sensitive to the transponder cost if all logs are to be marked.
2. *The cost for applying the transponders.* Automatic application can be done by the harvester, but the application is not allowed to delay the harvesting operation.

---

[1] A competence centre for applied wood technology research, see www.ltu.se/ske/tcn.

3. *No plastic is allowed in the transponders.* Sawmills sell the wood chips to pulpmills, and the pulpmills do not accept any plastic material in the wood chips.
4. *Small amount of metal.* Only a small amount of metal is allowed in the tranponders. One reason is that hard materials in the logs will damage the tools when sawing the logs. The other reason is that because of this problem, logs with metal are automatically put aside in the log sorting.
5. *Harsh conditions.* The transponders have to be applied into the log for protection and must also tolerate damp and cold conditions (-30°C).

The aim of the project was, as mentioned above, to make it possible to obtain feedback for optimization of the production. Based on the limitations due to cost, it was decided to aim initially at a system that makes it possible to trace a sub-sample of logs, around 0.1-0.5 % of all logs that are sawn. A sub-sample traceability system solution is less sensitive to the cost of the transponders and allows manual transponder application either in the forest or in the log sorting. The manual application is possible since the operators already today has to handle a small number of logs manually (around 0.3-0.7 %) for quality control purposes. Hence, the manually handled logs could easily be manually marked with a RFID transponder at the same time.

Due to the material limitations (no plastics), it was decided to use low-frequency (125 kilo Hertz) glass transponders of the same type as in Figure 4. A special tool was developed for manual application of this type of transponders, see Figure 8.

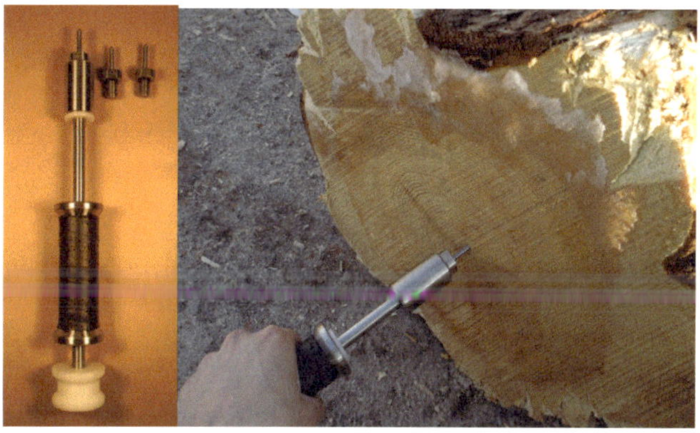

Fig. 8. Left: Tool for manual application of transponders. Right: Manual application of transponder to log.

In the sawmill, the transponders are to be read in the log sorting and in the sawline. Different antennas were designed specifically for these two positions, see Figure 9 and 10. When a transponder is read, the transponder id is sent to the process control system. In the log sorting, the log is then scanned by both X-ray and 3D scanners. Based on the scanner information the logs are sorted into different bins, each bin corresponds to logs aimed for specific products. A crucial step is to match the log transponder id to the scanner data for the same log. To match the correct id and scanner data is difficult, since the logs travel with 2.5 meters/second and the distance between logs can be down to 30 centimetres even during normal conditions. The correct transponder id also needs to be matched to the correct data in the sawline, but in the sawline the speed is lower and the distance between logs is longer.

Fig. 9. Antenna for RFID reading in the log sorting.

Fig. 10. Antenna for RFID reading in the sawline. The arrow is indicating the position of the antenna.

After sawing, the order of the boards is changed randomly before going through the green sorting. However, to get feedback from the result after sawing, i.e. the boards, traceability need to be obtained through the sawing. To obtain traceability the boards were colour marked and then the marked boards were identified in the green sorting before printing the matrix code. Hence, by combining the transponder id, colour marking id and the matrix id, traceability may be created from the forest to the secondary manufacturer.

5.3 Results from tests with RFID

The initial plan was to use 13.3 millimetres long transponders with 3.15 millimetres in diameter for the RFID application. However, the signal to noise ratio for the 13.3 millimetres transponder was to low. The low ratio was mainly caused by disturbances due to moving metal parts in the conveyor. A sufficient ratio was not possible to reach using the 13.3 millimetres transponder, so it was decided to use 22 millimetres long glass transponders with 4 millimetres in diameter. The 22 millimetres transponders are more expensive, but the main reason for wanting to avoid the longer transponders was that these were causing problems with the metal detection equipment used in the log sorting. To ensure that logs with metal do not reach the sawline the metal detection equipment sort out logs containing more metal than a certain value. The problem was that the 22 millimetres transponders contained to much metal and was sorted out by the metal detection equipment. To avoid that logs marked with the longer transponder were sorted out two different threshold levels for the metal detector were created, one level for logs with RFID transponders and a lower level for all other logs.

The read rate for the RFID application was tested by applying the longer transponders to 49 logs in a batch of totally 85 logs. The logs were run through the log sorting and then sawn during normal conditions. In the log sorting 49 transponders (100 %) were read and matched to the correct data, while 45 transponders (92 %) were read and matched to the correct data in the sawline. The 49 logs resulted in 196 centreboards and for 150 of these boards (77 %) the colour marking could be read and matched to the properties of the log.

The results from the read rate tests are promising. If 10 000 logs are marked with RFID transponders, this results in a database with connected log and board data for 7 700 logs, based on a 77 % read rate for both the RFID transponder and the colour marking. The RFID transponders costs around 1.5 EURO, resulting in 15 000 EURO for 7 700 traced logs, i.e. 1.9 EURO/traced log. For a comparison, the estimated cost for a test sawing where 150 logs were traced manually was 4 500 EURO, i.e. 30 EURO/traced log.

The traceability data received from the application could, for example, be used to identify the relation between relative knottiness in boards and knottiness in logs, see Figure 11. Hence, if a customer wants boards with a relative knottiness less than 0.3, Figure 11 shows that logs with a relative knottiness less than 45 produce boards that are suitable for that customer. By using traceability data, like the data in Figure 11, it is easier to optimize the production process, since the connection between process data and product data is available.

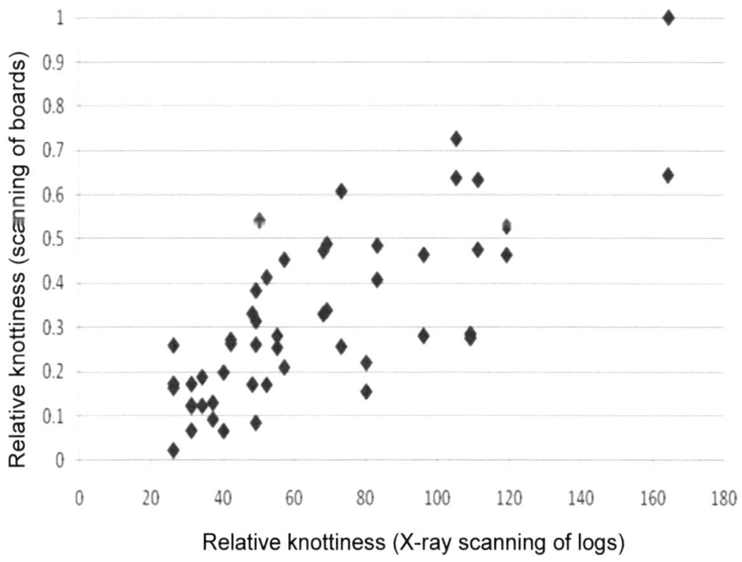

Fig. 11. Relative knottiness in boards as a function of relative knottiness in logs.

5.4 On-going and future development of RFID technology for the sawmill industry

The example shows a successful application of RFID for a sawmill industry. However, a wider use in the forest industry will require RFID transponders that are specifically developed for this application. That means a low cost transponder that can be automatically applied and that does not cause problems for neither pulp mills nor sawmills, i.e. such a transponder can not consist of plastic and can only contain a minimum amount of metal. Research work aiming at such a transponder is being done and a first prototype have been developed within the EU-project Indisputable key, for more information see Anonymous (2009).

6. An outlook on other RFID applications

There have been other ideas and uses of RFID to improve traceability in continuous process except from the two examples described. We here briefly describe some of these applications found in the literature.

RFID has been used in a number of mine to mill traceability projects around the world. La Rosa et al. (2007), for example, describes an application were hardened transponders and readers was used to track ore from the blast to the concentrator. The transponders were inserted in the blast holes or/and on the surface of the resulting muckpile. For each transponder the x-y-z position and identification number were linked. Tracking the transponder the source and properties of the ore being treated in a process step was accurately known at all times Also, tracking the ore with transponders made it possible to relate the metallurgical performance to original location in the mine and geometallurgical conditions. In the application, the measured survival rate for the hardened transponders after operations, such as blasting and crushing, was around 80 %. (La Rosa et al., 2007)

Lauf (2008) describes another application were RFID have been used in a coal fired power plant to track the quality of the input coal. Knowing the coal quality used in a power plant is important, since the process settings may then be optimized to the current coal quality and power plant efficiency can be improved. On-line analysis instruments for coal quality are very expensive, but analysis from the equipment is available in real time. Therefore, transponders with the same size as the average sized piece of coal were dropped into the product flow in connection to the analysis equipment of coal quality and the quality of the product was then linked to the transponder identification number and stored in a database. The movement of the transponders could then be tracked by a set of readers which forwarded the transponder identification number and time for detection to a database. The information in the database was used by the operators to continuously monitor the coal quality and optimize the plant settings. (Lauf, 2008)

Chen et al. (2005) have used RFID to collect data for modelling the flow pattern of pellets in a full scale test silo. Transponders were placed in the silo at different x-y-z positions and the position was linked to the identification number of the transponder. The transponders residence time were measured at the exit during discharge. To deduce the discharge flow pattern the residence times for the transponders at various positions was studied. The results from the tests showed that using RFID to study flow pattern is a reliable and cost effective technique that does not interfere with the silo's structure or operation. (Chen et al., 2005)

Moussavi et al (2005) present a novel traceability solution for meat products in an abattoir and boning hall. The solution comprise of a smart conveyor system integrating mechanical design, electronic architecture and RFID equipment. Besides improved traceability the solution resulted in better yield, less waste, and improvement in labour assignment. (Mousavi et al., 2005)

A more remote application is presented by Allan et al. (2006), they used RFID to trace the transport of pebbles in a mixed sand-and-gravel beach. To trace the movement of pebbles various transponders with similar characteristics as the pebbles were placed at five different locations. How the transponders had moved was then examined at various times by scanning the bottom with hand readers to see the new positions of the reader. The investigation demonstrated that RFID can yield important insights into the transport of gravel.

7. Conclusions, discussion and future research

The fragility of the transponders was beforehand thought as a limitation when considering various RFID applications. However, the described applications in continuous processes have shown that transponders can be designed to survive blasting, crushing, extreme temperatures and large shear forces. Therefore, we argue that the transponder fragility should not be seen as a limitation for most uses. However, designing transponders that survive in high temperature processes are still thought to be problematic or even impossible. In some continuous processes, for example the paper industry, the transponder may be a contaminating factor in subsequent process stages. However, if the whole process is kept in mind when choosing and developing the transponder the contamination can be avoided like in, for example, the sawmill process. The selection and development of a suitable transponder may be expensive and time consuming.

When transponders are dropped directly in the product flow the orientation between the reader antenna and the transponder antenna cannot be controlled. For some application all orientation may not be readable and some transponders may pass the reader without being read. By serially mount readers at different orientations in relation to the passage of the product flow, like in the pellets distribution chain, a higher fraction of transponders can be read. A low read rate implies more costs either in an increased demand of transponders or number of readers to be mounted. However, a low read rate does not imply that RFID has to be abandon as an application.

Installing a reader in some continuous process may demand that the process is stopped during the installation or that the installation must be done during a maintenance stop. Therefore, the installation may be costly or limited to a specific time and making changes to an installed reader may be both difficult and costly. Simulations may provide information that can be used to reduce the risk for the need to modify an installation and thereby the cost of an installation.

The production process in continuous processes is often highly automated and richly instrumented. An RFID application may affect or be affected by an existing instrument, and unexpected problems may occur similar to the problems with the metal detection equipment in the sawmill process. Hence, before an RFID application is installed it is recommended to perform a risk evaluation in an attempt to identify potential risks with the installation. To predict all potential risks for problems connected to an RFID application is complicated. Therefore, it could be wise to be cautious and initiate the application stepwise in an attempt to minimize the consequences of unexpected problems.

One factor limiting the use of RFID is the physical size of the transponders. Today the size of the transponders may limit the use of RFID to trace the movement to particles $\geq$ 10 millimetres. Another factor limiting the use is the read range, with longer read range the transponder position could be more thoroughly followed and new applications would be available. The third factor limiting the use of RFID is the transponder price. However, new transponders with longer read range, lower price and smaller sizes are continuously developed.

Even though we have found several RFID applications in continuous process to improve traceability we still think that this is a field that should receive more attention in contemporary research. The presented applications have resulted in a number of benefits beyond an improved traceability and the benefits have outweighed the costs for the described applications.

8. References

Allan, J. C., Hart, R. and Tranquili, J. V. (2006). The Use of Passive Integrated Transponder (PIT) Tags to Trace Cobble Transport in a Mixed Sand-and-gravel Beach on the High-energy Oregon Coast, USA. *Marine Geology*, Vol. 232, No. 1-2, page numbers (63-86), ISSN: 0025-3227.

Anonymous. (2009). "Indisputable Key." Retrieved the 7 of September, 2009, from http://www.indisputablekey.com/.

Chen, J. F., Rotter, J. M., Ooi, J. Y. and Zhong, Z. (2005). Flow Pattern Measurement in a Full Scale Silo Containing Iron Ore. *Chemical Engineering Science*, Vol. 60, No. 11, page numbers (3029-3041), ISSN: 0009-2509.

Chiorescu, S. (2003). *The forestry-wood chain: simulation technique, measurement accuracy, traceability concept, Doctoral thesis*, Wood Science and Technology, Luleå University of Technology, Skellefteå, Sweden.

Choi, B. S., Wan, B., Philyaw, S., Dhanasekharan, K. and Ring, T. A. (2004). Residence Time Distributions in a Stirred Tank: Comparison of CFD Predictions with Experiments. *Industrial and Engineering Chemistry Research*, Vol. 43, No. 20, page numbers (6548-6556), ISSN: 1520-5045.

De Andrade Lima, L. R. P. and Hodouin, D. (2005). Residence Time Distribution of an Industrial Mechanically Agitated Cyanidation Tank. *Minerals Engineering*, Vol. 18, No. 6, page numbers (613-621), ISSN: 0892-6875

Dennis, D. and Meredith, J. (2000). An Empirical Analysis of Process Industry Transformation Systems. *Management Science*, Vol. 46, No. 8, page numbers (1085-1099), ISSN: 0025-1909.

Duffin, M. (1995). Techniques Guidelines to TQC. *The TQM Magazine*, Vol. 7, No. 4, page numbers (35-41), ISSN: 0954-478X.

Dykstra, D. P., Kuru, G., Taylor, R., Nussbaum, R., Magrath, W. and Story, J. (2002). *Technologies for Wood Tracking: Verifying and Monitoring the Chain of Custody and Legal Compliance in the Timber Industry*

Fisk, G. and Chandran, R. (1975). Tracing and recalling products. *Harvard business review*, Vol. 53, No. November-December, page numbers (90-96), ISSN: 0017-8012.

Flapper, S. D. P., Fransoo, J. C., Broekmeulen, R. A. C. M. and Inderfurth, K. (2002). Planning and control of rework in the process industries: a review. *Production Planning & Control*, Vol. 13, No. 1, page numbers (26-34), ISSN: 0953-7287.

Flodin, J., Oja, J. and Grönlund, A. (2008). Fingerprint traceability of Sawn Products Using Industrial Measurement Systems for X-ray Log Scanning and Sawn Timber Surface Scanning. *Forest Product Journal*, Vol. 58, No. 11, page numbers (100-103), ISSN: 0015-7473.

Fransoo, J. C. and Rutten, W. G. M. M. (1993). A Typology of Production Control Situations in Process Industries. *International Journal of Operations & Production Management*, Vol. 14, No. 12, page numbers (47-57), ISSN: 0144-3577.

Hild, C., Sanders, D. and Cooper, T. (2000). Six Sigma on Continuous Processes: how and why it Differs. *Quality Engineering*, Vol. 13, No. 1, page numbers (1-9), ISSN: 0898-2112.

Jacobs, R. M. (1996). Product recall -A vendor/vendee nightmare. *Microelectronics and Reliability*, Vol. 36, No. 1, page numbers (101-103), ISSN: 0026-2714.

Jacobs, R. M. and Mundel, A. B. (1975). Quality Tasks In Product Recall. *Quality Progress*, Vol. 8, No. 6, page numbers ISSN: 0033-524X.

Kvarnström, B. and Bergquist, B. (2009). Improving Traceability in Continuous Processes Using Flow Simulations. *Submitted for publication.*

Kvarnström, B. and Oghazi, P. (2008). Methods for Traceability in Continuous Processes: Experience from an Iron Ore Refinement Process. *Minerals Engineering*, Vol. 21, No. 10, page numbers (720-730), ISSN: 0892-6875

Kvarnström, B. and Vanhatalo, E. (2010). Using RFID to Improve Traceability in Process Industry. *Journal of Manufacturing Technology Management*, ISSN: 1741-038x.

La Rosa, D., Valery, W., Wortley, M., Ozkocak, T. and Pike, M. (2007). The Use of Radio Frequency ID Tags to Track Ore in Mining Operations, *Proceedings of APCOM 2007: 33rd International Symposium on Application of Computers and Operations Research in the Mineral Industry, ISBN: 978-956-8504-06-9 Santiago, Chile* April 24 - 27, 2007

Lauf, D. (2008). "RFID Coal Tracking Technology." *Transponder News* Retrieved 20 of February 2008, from http://transpondernews.com/tcoalsrt.html.

Lelinski, D., Allen, J., Redden, L. and Weber, A. (2002). Analysis of the Residence Time Distribution in Large Flotation Machines. *Minerals Engineering*, Vol. 15, No. 7, page numbers (499-505), ISSN: 0892-6875

Lundqvist, S. O. and Kubulnieks, E. (1995). Improved Production and Product Information for Chemical Pulp Mill Operators, *Proceedings of 5th IFAC Symposium on Automated Systems Based on Human Skill -Joint Design of Technology and Organisation, Berlin, Germany*, September 26-28

Mahoney, F. X. and Thor, C. G. (1994). *The TQM Trilogy: Using ISO 9000, the Deming Prize, and the Baldrige Award to Establish a System for Total Quality Management* American Management Association, ISBN: 0-8144-5105-5, New York.

Mousavi, A., Sarhadi, M., Fawcett, S., Bowles, S. and York, M. (2005). Tracking and Traceability Solution Using a Novel Material Handling System. *Innovative Food Science & Emerging Technologies*, Vol. 6, No. 1, page numbers (91-105),

Oakland, J. S. (1995). *Total Quality Management Text with Cases*, Butterworth-Heinemann, ISBN: 0-7506-2124-9, Oxford.

Oghazi, P., Pålsson, B. and Tano, K. (2009). Applying traceability to grinding circuits by using Particle Texture Analysis (PTA). *Minerals Engineering*, Vol. 22, No. 7-8, page numbers (710-718), ISSN: 0892-6875

Skoglund, T. and Dejmek, P. (2007). Fuzzy Traceability: A Process Simulation Derived Extension of the Traceability Concept in Continuous Food Processing. *Food and bioproducts processing*, Vol. 85, No. 4, page numbers (354-359), ISSN: 0960-3085.

Töyrylä, I. (1999). *Realising the Potential of Traceability -A Case Study Research on Usage and Impacts of Product Traceability, Doctorial degree,* Industrial Management, Helsinki University of Technology, Helsinki.

Uusijärvi, R. (2003). *Linking raw material characteristics with Industrial Needs for Environmentally Sustainable and Efficient Transformation processes (LINESET).* Research Report no. P 0309034, QLRT-1999-01476 Final Report, SP Technical Research Institute of Sweden, Wood Technology, (SP Trätek).

Vanhatalo, E., Kvarnström, B., Bergquist, B. and Vännman, K. (2009). A Method to Determine Transition Time for Experiments in Dynamic Processes. *Submitted for publication.*

Yianatos, J., Bergh, L., Condori, P. and Aguilera, J. (2001). Hydrodynamic and Metallurgical Characterization of Industrial Flotation Banks for Control Purposes. *Minerals Engineering,* Vol. 14, No. 9, page numbers (1033-1046), ISSN: 0892-6875

6

A Case Study of an RFID-based System for Pilgrims Identification and Tracking

Mohamed Mohandes
King Fahd University of Petroleum & Minerals
Kingdom of Saudi Arabia

1. Introduction

RFID represents an ideal solution for crowd control and personal identification at Pilgrimage season (Hajj). It provides means to save lives and improve services at an affordable cost. This Chapter describes a developed prototype that uses a passive RFID wristband tag for identification of pilgrims in the holy areas during Hajj. Upon presentation of the tag to a handheld reader all information related to the pilgrim holding the tag will be displayed on the screen of the reader. This way if a pilgrim gets sick, lost, or dies he can be easily identified through the tag on the spot. Officers around the holy area will carry a light-weight handheld reader that can get all information from the tag without the need to burden the already overloaded communication network during the Hajj season.
Hajj (Pilgrimage) is undoubtedly the most crowded gathering of Muslims on earth. It has unique characteristics with regard to the people who attend it (pilgrims), the place they meet in, and the kind of rituals they perform. These characteristics result in a set of challenges to the authorities in controlling the crowd, and identifying the personalities. What makes it even more challenging is that the whole crowd makes the same movements at the same times doing essentially the same thing. Muslims are obliged to perform Hajj once in their lifetime. Many of them love to perform it more than once. It is performed on specific days of the year (8th- 13th of the 12th Hijri month) in designated boundaries around Makkah City in Saudi Arabia. The authorities for Hajj try their best to limit the crowd flooding to the area by assigning quotas for pilgrims from each country. Yet the number of pilgrims exceeds 2.5 millions every year and is in the rise. Additional 4 million visitors who come to the holy areas every year at other than Hajj times may benefit from developed RFID systems as well. It is expected that this number will reach 10 million visitors in the near future.
While it is a great spiritual experience for all the pilgrims, at the same time it poses great challenges of all sorts for the authorities responsible for facilitating the Hajj. Despite all that is done to facilitate the Hajj, the following are some of the common difficulties faced by the pilgrims and the authorities alike:

- Identification of pilgrims (lost, dead, or injured)
- Medical Emergencies
- Guiding lost pilgrims to their camps.
- Loss of identity documents and money
- Crowd control

In the last few years RFID systems have experienced a surge of deployment (due to the drop in cost). Inventory tracking is now necessary on an unprecedented scale to support growing consumer markets at low operating costs. Furthermore, the components used to build the tags and readers have become more sophisticated as they provide greater functionality, longer reading range, and higher speed of data transfer (Finkenzeller, 1999). As a result, a number of influential organizations such as Wal-Mart, GE, U.S. Department of Immigration, and the U.S. Department of Defense, recognized the capabilities of RFID to improve operational efficiencies (Roberti, 2004). Additionally, it has been used successfully during Beijing Olympic 2008, where 16 million RFID embedded tickets have been used for admission to the games (Chhabra, 2008).

The idea of utilizing RFID for Hajj services was introduced and passed through several implementations and discussions with Hajj officials that have been concluded with a prototype using passive RFID technology (Mohandes, 2009). The developed prototype was tested on 1000 pilgrims from the country of Ivory Coast in collaboration with officials from Hajj Ministry and Development Commission of the Holy cities of Makkah and Madinah. The results of the experiment have convinced Hajj authorities to utilize this technology for all pilgrims in the near future. Additionally, the possibility of tracking pilgrims using active RFID system has been explored. Due to the small number of working tags, experiment was performed only on 6 persons. Results of the latter experiment have shown that the system requires further development and needs to be validated with a larger number of tags.

2. Challenges during Hajj

While it is a great spiritual experience for all the pilgrims, at the same time it poses great challenges of all sorts for the authorities responsible for facilitating the Hajj. The following describes some of the unique characteristics about Hajj:

- **Large Number of Elderly Pilgrims**
  Many Muslims decide to perform Hajj when they are old. Moreover, many countries, in attempting to adhere to the quota, give preference to the elderly. Added to the crowd and exhaustion they face, a significant number of them are exposed to falling sick or even losing their lives.
- **Language Barrier**
  Pilgrims in Hajj season are, literally, coming from all countries on earth, with no exception. They speak different languages. It is a serious issue when it comes to helping the lost or treating the sick.
- **Dress Code**
  The dress of the pilgrim consists of two pieces of garments, one is rolled around the lower part of the body and the other is put on the shoulder. There are no pockets to carry the belongings. Although pilgrims are allowed to put on a belt with small pockets, most of them do not prefer to carry valuable documents because of the fear of losing them in the crowd.
- **Identification of pilgrims (dead or injured)**
  Due to the crowd, age factor and sometimes unfortunate accidents, the death toll in Hajj season is on the high side. Many of the dead are on their own, carrying no documents. Every year at the end of Hajj season, authority are confronted with tens of bodies of pilgrims that are never identified or claimed. This problem is not easy to solve as it is

not known when and if any of these bodies will be claimed in the future. Moreover, it puts a moral obligation on Hajj authorities to identify and inform the families of these pilgrims of the sad demise of their loved ones. Pilgrims are considered as guests of the Kingdom until their safe return home.

- **Knowing the medical record of patients**
  Tens of thousand of pilgrims are brought to hospitals and clinics for emergency treatment. Doctors always complain that would they knew the medical record of the patient they would be more efficient in treatment and utilization of the limited resources available during this period. Unfortunately, this is never the case. Even worse, due to the language barrier they may not be able to do basic investigations with the patient to find any information that may help them in the diagnosis.
- **Guiding Lost Pilgrims to their camps**
  Large numbers of pilgrims report on the hour to Help Kiosks to guide them to their camps. It is not always easy to find a speaker of the lost pilgrim's native language. Without papers showing their location and contact numbers, and sometimes without a common language to communicate, it becomes almost impossible to help them. This puts a tremendous pressure on the security officers who are taking care of pilgrims.
- **Loss of Important Documents**
  With the special cloth of Ihram, it is highly likely that the pilgrim may lose some of his identification papers, like Passport, Residence Permit, and national ID in addition to money or credit cards. With lost passports the return of pilgrims to their home countries is delayed which imposes additional burden not only on the pilgrims but also on the authorities.
- **Crowd Control**
  Crowd control has become a major problem with many pilgrims dying in accidents and stampedes in congested places during the Hajj, especially, in Mina. These accidents happen due to insufficient planning and management of available space as well as due to crowd behavior.

Fig. 1. Pilgrims in Makkah during Hajj

3. RFID technology

An RFID system consists of the following three components as shown in Figure 2:

- RFID tag or transponder
- RFID reader or transceiver with a scanning antenna

Data processing subsystem that can be embedded in a handheld reader.

An RFID reader consists of an antenna, transceiver and decoder, which sends periodic signals to inquire about any tag in its vicinity. On receiving any signal from a tag it passes on that information to its embedded data processor.

The scanning antenna puts out radio-frequency signals in a relatively short range. The RF radiation does two things; it provides a means of communication with the transponder tag (the RFID chip) and (in the case of passive RFID tags) it provides it with the necessary energy to function. This is an absolutely key part of the technology; RFID tags do not need to contain batteries, and can therefore be made simple, cheap and remain usable for very long periods of time. The scanning antennas can be permanently affixed to a surface; handheld antennas are also available. They can take whatever shape needed; as shown in Figure 3.

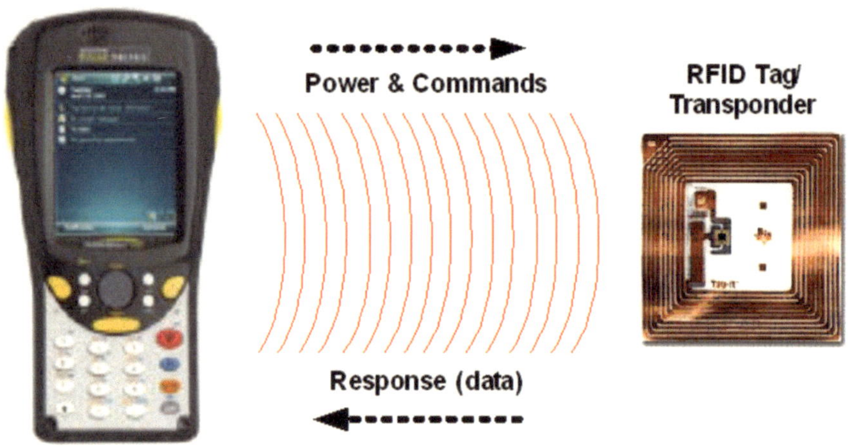

Fig. 2. A Typical RFID System with Handheld Reader

When an RFID tag passes through the field of the scanning antenna, it detects the activation signal from the antenna. That "wakes up" the RFID chip, and it transmits the information on its microchip to be picked up by the scanning antenna.

Fig. 3. Different forms of Handheld RFID readers

3.1 RFID tags

An RFID tag is composed of an antenna, a wireless transducer and an encapsulating material as shown in Figure 4. These tags can be either active or passive. While the active tags have on-chip power, passive tags use the power induced by the magnetic field of the RFID reader. Thus, passive tags are cheaper but with lower range (<10m) and more sensitive to regulatory and environmental constraints, as compared to active tags.

The vast majority of RFID tags or transponders use a silicon microchip to store a unique serial number and usually some additional information. Passive RFID tags do not have a transmitter; they simply reflect back energy (radio waves) coming from the reader antenna. Passive RFID tags do not require batteries, and can be much smaller and have a virtually unlimited life span.

Active tags have their own transmitter and power source, usually—but not always—a battery (active tags could draw energy from the sun or other sources). They broadcast a signal to transmit the information stored on the microchip. Wristband tags for human identification and access control as shown in Figure 5 have been used in this pilot implementation. Alternatively, it is possible to have very small tags implanted in the pilgrim's hand. The RFID chip used for this purpose is a glass capsule measuring 1.3mm by 1mm as shown in Figure 6 (VeriChip, 2009).

3.2 Types of RFID systems

Operating frequency is the determining factor for the type of application an RFID system is best suited for. These include high frequency (850-950 MHz and 2.4-5 GHz), intermediate frequency (10-15 MHz) and low frequency (100-500 kHz).

- High-frequency RFID systems are suitable for applications requiring a longer read range such as supply chain, inventory, assembly lines, toll-collection systems and railroad car and container tracking.
- Intermediate-frequency RFID systems are just now beginning to emerge in the financial transaction processing areas of smart card use.
- Low-frequency systems have short reading ranges and lower system costs. They are most commonly used in security access, asset tracking, and animal identification applications.

As you move up in frequency, tag and reader costs move up as well. Tags at 125 kHz operating frequency have ICs costing tens of cents as compared to 2.4 GHz tag ICs costing several dollars.

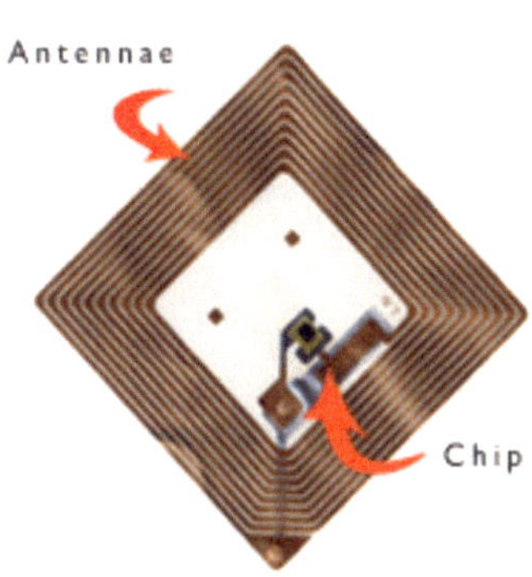

Fig. 4. Components of an RFID tag

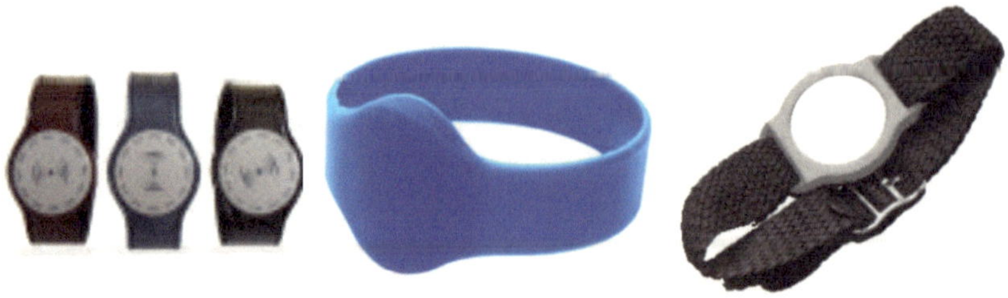

Fig. 5. Wristband RFID tags

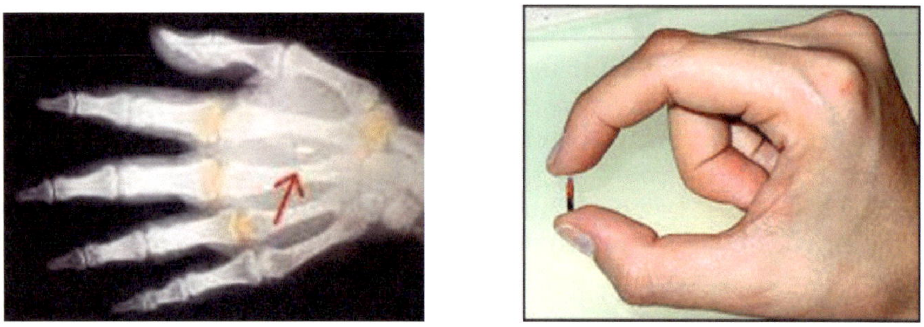

Fig. 6. RFID Chip for Implanting in Human Body

3.3 Applications for RFID

Applications fall into two principal categories: firstly, short range applications where the reader and tag must be in close proximity (such as in access control) and secondly, medium to long range applications, where the distance may be greater (such as reading across a distribution centre dock door). A sample of applications includes:

- Security, ticketing and access control
- Supply chain logistics & retail
- Asset tracking and traceability
- Lifetime item identification
- Medical applications
- Animal and specimen identification
- Airline baggage handling.
- Vehicle access control

4. RFID solution for Hajj

RFID offers an effective, convenient and economical solution to the Hajj problems described above. A wristband RFID tag that can be worn at all times can be provided to each pilgrim. This tag when presented to an RFID reader the information stored in the tag can be retrieved and displayed on a handheld portable unit. The following is some of the information that can be stored in a tag:

- Personal details - like name, address, blood type, nationality, etc.
- Medical condition
- Contact information of the pilgrim's Hajj group
- E-purse that can be loaded with optional amount of money

RFID may help in crowd control by allowing the authorities to estimate the number of people at specific locations where there is a risk of accidents taking place. Combined with the efforts of the security personnel it may be possible to ease the flow of pilgrims at key points resulting in proper utilization of space. RFID readers can be installed at key entry and exit points to ping wristband RFID tags and thus get fairly accurate estimate of people passing those points. This will help the security personnel in controlling the entry of pilgrims in high risk areas.

The utilization of the wristband can allow the pilgrims to leave their IDs on secured places like hotels safe deposit boxes. Moreover, if an ID is lost, it is possible to retrieve the information needed to produce a replacement of the lost ID with the presence of wristband tag. Information stored in the tag about the medical condition of the pilgrim can be very useful in case of a medical emergency. Having access to the medical condition of the patient in the emergency room from the tag he holds on his wrist may help safe the life of the patient and improve efficiency of the medical team. The tags will also help in guiding the pilgrims who get lost - besides helping in the immediate identification of pilgrims who die during Hajj.

With the new rules that each local or external pilgrim should obtain a permission to perform Hajj, the permission certificate can be the wristband itself. The presence of the tag can be detected at a distance and does not require visual reading by the security officers. For example, the officer can read tags from a distance and find out if all passengers of a car have permission to perform Hajj. This is because RFID tags do not require line of sight or contact with the reader for the information to be transferred. This would increase efficiency and deter violators in a way to achieve the objectives of Hajj authorities to control the number of pilgrims every year and make it easier for authorized pilgrims to perform their once in life time Hajj.

Some types of RFID wristbands can't be unlocked once fixed and requires cutting the band. This will guarantee that the wristband will not get lost or disposed. There is a possibility to collect the wristbands upon finishing Hajj activities. The collected RFID tags could be re-programmed and installed in new wristbands to be used for upcoming Hajj seasons. The wristband tag can be a real working watch; this should remove the fear and sensitivity of pilgrims and their reluctance to wear such tags. Other paper wristbands did not find the full cooperation from pilgrims in the past.

5. Development of the prototype system

The developed prototype system consists of the following:

5.1 WorkAbout Pro S Hand-Held computer

The WorkAbout Pro (Teklogix, 2009) is a flexible, programmable and expandable portable data collection terminal, designed for a range of data collection needs. It is available with either Windows Mobile 2003 SE or Windows CE .NET configurations. The WorkAbout Pro delivers the performance, ruggedness and durability required for mobile data collection in an ergonomic package.

The device used in this prototype is WorkAbout Pro S having Windows Mobile 2003 as the operating system and a built-in RFID reader that supports worldwide RFID standards in LF (125 - 134.2 KHz) or HF (13.56MHz) technologies. Based on industry standard Microsoft and Intel XScale architecture, the Workabout Pro ensures seamless application development and integration capabilities. Compact flash and SD/MMC memory expansion slots combined with USB, and a 100-pin high speed expansion interface, to provide flexible radio and peripheral connectivity.

5.2 Psion Teklogix Mobile Devices SDK

The Psion Teklogix Mobile Devices SDK (Psion, 2009) for Windows Mobile 2003 SE, provides support for programming in .NET compact framework, C/C++ and Java. Developers are able to write flexible mobile applications to take advantage of a variety of scanners and drivers, all from the single platform. Once developed, an application can be deployed, without modification, on any Psion Teklogix mobile computer. The Mobile devices SDK provides a consistent set of APIs for all the Psion Teklogix Windows CE and Windows mobile based computers. Applications built using this release of the mobile devices SDK are expected to be fully compliant with future software releases for the Psion Teklogix handheld computers.

5.3 Development tools

- **Microsoft Visual Studio 2005**
  Microsoft Visual Studio (Microsoft, 2009), is the main Integrated Development Environment (IDE) from Microsoft. It can be used to develop console and GUI applications along with Windows forms applications, web sites, web applications, and web services in both native code as well as managed code for all platforms supported by Microsoft Windows, Windows mobile, .NET framework, and .NET compact framework.
- **SQL Compact Edition**
  SQL Server Compact (Microsoft, 2005) is a free and easy-to-use embedded database engine that lets developers build robust mobile applications that run on all Windows platforms including Windows XP, Vista, Pocket PC, and Smartphone.
- **Mughamrat Windows Mobile Arabic Layer**
  In the Middle East market, the Arabic language support is mandatory. Hence, building Arabic C# .NET applications, running on a Pocket PC or on Windows CE is a challenging task due to the many problems related to the Arabic language specifications that developers have to cope with. To avoid these problems, Mughamrat has developed a software development kit for developing application with an Arabic interface for mobile devices called WinCE Arabic Layer, WAL (Mughamrat, 2009)

5.4 Pilgrim information system

This software is designed to read and write pilgrims information from the data source using passive RFID tags. The purpose of this software requirement specification is to describe all the functional requirements, constraints, dependencies and the way the requirements be fulfilled. Through the use of this software the user will be able to save user information to data source, read user information using hand held device, and assign RFID tags to users. The following provide details about the application architecture and usage.

- **User Info Writer**

The application allows the user to input pilgrim information and assign them a tag ID. The application has been developed using C# and .NET framework. There is only one class in this application that takes care of the required functionality. The userInfo class provides all the functions needed to store the pilgrim's information on the data source which is an SQL compact edition database. The class diagram is shown in Figure 7. The following functions are implemented for this class.

- *private void add_image_Click(object sender, EventArgs e)*
  This function is executed when the user wants to add the image for any pilgrim.
- *private void Btn_Save_Click(object sender, EventArgs e)*
  This function is called when the user saves the information of any pilgrim
- *private int SaveImage(byte[ ] userImage)*
  This function is called internally from the Btn_Save_Click function and creates the connection to the database and executes to the query required to create a new entry in the database.

Fig. 7. UserInfo Class Diagram

- **User Interface Screens**
  Figure 8 shows the main screen of the application. It allows the user to input all the required information for the pilgrim. User can search for the required image by clicking the Browse button. The screen shown in Figure 9 allows the user to search for the image of the pilgrim from the PC. Once all the information is filled in, the user clicks the Save button and all the information along with the image and RFID tag number are saved to the database.
- **Software for RFID Tag Reader**
  This application has been developed to be deployed on a Psion Teklogix Workabout Pro S that has a built in RFID reader and Windows Mobile 2003 as the operating system. The application will allow the user to scan any RFID tag to get the tag number and then retrieve the information related to the tag from the data source that is deployed along with the software. There is only one class in this application that takes care of the required functionality. The class diagram is shown in Figure 10. The functions that are developed for this class are:
  - *private void fTagSYSReader_Load(object sender,EventArgs e)*
    This function is excuted when the application is started and is responsible for connecting to and starting the RFID reader. The RFID reader is set to work on COM

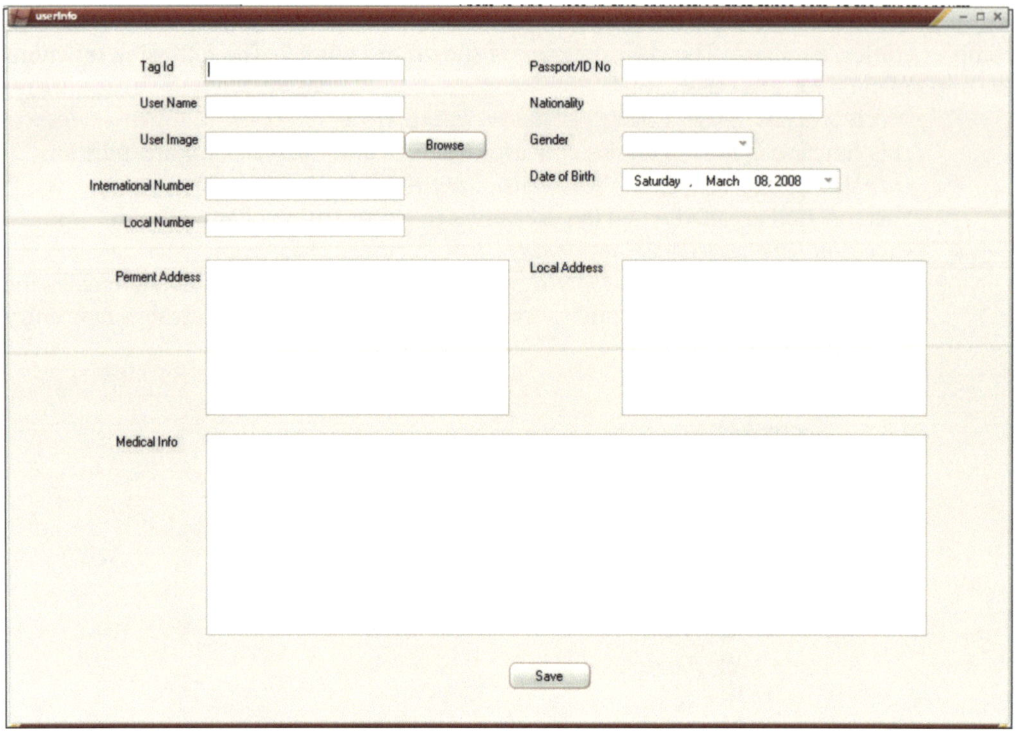

Fig. 8. Main Screen of the Application

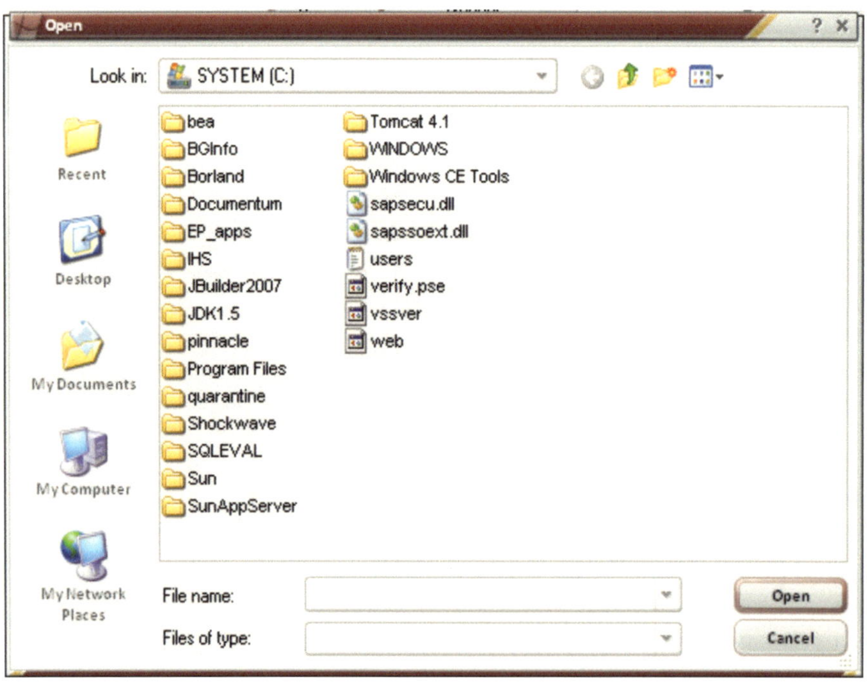

Fig. 9. Image Search Window

Port=1 and the baudRate = 9600. If there are any errors during this operation the appliction will highlight them by showing an error message. After the connection is made this function activates the timer which keeps the RFID reader in a continuous read mode.

- *private void fTagSYSReader_Closing(object sender,CancelEventArgs e)*
  This function is excuted when closing the application. This is like a clean up function and closes all the connections the appliction has setup.
- *private void timer_Tick(object sender, EventArgs e)*
  This function is executed when the timer is activated and everytime the timer expires. The timer is a very useful .Net control that allows the application to excute the same code based on a time interval. The time interval for our appliction is set to 10 milliseconds. So after every 10 milliseconds the code in this function is excuted. This function reads any data availbale on the tag reader, after getting the tag it connects to the database and queries the database to get all the required information for this tag ID and displays the information to the user.

Fig. 10. RfidTagReader Class Diagram

- **Database Design**
  There is only one database table so that the size remains small and the database doesn't have to store any relation information. tag_id is the primary key and is the tag number of the RFID. The fields used in the Database are given in Table 1.
- **Handheld Reader Screens**
  The first screen you see when you start the application is shown in Figure 11. It shows the tag ID when it is scanned. If the tag is assigned to a pilgrim in the database it will show the pilgrim's name and picture. At the bottom of the screen there are four tabs; each one shows different information regarding the pilgrim. The second screen is shown in Figure 12. You get this screen when you click on the General Info tab. This screen shows the pilgrim's gender, nationality, passport/ID number, local and international phone numbers and blood group. The third screen is shown in Figure 13. You get this screen by clicking the Address tab and it shows the Current and Permanent addresses. Figure 14 shows the last screen which you get by clicking the Medical tab and it shows the medical information of the pilgrim.

Fig. 11. Handheld Reader Screen 1

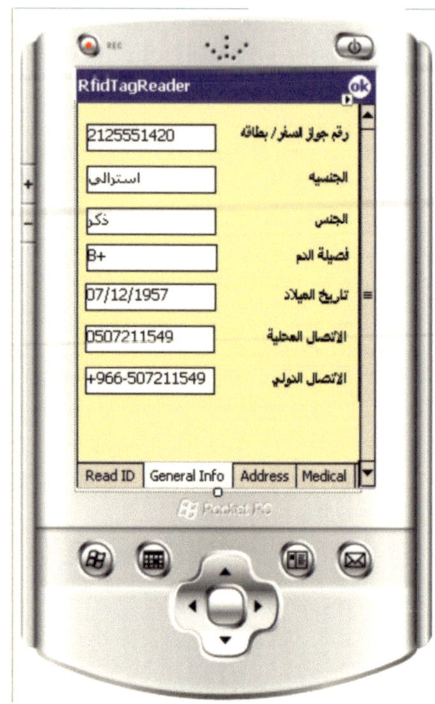

Fig. 12. Handheld Reader Screen 2

Fig. 13. Handheld Reader Screen 3

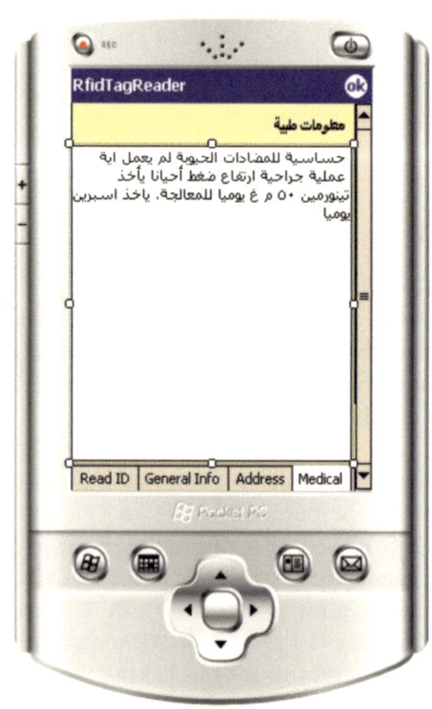

Fig. 14. Handheld Reader Screen 4

Column Name	Date Type	Length
user_name	nvarchar	200
tag_id (primary key)	nvarchar	100
user_pic	varbinary	8000
user_id_no	nvarchar	100
user_birth_date	datetime	8
user_gender	nvarchar	8
user_perm_address	ntext	16
user_local_address	ntext	16
user_med_info	ntext	16
user_nationality	nvarchar	100
user_inter_ph	nvarchar	20
user_local_ph	nvarchar	20

Table 1. The fields used in the Database

6. Pilot project for Pilgrim identification

A pilot project for the developed Pilgrim Identification is implemented to demonstrate the advantages of using RFID technology during the Hajj. There were 1,000 tags and a single reader. Pilgrims from the country of Ivory Coast volunteered for the pilot (see Figure 15 &16). However, the group consisted of about 4,000 pilgrims. The elderly 1,000 were selected to participate on the pilot implementation. An IT engineer accompanying the group had a complete database that includes picture, name, blood type, date of birth, and address in their home country. The medical team of the group had the medical condition of all pilgrims. Therefore, programming and distributing the tags was a simple process.

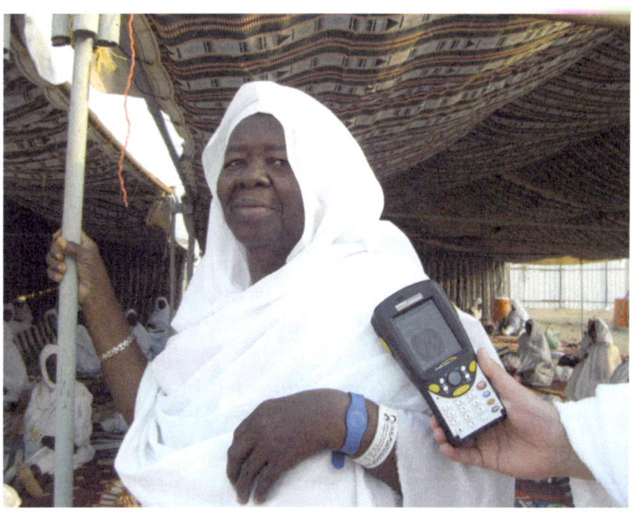

Fig. 15. Information of a pilgrim is displayed on the handheld reader as it is brought close to the tag.

Fig. 16. Some Participants of the pilot implementation

Due to the fact that only one reader is available and not all pilgrims in the group have wristband tags limited the potential of the applications. Nonetheless, scenarios were created to show some of the possible benefits of the system. In particular, the pilot study has shown that the RFID system would help identification of pilgrims at check points, as no need for officer to read Hajj permit to check its validity for every person. Checking the wristband tags by the reader takes a small fraction of the time it takes to authenticate the paper permit. At the entrance of the hotel, pilgrims with the blue wristband tags did not need to be stopped as it was clear from far that they belong to the group. The same was observed at the entrance of the camp in Mina, Arafat, and Muzdalipha. Queues for food and access to the busses for transportation through out the holy areas are other occasions where RFID wristband tags would improve efficiency in checking that only authorized pilgrims use these services. Automation is necessitated further by the fact that the rituals of Hajj are to be performed by pilgrims at the same instant of times. No manual system would perform this authorization process with ease and satisfaction. From the pilot, it was clear that participating pilgrims were identified from the unique blue wristband even without RFID reader. This indicated that on the full implementation, it is better if wristband with unique colour is used for every group. Even better if the wristband looks like the flag of the country from where the group of pilgrims came from. This will help drastically in guiding lost pilgrims to their camps. Moreover, if tags are programmed at the country of origin before departure then information may be verified from the tag and the computer of immigration centre, which improves efficiency and reduces long waiting queues at the airport.

7. Active RFID wristband tags for pilgrim tracking

Every year a significant number of violators pretend that they want to perform Hajj while their real intention is to find jobs and stay in the country illegally. The Hajj area is limited to few squared kilometers. Therefore, if a network of RFID readers is distributed in the holy area and an active RFID wristband is used for each pilgrim, then when a pretend pilgrim attempts to leave the area or cut the wristband an alarm is sent to security officers indicating a violator. Moreover, although short range tags proved useful for identification (Tuttle, 1997), a longer range would improve efficiency. At check points, officers need to check the Hajj permit of each pilgrim. Pilgrims enter the holy area on cars or busses. Short range RFID system would require officers to bring the reader close to the tags to be able to read it. A longer range would allow pilgrims permits to be checked while pilgrims are seated in their vehicles without the need for officers to come close to read the wristband tag. Additionally, statistics about distribution of pilgrims throughout the holy area help provide better services in future Hajj seasons. Long range RFID system would provide such statistics without creating bottlenecks in the area. Moreover, sometimes there is a need to locate a pilgrim among the crowd. Long range RFID system with a network of readers help locate a particular pilgrim when needed.

To develop a solution for these tasks, an off the shelf development kit is acquired from ActiveWave. Its specifications (ActiveWave, 2009) matched our requirements reasonably well. The wristband has a replaceable 3V battery that lasts from one to three years. It's read range for receiving is up to 30 meters while for transmission is up to 85 meters. Its user memory is up to 256 kbits, which is enough to store a pilgrim's information with his picture and a finger print if needed. Anti-collision capability is also available. It sends an alarm if removed or cut. The tag is shown in Figure 17.

The reader has three basic modes of operation: program, monitor, and call. When in programming mode, the reader configures the tags. When in monitoring mode, the reader listens to all tag activity and relays this information back to the host in real-time. When in call mode, the reader wakes up specific tags, specific groups of tags, or all tags within range. The reader is shown in Figure 18.

Experiment for Tracking System

A development kit has been acquired that includes a reader and about 24 wristband tags. Upon testing the system, 18 tags were found not functional. The remaining 6 were programmed to track 6 pilgrims. The reader is placed on the ground and pilgrims were asked to start moving from about 100 meters towards the reader. It was noticed that at about 30 meters the pilgrims were detected, however, if a pilgrim occludes another, both will not detected by the reader. The same process is observed when more pilgrims stand close to each others. The detection range is observed from all directions consistently. For the crowd in Hajj, the system as it was tested is not useful for the intended purpose of tracking. More refinement of the used frequencies, location of readers, and antennas design is needed.

Further experiment was performed to evaluate the system on detecting passengers with active wristband tags on moving vehicles. Therefore, 5 passengers with active wristband tags are asked to ride on a car. The car moved on different speeds while the reader is fixed on the side of the road. All 5 tags are detected only when the passengers raised their hands so the tags are seen from the window of the car. No tags were detected if passengers put their hands on normal positions. It is possible to ask pilgrims to raise their hands when passing through a check point to improve efficiency and remove bottlenecks on the road. However, better system and further investigations are needed for tracking pilgrims

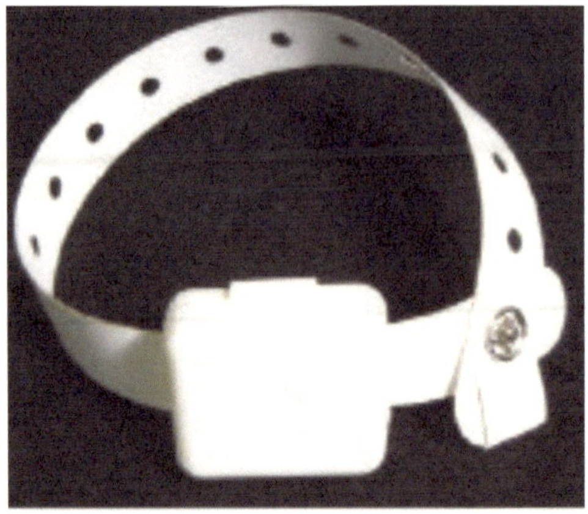

Fig. 17. Active RFID Wristband Tag

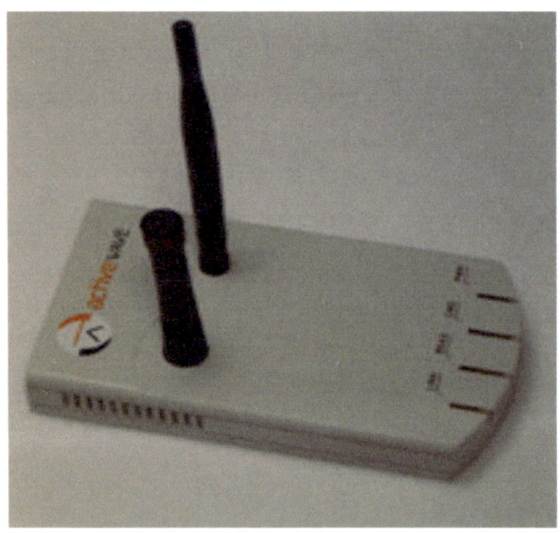

Fig. 18. The ActiveWave RFID reader

8. Conclusions and recommendations

Hajj season poses many challenges for the authorities. The main challenges have been reviewed in this work based on official reports, interviews with key officials and the authors' personal experience. Using advanced technologies is very promising in tackling these challenges. In this project, we have demonstrated the use of RFID technology to ease some of these challenges

A pilot system was carried out to experiment the system during Hajj season. The prototype RFID-based Pilgrim Identification System was tested with a group of 1000 pilgrims from Ivory Coast. This experiment proved to be very successful in demonstrating the benefits of the system. In particular, it demonstrated the effectiveness of RFID system in removing bottlenecks of the traditional authentication system. This is due to the fact that checking a wristband tag by the reader takes a small fraction of the time it takes to authenticate the paper permit.

It must be noted that the cooperation of pilgrims is very vital for the success of the system. The experiment has shown that pilgrims happily participated after enough explanation was given about the objectives of such a system and the improved services it leads to for future pilgrims. Moreover, it was shown that collecting the wristband tags to use them in future seasons is not practical, as most pilgrims wanted to keep the tag as a souvenir from Hajj. Pilgrims should be educated in their own countries before coming for Hajj about the system and the importance of keeping the tags as a Hajj permit and its loss can cause them difficulties during their Hajj.

Distribution of pre-programmed tags in each pilgrim's home country makes the job of immigration officers in Jeddah easier and improves efficiency at the airport as information needs only be verified from the tags rather than inputting the information while pilgrim is waiting.

In the future, applications such as e-purse may be added to the tag, so that all what the pilgrim need to carry is his wristband tag. This saves pilgrims the worry that they may loose money or official documents.

Although 1000 pilgrims is a relatively large number for the current project, it only makes 0.04% from total pilgrims. Moreover, having only 1000 tags and one reader limited the full benefits of the RFID system for such application. Especially that the group of Ivory Coast consists of about 4000 pilgrims. Therefore, it is recommended that the experiment be performed on an entire contingent of pilgrims from a country such as Malaysia which has about 14,000 pilgrims coming every year. This will help in validating the system with a relatively good sample size before deploying the system for more than 2.5 million pilgrims.

The use of active RFID system to track pilgrims has been investigated using a developed prototype. The system has proved to be useful for detection of Hajj permits while pilgrims are on moving vehicles. More investigation is needed for this system. Improvement of antennas design, selection of readers' location, and communication frequency are also to be tackled in future work.

9. Acknowledgement

This work has been conducted under KFUPM Funded Project No. IN060340. The author is grateful to KFUPM for the support. The author acknowledges the help and support of Ahmed Abul Hussain, Maan Kousa, and Saad Al Shahrani. The author thanks the pilgrims of Ivory Coast for their participation on the pilot implementation, particularly, the help of the IT Engineer with the group is highly acknowledged.

10. References

ActiveWave Incorporation, (2009), http://www.activewaveinc.com

Chhabra, G. (2008), How RFID will benefit from Beijing Olympics 2008, *The RFID Weblog* http://www.rfid-weblog.com/50226711/how_rfid_will_benefit_from_beijing_olympics_2008.php

Finkenzeller, K. (1999), *RFID Handbook, Radio-Frequency Identification Fundamentals and Applications*, John Wiley, NY ISBN-13: 978-0471988519

Microsoft Developer Network (2009), http://msdn.microsoft.com/en-us/default.aspx

Microsoft SQL Server (2005), http://www.microsoft.com/sql/editions/compact/default.mspx

Mohandes, M., (2009), RFID-based System for Pilgrims Identification & Tracking, *The Applied Computational Electromagnetics Society Journal (ACES)*, in press, 2009.

Mughamrat Company (2009), WindowsCE Arabic Layer (WAL), http://www.mughamrat.com/ProductsSolutions/MobilityArabizationProducts/WindowsCEArabicLayerWAL/tabid/67/Default.aspx

Psion Teklogix Mobile Devices SDK (2009), http://www.psionteklogix.com/us/index.htm

Roberti, M. (2004), US DOD Final RFID Policy, *RFID Journal*, http://www.rfidjournal.com/article/view/1080/1/1

Tuttle, J.R. (1997), Traditional and Emerging Technologies and Applications in the Radio Frequency Identification (RFID) Industry, *1997 IEEE Radio Frequency Integrated Circuits Symposium*, Denver, USA, June 1997, IEEE.

VeriChip, (2009), VeriChip Corporation. http://www.verichipcorp.com/index.html

Workabout Pro Hand-Held Computer (2009) http://www.psionteklogix.com/us/products/handheld/workaboutpro.htm

7

Implementation Protocol Utilising Radio Frequency Identification (RFID) and Biometric Identifiers; In the Context of Irish Bovine Traceability

Conor Shanahan, Gashaw Ayalew, Francis Butler, Shane Ward and Kevin McDonnell
Biosystems Engineering, University College Dublin, Belfield, Dublin 4 Ireland

1. Introduction

1.1 Irish cattle traceability

The Irish Department of Agriculture, Fisheries and Food (DAFF) is the authority responsible for the implementation of EC 1760/2000, which outlines the requirements for the identification and registration of bovine animals. Since the mid 1990's computerised databases have been established specifically the Calf Birth Registration System and the Cattle Movement Monitoring System (CMMS) (DAFF, 2003). The CMMS was initiated to record all information on births, movement, deaths and disposals. Computer equipment linked to the central database was installed in abattoirs and export points to electronically record all animal movements to and/or from these premises. As the CMMS relies on paper records for notification of certain events such as, births (herd keepers have seven days to notify the Calf Birth Registration System) and on-farm deaths, it cannot be said that it is totally accurate at a given point in time. The figures from the official CMMS statistics reports published each year reflect this, out of which a summary of the herd population statistics for recent years can be seen in Table 1. The start of January figures are calculated by taking the population figure at the end of December for the year in question from the CMMS database and adding all the disposals (cattle slaughtered, exported and on-farm deaths) and subtracting all the births and imports that have been recorded on the CMMS for that year.

	Year	Pop.	Year	Pop.	Year	Pop.	Year	Pop.
End of Dec	2003	6,589,974	2004	6,501,788	2005	6,532,706	2006	6,321,823
Start of Jan	2004	6,502,322	2005	6,489,962	2006	6,464,038	2007	6,303,313
Surplus		**87,652**		**11,826**		**68,668**		**18,510**

Table 1. CMMS cattle population data 2007 – 2003 (DAFF, 2004, DAFF, 2005, DAFF, 2006a, DAFF, 2007, DAFF, 2008)

At the current point in time barcodes are the data carriers used on cattle eartags in Ireland. However, there are issues with the readability of barcodes, of which Stanford et al., (2001) list a number of factors that can affect the read rates of eartag barcodes:

- quality of readers/scanners (ruggedness);
- direct sunlight;
- environmental conditions (i.e. cold, rain, etc);
- dirt/manure contamination of tags;
- quality and size of barcodes; and,
- the contrast between barcode and the colour of the tag.

As a result of this the identity of bovines is not being exclusively electronically recorded at the point of slaughter as called for by EC/1760/2000. Shanahan et al., (2007) have developed a genetic algorithm that can recover up to 42.4 % of damaged GS1 Code128C barcode contents, however, this still requires the barcode check digit to be recovered.

1.2 Radio frequency identification (RFID)

RFID tags are becoming more ubiquitous in the modern day world; they are finding more applications daily. Some current uses are theft prevention, automatic toll collection, building access control and animal identification (Landt, 2005). A basic RFID system is comprised of tags and readers. RFID tags each programmed with a unique identification number, may be passive (no on-board power source), semi-active (containing on-board source, for on-board power consumption) or active tags (containing on-board batteries, used for all power requirements including reader communication). The readers are comprised of an antenna, transceiver, decoder (Domdouzis et al., 2007). The reader emits a radio signal via its antenna and if there is a tag in the activation field that responds with an adequate signal strength, it picks up the signal from the RFID tag, translates the data into binary format and forwards data to a connected computer or displays it (Domdouzis et al., 2007). RFID systems operate at one of a number of different frequencies which are shown in Table 2.

	Low Frequency (LF)	High Frequency (HF)	Ultra High Frequency (UHF)	Microwave
Frequency Range	120-139 kHz	13.56 MHz	860-960 MHz	2.45 GHz
Read Range	< 0.5 m	~ 1 m	~ 4 – 5 m	~ 10 m
Main Application	Animal identification	Smart cards	Item level tracking	Electronic car toll collection

Table 2. RFID standard frequencies and related applications (The Institution of Engineering and Technology, 2005)

The ISO (International Standards Organisation) have published standards, ISO 11784 / 11785, for the use of RFID tags for animal identification (Kampers et al., 1999). These standards relate to 64 bit passive RFID tags operating at a low frequency (LF) range of 124 kHz – 139 kHz. Low frequency RFID tags are suitable for use in animal identification as electromagnetic waves at this frequency of operation suffer lower levels of attenuation due to environmental factors such as water, however, there is a reduction in read range compared to RFID tags that operate at a higher frequencies (Gandino et al., 2007). The code structure for ISO 11784 can be seen in Table 3. At the moment the ISO is working on a new standard, the ISO 14223, for use in animal identification which as of yet has not been published. In the new standard the ISO will define the code structure for advanced RFID tags, which can have a lager memory than 64 bits, such a tag will be identified by a binary flag '**1**' at its 16th bit to indicate it has a larger memory capacity. The advanced tags will have features such as: authentication (for greater security), multipage memory (pages will have predefined number of bits, and can be read-only, write once/read many times or random access memory) and on-board sensors which enables them to record environmental conditions (Jansen & Eradus, 1999).

Bit no.	Information	Numeric content	Numeric length	Values
1	Animal	0 – 1	1	**1** (animal flag)
2-15	Reserved field	00,000 – 16,383	5	**00000**
16	Additional data block flag	0 – 1	1	**0** (no additional data)
17-26	Country code (ISO 3166)	0000 – 999	3	**372** (Ireland)
27-64	Individual identification code	000,000,000,000 – 274,877,906,943	12	**000,000,000,000 – 274,877,906,943**

Table 3. ISO 11784, 64 bit code structure for animal identification (Kampers et al., 1999)

1.3 EPCglobal network

The EPCglobal Network originally developed at the Auto-ID centre in Massachusetts Institute Technology, now has its standards maintained by EPCglobal Inc. (Min Kyu et al., 2006). The EPCglobal Inc., seeks to facilitate the realisation of an "internet of things" by producing standards for the transfer of data relating to items tagged with RFIDs. Central to the strategy is the use of an Electronic Product Code (EPC), which is a globally unique number, carried by an RFID tag that can be used to identify the product to which it is attached at item or pallet level. The use of the EPCglobal Network will help reduce potential discrepancies inherent in the current CMMS database as exemplified by the figures in Table 1. GS1 (Global Standards Agency, formed by a merger of Uniform Code Council (UCC) and European Article Numbering (EAN) International) is the parent company of EPCglobal Inc. and allows the use of GS1 codes such as SGTIN (serialised global trade item number) and SGLN (serialised global location number, used to identify individual trade units which would be a bovine in this instance) in the formation of EPCs (EPCglobal Inc., 2007a).

The EPCglobal Network aims to share product traceability data between different partners in the supply chain, through the EPCIS (electronic product code information service). The EPCIS encompasses both interfaces for data exchange and specifications of the data itself, which may be either: *static data,* which will not change over the life of the object; or *transactional data,* which will change as the object travels through the supply chain (EPCglobal Inc., 2007a).

The EPCglobal Network locates information about specific EPCs by the use of Object Name Service (ONS) that is analogous to the Internet Domain Name Service (DNS) (Min Kyu et al., 2006). If information is required about a particular EPC (for instance, an SGTIN), the enquirer will contact the root ONS controlled by EPCglobal which will use the company prefix, also known as the EPC Manager Number, to point towards the local ONS and the related EPC, which in turn will provide a pointer to the EPCIS service in question (EPCglobal Inc., 2007a). A graphical representation of this process can be seen in Figure 1. A manager number is assigned to an organisation who subscribes to the EPCglobal Network by the issuing agency (GS1); in this case it will be used to identify DAFF as the manager. The manager number gives an organisation the right to use the EPC namespace based on the GS1 family of codes; also it ensures the uniqueness of the serial number that follows.

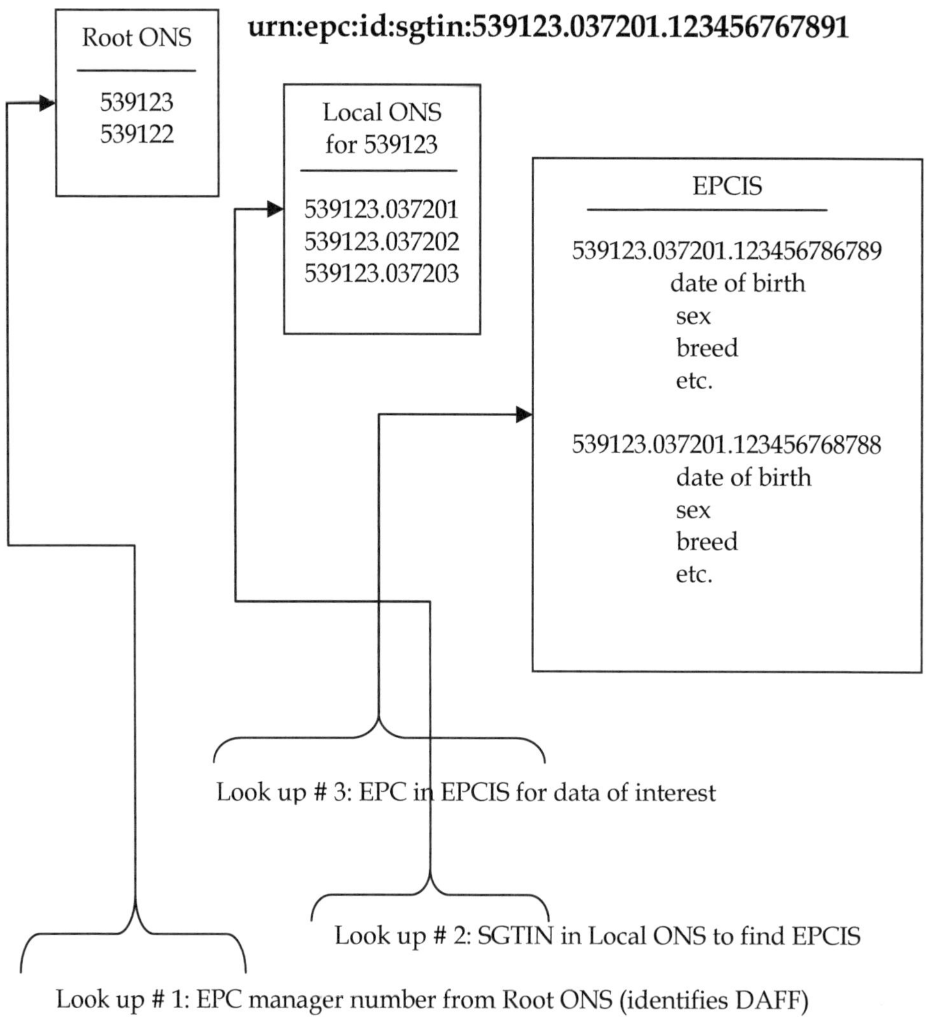

Fig. 1. Illustration of an EPCIS query for the SGTIN 539123.37201.12346786789 (adapted from EPCglobal Inc., 2007a)

1.4 Biometric identification

Biometrics has a great potential as a tool for identification or the verification of identity of individual animals. A physiological characteristic may be used as a biometric identifier as long as it satisfies the following (Jain et al., 2004):

- universality - the characteristic must be common to the population;
- distinctiveness - there should be sufficient difference in the characteristic among the population;
- permanence - the characteristic should not vary over time; and,
- collectability - the characteristic must be able to be measured quantitatively.

Various methods may be applied such as DNA testing (Stanford et al., 2001), muzzle prints (Barry et al., 2007) or retinal imaging (Barcos, 2001; Barry et al., 2007; Stanford et al., 2001). With regards to cattle, DNA testing has been developed into commercial applications in Ireland such as the identiGEN Ltd. system (Cunningham & Meghen, 2001). However, the disadvantage of using DNA testing is that it cannot provide instant results and is costly for testing large numbers of animals (Stanford et al., 2001), Dalvit et al., (2001) also points out that this technique will not be able to differentiate between monozygotic twins and cloned animals. Muzzle printing has been examined by Barry et al., (2007), who noted difficulties in obtaining satisfactory prints using ink, they also experimented with taking gray-level digital images but concluded that further study was necessary to be carried out on the automation of image capture and an increase of the testing population before this method of biometric identification could be recommended as a feasible method of identity verification.

Retinal imaging of cattle has been examined by Allen et al., (2008) where they reported 100 % accuracy in the matching of retinal images from their tests, with 98.3 % matched using a computer algorithm and the remaining 1.7 % visually. They completed their trials using the Optibrand™ system, which comprises of a handheld device for the capture of retinal images. Retinal images for both eyes of the cow can be captured in under two minutes with the use of a cattle crush to restrain the animal (Allen et al., 2008). Gonzales-Barron et al., (2008) carried out a similar study with the Optibrand™ system on sheep, where they proposed using retinal images from both eyes to confirm identification and found that this method provided a matching rate of 100 %. Retinal imaging, although a highly accurate method of biometric identification, has some drawbacks: operators require training periods; animals have to be restrained; and if imaging is required to be done outside, a shade will have to be provided to reduce the narrowing of the animal's iris which can affect the quality of the image obtained (Allen et al., 2008; Gonzales-Barron et al., 2008).

The main advantage of retinal imaging over other types of biometric identification is the near real-time results; images obtained for the Optireader™ can be compared visually to a printout or an online database beside the animal. Work completed by Rusk et al., (2006) summarised in Table 4, demonstrates the reliability of visual matching of retinal scans. The subjects who completed the visual matching had not viewed retinal images before, implying no need for previous experience. In Table 4 the final column shows the total verified after being sent to the Optibrand technicians for matching using electronic methods. In the same study, Rusk et al., (2006), had the test subjects view 156 ovine retinal images, 100 % of which were visually matched.

As the anticipated number of images and their matching is an enormous task, the BioTrack database will constitute an essential component of the traceability infrastructure. This data will hold information on the origin and identity of cattle and will be used in conjunction with the CMMS database for the verification of cattle identity but will operate independently of the CMMS database. In cases where the existing contents of a barcode are unrecoverable, the matching of retinal scans would be sufficient for the electronic identity verification of the animal. However, when phased in the use of RFID tags as data carriers would improve this situation as RFID tags can operate with a high level of accuracy in adverse environmental conditions more so than the incumbent barcode (Huber et al., 2007). In conjunction with the BioTrack database the use of RFID will improve the electronic recording of animal identity at the point of slaughter.

Animals enrolled	Re-imaged	Visual match %	Verified overall %
70	45	100	100
88	66	100	100
163	84	88.8	91.7
78	57	91.2	96.4
92	65	95.4	98.5
	317	**95.08**	**97.32**

Table 4. Visual matching of retinal images adapted from Rusk et al., (2006)

This study will detail two alternative code structures for Irish bovine identification that can be accommodated on ISO 11784 compliant tags and can be integrated into the EPCglobal Network by the use of a middleware, thus allowing DAFF to become the custodian of real-time traceability data including identity verification through the use of biometric identifiers. This study will also suggest an implementation strategy for a web based BioTrack database, for cattle retinal scans.

2. Irish cattle identification: eartag number, ISO 11784, EPC and compatibility

The identification number used in Ireland for cattle is a 12-digit numeric code encoded in GS1 Code128C barcode. In addition, the eartag also displays a 2 letter country code, which is not encoded on the eartag barcode. The numeric code is comprised of; a two digit region code, representing one of 29 separate regions in Ireland each controlled by a District Veterinary Office (DVO); a 5 digit herd code, each representing a herd keeper; a check digit, the algorithm of which is controlled by DAFF and is not in the public domain; and finally, a 4 digit animal identifier. An example of an eartag is shown in Figure 2.

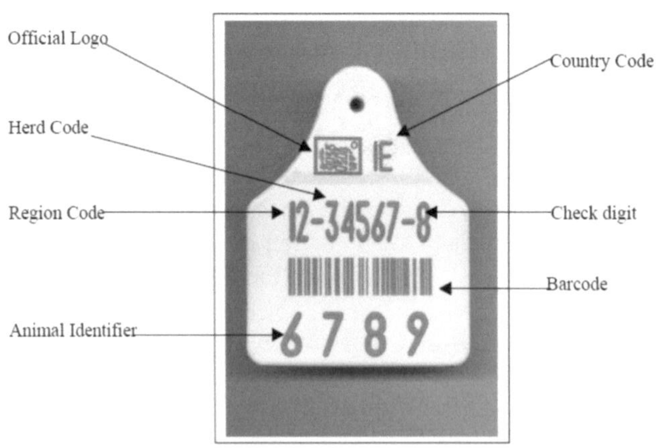

Fig. 2. An example of an Irish cattle eartag issued by DAFF (DAFF, 2006b)

In its current format, the identification number used in Ireland for cattle cannot be represented on either an ISO 11784 compliant RFID tag or as an EPC, due to the fact that the twelve digit serial number field in both does not have a wide enough range to accommodate the identification number. From the point of view of the integrity of the traceability system, it is necessary to have a single constant identification number that can be used on visual eartags, the electronic RFID tags and as an EPC. It is a legal requirement, that both eartags shall have the same unique identification code which can identify the individual animal and the holding where it was born (EC, 2000). In this section two alternative code structures are proposed that can be expressed as an EPC and comply with ISO 11784 and at the same time satisfying the requirements of EC 1760/2000.

EPCglobal Inc. have published standards on the data content of Class-1 Generation-2 passive UHF RFID tags that operate at a frequency range of 860 - 960 MHz (EPCglobal Inc., 2008). LF RFID tags are used for electronic identification of animals, with ISO 11784 being the standard. As stated earlier UHF RFID tags are not generally considered suitable for use in animal identification, a study by Sundermann and Pugh, (2008) indicated that UHF tags may have some potential but they recommended further study with larger test numbers. Tables 5 and 6 detail two alternative proposed code structures for an EPC based upon the SGTIN code structure. It is important to note that the EPC serial number field is of the same length as the ISO 11784 serial number field, both accommodating 12 digits.

There are three important fields in the SGTIN code structure. The first field is the manager number assigned to DAFF to identify them as custodians of the EPC in question. The item reference is the second important field and is used by the EPC manager (DAFF in this case) to further classify the EPC into different product classes. In Table 5 the number '037201' is suggested as the item reference, where the leading '**0**' is used to expand the field to six digits; the '372' is the ISO 3166 Country Code for Ireland; and the '01' refers to a bovine. This field can be used to identify other animals that will require tracing, for example '02' may be assigned to identify sheep, '03' for goats etc. The final field is the serial number field which is a 38 bit field that can be used to represent a 12 digit number with a maximum value of 274,877,906,943. This field will hold the identification number of the individual bovine; however, it is not large enough to contain the eartag identification number in its current format. As stated earlier there are 29 separate regions in Ireland, these regions are encoded by using the set of numbers from 11 to 39 inclusive. The first two digits of the cattle identification number are comprised of the region code. The cattle identification number could not be accommodated by either ISO 11784 or EPC in its current format, because the highest representation available to the first two digits in the serial number field is '27'. A proposed solution, shown in Table 5, is to omit the check digit from the ISO 11784 compliant tag and the EPC, thereby decreasing the length of the cattle identification number to eleven digits thus allowing adequate space to accommodate the identification number in both standards. As there is an inbuilt error check mechanism that exists in electronic tags (Kampers et al., 1999), the omission of the check digit from the eartag number will not seriously affect its integrity.

An alternative structure for a SGTIN is presented in Table 6. In this structure the region code is omitted completely. As of the end of 2007, there were 112,931 separate herds in Ireland (DAFF, 2008). The current identification number carried on eartags has a five digit herd code, this is not sufficient on its own to uniquely identify the holding of origin, (as required by EC/1760/2000) but it is used in conjunction with the region code (2 digits), meaning in total seven digits are used to uniquely identify the holding of origin. If the herd code was increased to six digits, this would provide adequate namespace to uniquely identify the holding of origin, without the need for a region code. The numeric representation of which could be as high as 274,877; more than twice the number of herds currently in Ireland. This would also enable the animal identifier to be increased to five digits utilising the extra digit freed by the omission of the region code. An expansion to five digits for the animal identifier may prove necessary to ensure adequate namespace so that the EPC is a globally unique number over the lifespan of a bovine, which is typically 20 – 25 years (Kernan, GS1 Ireland, personal communication). In this proposed structure the check digit could be retained.

Field description	Bits	Allowed Digits	Values	Meaning
Header	8	n/a	0011 0000	Identifies the following code as a SGTIN.
Filter value	3	1	010	Standard trade item grouping.
Partition	3	1	6	Determines the length of next three fields.
Company prefix	20	6	539123	DAFF assigned prefix from GS1.
Indicator digit	4	1	0	Default.
Item Reference	20	6	037201	International region code '372' ISO-3166 & Bovine identifier '01'.
Serial number	38	12	0 – 274,877,906,943	Region code (2) & Herd code (5) & Individual animal identifier (4).

Table 5. Proposed EPC SGTIN code structure for Irish bovine identification

For both examples given above the ISO 11784 encoding will be the same as the serial number field of the EPC, this serial number would then be used on the visual eartag to uniquely identify the animal and satisfy the requirements of EC/1760/2000.

Field description	Bits	Allowed Digits	Values	Meaning
Header	8	n/a	0011 0000	Identifies the following code as a SGTIN.
Filter value	3	1	010	Standard trade item grouping.
Partition	3	1	6	Determines the length of next three fields.
Company prefix	20	6	539123	DAFF assigned prefix from GS1.
Indicator digit	4	1	0	Default.
Item Reference	20	6	037201	International region code '372' ISO-3166 & Bovine identifier '01'.
Serial number	38	12	0 – 274,877,906,943	Herd code (6) & Check digit (1) & Individual animal identifier (5).

Table 6. Alternative proposed EPC SGTIN code structure for Irish bovine identification

Within the GS1 encoding structure a GLN (global location number) is the identification key for physical locations. A GLN is a 13 digit number, consisting of a company prefix, location reference and a check digit (EPCglobal Inc., 2008). Within the Irish beef traceability context a GLN will be used to identify individual holdings, for this purpose it would be desirable to maintain the 6 digit herd code as the location reference as proposed in the code structure outlined in Table 6. The code structure of a SGLN (serialised global location number) is similar to the SGTIN, the main differences being the serial number field will not be utilised and the decoding of the SGLN involves calculating a check digit that is a component of the GLN. Table 7 shows the code structure for a 96 - bit SGLN, based on the 6 digit herd code as proposed in Table 6 for the SGTIN.
In order to extract a GLN from a SGLN it is necessary to calculate the check digit, this is achieved by concatenating the company prefix and the location reference into a string represented by (d_1 d_2...d_{12}). The check digit (d_{13}) is calculated using the formula in Eqn. 1 (EPCglobal Inc, 2008).

$$d_{13} = (-3(d_2 + d_4 + d_6 + d_8 + d_{10} + d_{12}) - (d_1 + d_3 + d_5 + d_7 + d_9 + d_{11})) \bmod 10 \qquad (1)$$

Once calculated the check digit, d_{13}, is concatenated to the company prefix and the location reference to give the GLN, represented by (d_1 d_2 ... d_{13}). For the SGLN in Table 7 the GLN would be displayed as the decimal 5391231586520.

Field description	Bits	Allowed Digits	Values	Meaning
Header	8	n/a	0011 0010	Identifies SGLN.
Filter value	3	1	001	Physical location
Partition	3	1	6	Determines the length of next two fields.
Company prefix	20	6	539123	DAFF assigned prefix from GS1.
Location reference	20	6	158652	6 digit Herd code
Extension component	41	15	0 – 999,999,999,999	Not applicable to this example

Table 7. Proposed EPC 96-bit SGLN code structure for Irish herd identification

Subscribers to the EPCglobal Network often have different requirements / business processes to be accommodated. To facilitate this, a middleware is often required to translate RFID data into appropriate information for specific purposes. To this end, we have developed software to translate cattle eartag numbers into the suggested ISO 11784 and EPC encodings and formulate a GLN from the company prefix and the 6 digit herd code, an example of the transformation of an identification number into EPC and ISO formats can be seen in Figure 3. The EPCglobal Network is system independent; that is, it is concerned with the transfer of information in a standard format that can be read by users no matter what internal software they are operating (EPCglobal Inc., 2007b). The manner in which the information is translated into the required standard can be defined by individual organisations through the use of middleware. The advantage to this approach is that it gives organisations the freedom to tailor the software to their specific needs. In the case of DAFF this would mean that instead of purchasing / developing an entirely new software package from the ground up, they could employ middleware to translate the traceability information already stored in the CMMS database into EPC compliant format.

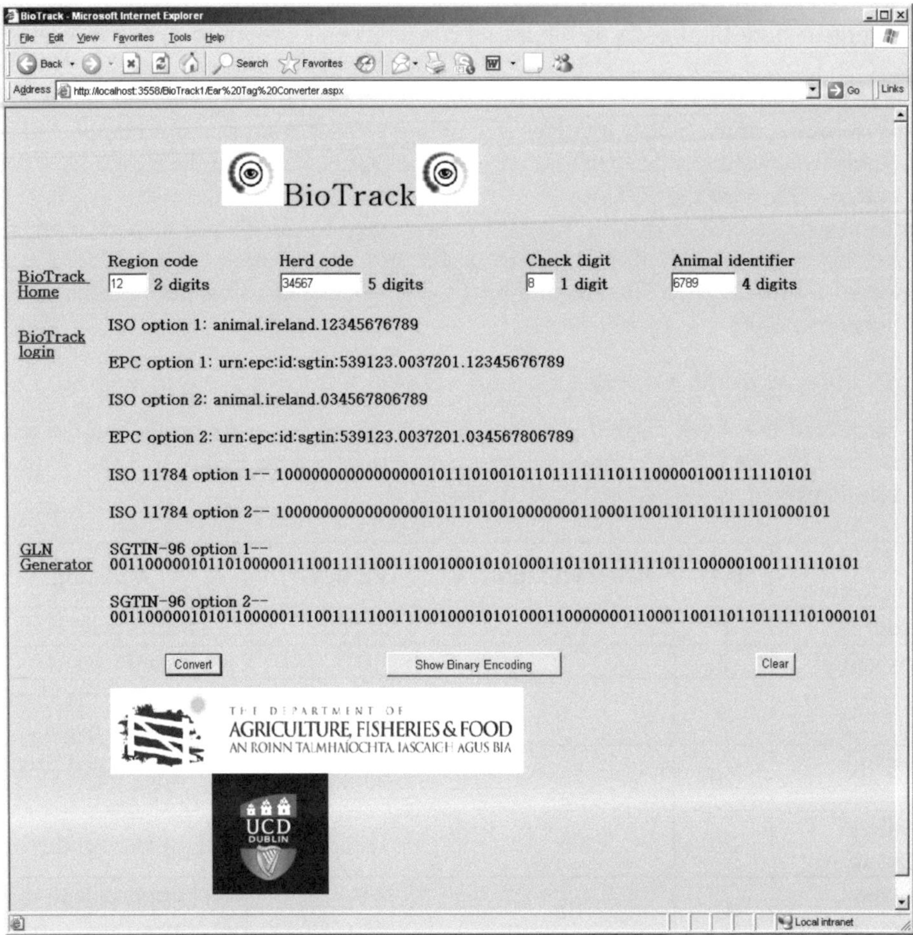

Fig. 3. Example of web based middleware for the transformation of Cattle identification number to ISO 11784 and EPC

3. A model for phased implementation of the BioTrack database

The provision of identity verification at the point of slaughter is the main aim of BioTrack - the ability to know for sure that animal A is in fact animal A, to validate all the traceability information associated with the given animal for the pre-slaughter supply chain. The contents of the BioTrack database have been outlined by Shanahan et al., *2009*:

- eartag number
- retinal scan right
- retinal scan left
- eartag image
- herd details (name/address)

- date of birth
- GPS location
- scan date/time stamp
- device ID
- operator ID

An example of a record stored in the BioTrack database and the visual display in a web browser can be seen in Figure 4.

The implementation of BioTrack should be completed on a phased basis. Optireaders™ should be made available where there is an existing link to the CMMS database, namely; marts, export points and slaughtering locations. Once they are located in these areas they can be used for the capture and verification of cattle identity. Of a total of 2,519,885 cattle movements which occurred in 2007, 1,625,290 of these were carried out at marts (DAFF, 2008), this represents approximately 26 % of the national herd (this figure does not take into account animals that were sold at mart multiple times during 2007). There are a number of advantages to installing biometric capture points at marts:

- farmers can become familiar with the technology;
- there is an existing link to the CMMS database for the upload of data; and,
- large proportion of the cattle being traded at marts are being sold for fattening purposes which means that they will ultimately end up going for slaughter (where the BioTrack database can be consulted for identity verification).

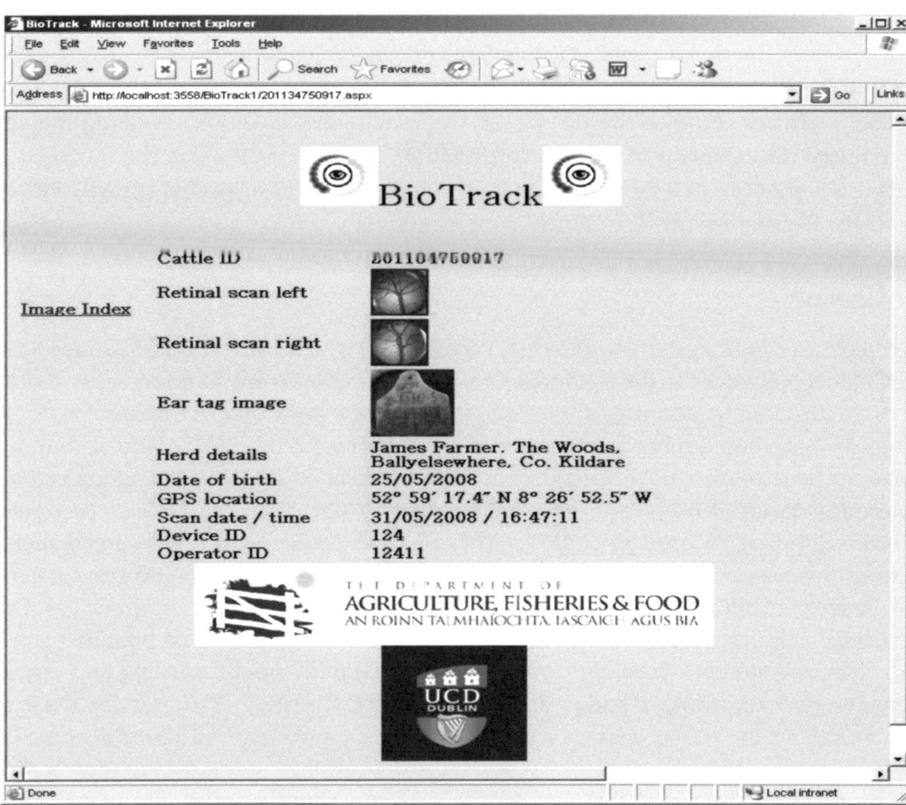

Fig. 4. Display of a BioTrack record in a web browser

Many large retailers and beef processors carryout direct trading with the herd keepers that supply their cattle, and this is an area that can be targeted for an initial capture of biometric identifiers. A total of 1,694,488 cattle were slaughtered in factories, this being 75 % of all cattle disposals in 2007 and 27 % of the national herd. On-farm deaths accounted for 12 % of disposals, the remainder of disposals were export (10 %), and local authority licensed abattoirs (3 %) (DAFF, 2008). As the statistics show if retailers and processors made it a requirement that their supplier's record a biometric identifier from their animals, the direct trading section of the supply chain would constitute a large proportion of the beef destined for the consumer's plate, and would have adopted biometric identity verification.

Aside from biometric collection at marts and animals destined for the commercial sector DAFF will have to implement a process for biometric collection of the national herd. This would be the responsibility of DVOs. Veterinary officers of the DVO currently carry out a number of farm inspections each year, the purpose of which is to ensure animal health and that housing conditions are of an adequate standard (DVO, County Dublin, personal communication). It would be possible for the veterinary officers to record the retinal images and identification numbers of all cattle on the farm during these inspections, which would also allow the DVO personnel to accumulate experience while gathering biometric identifiers. It would be unrealistic to expect that the BioTrack database could be populated with the retinal images of the 6 million plus Irish herd at once; however, efforts should be focused on a phased recording of biometrics for beef cattle as they are most likely to be destined for the consumer's plate.

Once the retinal image has been captured it will be up-loaded to the BioTrack database which will be under the control of DAFF. The BioTrack database will be linked to the EPCglobal Network through the use of the ONS, allowing stakeholders along the supply chain to query traceability and identity information. It is envisioned that there will be a flag indicating whether or not a retinal image has been captured for a specific animal, which will be displayed if a request for EPC information is received.

4. Discussion

While there are advantages to employing a traceability system based on RFID tags utilising the EPCglobal Network for the exchange of information, there are some considerations that have to be taken into account. Currently the average herd size in Ireland is 55 head, however 41 % of herds have less than 25 head of cattle (DAFF, 2008); and it may not be practical for herd keepers of this magnitude to implement RFID systems. Taking economics of scale into account it may be more advantageous to start with the 24 % of herd keepers that have more that 75 head of cattle (DAFF, 2008), it could be safely assumed that herd keepers of this size already have some farm management software to aid production and would be more comfortable with technological advancements in farming practices. Once a herd is fitted with electronic eartags there are other values that can be obtained from the system; such as automatic feed distribution, individual milk yield recording and automatic live weight gain recording (Eradus and Jansen, 1999; Rossing, 1976). Under the current system in Ireland the cost of eartags for cattle (currently priced at € 2.15 for a single eartag and € 2.94 for pair of eartags as shown in Figure 2 (Eurotags, personal communication) is borne by the herd keeper. RFID enabled eartags are slightly more expensive - costing approximately € 3 for a single tag (Eurotags, personal communication). A survey of

American electronic tag suppliers (7 quotes) gave an average price of € 1.73 (exchange rate of 0.635 as at 09/07/2008) for ISO 11784 compliant RFID eartags (This study).
There are now mobile phones and personal digital assistants (PDAs) with in-built RFID / barcode readers on the market, making it possible for herd keepers to record the identification numbers of cattle and through a system described by Min Kyu et al., (2006), enabling communication through the EPCglobal Network via a mobile phone network which could facilitate requests for movement authorisation from the CMMS, which would streamline the process and make redundant the need to apply in writing for cattle movement authorisation.

5. Conclusion

The accurate and timely identification of cattle is a necessity if full chain traceability from farm to fork is to be achieved. With current technology such as RFID cattle tags, cattle identification numbers can be captured automatically and shared along the supply chain through the use of the EPCglobal Network, which would rely upon the traceability infrastructure already maintained by DAFF. The use of retinal images as a biometric, stored on a BioTrack database to verify identity would provide a system check that would be virtually fraud-proof. While such a system may be costly to implement it is suggested that larger herd keepers and suppliers to major retailers and processors be the first to adopt the RFID tagging and biometric capture, while marts and commercial slaughterhouses can be the first premises to install biometric identity verification systems. A system such as this would be able to identify cattle whose eartag has been tampered with; in the case of a retinal image not matching to the identification number on the eartag, it would be an indication that fraudulent activity may have occurred and such an animal should not be allowed to go to slaughter and an investigation initiated by the district veterinary officer, who is required under Irish law to be present at all cattle slaughtering. The introduction of BioTrack would provide a mechanism for source and identity verification of Irish beef products, the utilisation of the EPCglobal Network would also ensure that trading partners around the globe will have confidence in the traceability infrastructure commensurate with the high standards of production employed in Ireland thus adding value to the beef sector.

6. References

Allen, A.; Golden, B.; Taylor, M.; Patterson, D.; Henriksen, D. & Skuce, R., (2008). Valuation of retinal imaging technology for the biometric identification of bovine animals in Northern Ireland. *Livestock Science*, 116, 1-3, pp. 42-52, ISSN 1871-1413

Barcos, L. O., (2001). Recent developments in animal identification and the traceability of animal products in international trade. *Rev. sci. tech. Off. int. Epiz.*, 20, 2, pp. 640-651

Barry, B.; Gonzales-Barron, U. A.; McDonnell, K.; Butler, F. & Ward, S., (2007) Using muzzle pattern recognition as a biometric approach for cattle identification. *Transactions of the ASABE*, 50, 3, pp. 1073-1080, ISSN 0001-2351

Cunningham, E. P. & Meghen, C. M., (2001) Biological Identification systems: genetic markers. *Rev. sci. tech. Off. int. Epiz.*, 20, 2, pp. 491-499

Dalvit, C.; De Marchi, M. & Cassandro, M., (2007) Genetic traceability of livestock products: A review. *Meat Science*, 77, 4, pp. 437-449, ISSN 0309-1740

Department of Agriculture, Fisheries and Food (DAFF), Ireland, (2003) *CMMS Statistics Report 2002*. Available online at: http://www.agriculture.gov.ie/media/migration/animalhealthwelfare/animalidentification/cattlemovementmonitoringsystem/cmmsstatsrept.pdf Accessed on 15/07/2008.

Department of Agriculture, Fisheries and Food (DAFF), Ireland, (2004) *CMMS Statistics Report 2003*. Available online at: http://www.agriculture.gov.ie/media/migration/animalhealthwelfare/animalidentification/cattlemovementmonitoringsystem/cmms_stats.pdf Accessed on 15/06/2008.

Department of Agriculture, Fisheries and Food (DAFF), Ireland, (2005) *CMMS Statistics Report 2004*. Available online at: http://www.agriculture.gov.ie/media/migration/animalhealthwelfare/animalidentification/cattlemovementmonitoringsystem/CMMSstatsreport2004.pdf Accessed on 15/06/2008.

Department of Agriculture, Fisheries and Food (DAFF), Ireland, (2006a) *CMMS Statistics Report 2005*. Available online at: http://www.agriculture.gov.ie/media/migration/animalhealthwelfare/animalidentification/cattlemovementmonitoringsystem/CMMSstatsreport2005.pdf Accessed on 15/06/2008.

Department of Agriculture, Fisheries and Food (DAFF), Ireland, (2006b) *Competent authority, model of passport, eartags and holding register*. Available online at: http://www.agriculture.gov.ie/media/migration/animalhealthwelfare/animalidentification/cattlemovementmonitoringsystem/ireland.pdf Accessed on 01/09/06.

Department of Agriculture, Fisheries and Food (DAFF), Ireland, (2007) *CMMS Statistics Report 2006*. Available online at: http://www.agriculture.gov.ie/media/migration/animalhealthwelfare/animalidentification/cattlemovementmonitoringsystem/StatisticsReportInside2006.pdf Accessed on 07/02/2008.

Department of Agriculture, Fisheries and Food (DAFF), Ireland, (2008) *CMMS Statistics Report 2007*. Available online at: http://www.agriculture.gov.ie/media/migration/animalhealthwelfare/animalidentification/cattlemovementmonitoringsystem/CMMSreport2007.pdf Accessed on 07/07/2008

District Veterinary Office, (2008), Dublin, Personal communication.

Domdouzis, K.; Kumar, B. & Anumba, C., (2007) Radio-Frequency Identification (RFID) applications: A brief introduction. *Advanced Engineering Informatics*, 21, 4, pp. 350-355, ISSN 1474-0346

EPCglobal Inc., (2007a) *EPC Information Services (EPCIS) Version 1.0.1*. Available online at: http://www.epcglobalinc.org/standards/epcis/epcis_1_0_1-standard-20070921.pdf Accessed on 25/07/2008

EPCglobal Inc., (2007b) The *EPCglobal Architecture Framework*. Available online at: http://www.epcglobalinc.org/standards/architecture/architecture_1_2-framework-20070910.pdf Accessed on 15/07/2008

EPCglobal Inc., (2008) EPCglobal *tag data standards version 1.4*. Available online at: http://www.epcglobalinc.org/standards/tds/tds_1_4-standard-20080611.pdf Accessed on 03/07/2008

European Commission, (2000) *Regulation (EC) No 1760/2000 of the European Parliament and of the Council of 17 July 2000 establishing a system for the identification and registration of bovine animals and regarding the labelling of beef and beef products and repealing Council Regulation (EC) No 820/97*. Available online at: http://eur-lex.europa.eu

/LexUriServ/LexUriServ.do?uri=OJ:L:2000:204:0001:0010:EN:PDF Accessed on 15/01/2008

Eurotags (2008), Mullinahone Co-op, Ireland, Personal communication.

Gandino, F.; Montrucchio, B.; Rebaudengo, M. & Sanchez, E. R., (2007) Analysis of an RFID-based Information System for Tracking and Tracing in an Agri-Food chain. *Proceedings of 2007 1st Annual RFID Eurasia*, pp. 1-6, ISBN 978-975-01566-0-1, Istanbul, Sept. 2007

Gonzales-Barron, U.; Corkery, G.; Barry, B.; Butler, F.; McDonnell, K. & Ward, S., (2008) Assessment of retinal recognition technology as a biometric method for sheep identification. *Computers and Electronics in Agriculture*, 60, 2, pp. 156-166, ISSN 0168-1699

Huber, N.; Michael, K. & McCathie, L., (2007) Barriers to RFID Adoption in the Supply Chain. *Proceedings of 2007 1st Annual RFID Eurasia*, pp. 1-6, ISBN 978-975-01566-0-1, Istanbul, Sept. 2007

Jain, A. K.; Ross, A. & Prabhakar, S., (2004) An introduction to biometric recognition. *Circuits and Systems for Video Technology, IEEE Transactions on*, 14, 1, pp. 4-20, ISSN 1051-8215

Jansen, M. B. & Eradus, W., (1999) Future developments on devices for animal radiofrequency identification. *Computers and Electronics in Agriculture*, 24, 1-2, pp. 109-117, ISSN 0168-1699

Kampers, F. W. H.; Rossing, W. & Eradus, W. J., (1999) The ISO standard for radiofrequency identification of animals. *Computers and Electronics in Agriculture*, 24, 1-2, pp. 27-43, ISSN 0168-1699

Kernan, B., (2008), GS1 Ireland, Personal communication.

Landt, J., (2005) The history of RFID. *Potentials, IEEE*, 24, 4, pp. 8-11, ISSN 0278-6648

Min Kyu, H.; Il Woo, P.; Byung Hee, L. & Jin Pyo, H., (2006) A Framework for Seamless Information Retrieval between an EPC Network and a Mobile RFID Network. *Proceedings of Computer and Information Technology, 2006. The Sixth IEEE International Conference on*, pp. 98-98, ISBN 0-7695-2687-X, Seoul, Sept. 2006

Rusk, C. P.; Blomeke, C. R.; Balschweid, M. A.; Elliott, S. J. & Baker, D., (2006) An evaluation of retinal imaging technology for 4-H beef and sheep identification. *Journal of Extension*, 44, 5, ISSN 1077-5315

Rossing, W. (1976) Cow identification for individual feeding in or outside the milking parlour. *Proceedings of the Symposium on Animal Identification Systems and their Applications*, Wageningen, 1976

Shanahan, C.; Kernan, B.; Ayalew, G.; McDonnell, K.; Butler, F. & Ward, S., (2009) A framework for beef traceability from farm to slaughter using global standards: an Irish perspective. *Computers and Electronics in Agriculture*, 66, 1, pp. 62-69, ISSN 0168-1699

Shanahan, C.; Ayalew, G.; McDonnell, K.; Butler, F. &Ward, S. (2007), The use of genetic algorithms for data recovery from damaged barcodes. *Proceedings of the 37th Annual Research Conference Food, Nutrition & Consumer Sciences*, pp 100-101, ISBN 978-0-9556109-1-2, University College Cork, Sept. 2007

Stanford, K.; Stitt, J.; Kellar, J. & McAllister, T., (2001) Traceability in cattle and small ruminants in Canada. *Rev. sci. tech. Off. int. Epiz.*, 20, 2, pp. 510-522

Sundermann, E. & Pugh, G., (2008) The application of UHF RFID technology for animal ear tagging. The New Zealand RFID Pathfinder Group Inc. Available online at: http://www.rfid-pathfinder.org.nz/images/pdf/report-uhf-animal-tag-trials-july08.pdf Accessed on 10/08/2008

The Institution of Engineering and Technology, (2005) Radio Frequency Identification device technology, a factfile provided by the Institution of Engineering and Technology. Available online at: http://www.theiet.org/factfiles/it/rfid.cfm?type=pdf Accessed on 08/04/2008

8

Improving on Passenger and Baggage Processes at Airports with RFID

Katalin Emese Bite
Budapest University of Technology and Economics
Faculty of Transportation Engineering, Department of Transport Economics
Hungary

1. Introduction

Today's airports are overcrowded. The queues are long, passengers don't have time to spend it on the airport queuing, but security restrictions must be kept. Everyone would like to lower the high cost wherever it is possible. Such an area is the amount of costs generated by the baggage loss within the air travel. Another factor is the delay of flights, which can be generated by passengers late at the boarding or even not appearing. The aircraft can only take off if all the checked-in baggage has its owner on board. If not, the baggage has to be offloaded.

The costs generated by baggage loss are very high for both the airlines and the airports. The application of RFID technology would reduce these costs extremely. Today's implementation and chip prices are very high but with time it will decrease. The average industry cost per mishandled baggage is US$100. Approximately 1% of the 1.7 billion bags that passes through the system every per year is mishandled and RFID is an ideal candidate to reduce these losses. Upon full implementation, RFID would save the industry US$760 million annually.

2. Airport passenger and baggage reconciliation systems in use

After arriving at the airport, the traveller enters the terminal building at the departure hall. There the passenger checks-in himself and his baggage, which will be part of the Departure Control System (DCS). The DCS after entering all the necessary data will print a Boarding Pass and the long Baggage Tag (BagTag) with a barcode. The Boarding Pass is printed to inform the passenger of the flight number, boarding time, boarding gate number and seat number, and it is used to identify the passenger at the security and immigration check and boarding gate too. The barcode of the checked-in baggage serves the identification until the final destination. The longer part of this BagTag is put on the checked-in baggage. The passenger receives the smaller slip that contains the same barcode as the checked-in baggage. In case of baggage loss the airline is able to identify and find out where the baggage has been lost. Without the passenger having this receipt the airline is not obliged to find the lost luggage and compensate the passenger.

In recent years industrial deployments have changed the previous infrastructure of the departure hall. The operation of the check-in system has not changed much, but for

lowering the cost, the used tools (check-in desks, boarding card) have changed. The operation became more automatic and the passengers are more independent.
Currently on many airports there are different facilities available:

1. Traditional check-in desks with an agent: serving mostly the business, frequent flyer and the through check-in passengers.
2. Self check-in kiosks: where the passenger has to check-in himself, following the indications of the touch-screen kiosks. The passenger has to provide the requested data and can print his own boarding pass and baggage tag and then continue to the Baggage Drop to weight and drop off the checked-in baggage. When self check-in kiosks are introduced, an agent can help the passengers.

Fig. 1. Self Check - in kiosk

3. Portable Agent Workstations, Mobile Check-in device: agents circulate around the check-in area looking for customers for checking them in with a hand-held personal computer. These agents can also print the boarding pass and baggage tag, and then the passenger can to continue to drop off its luggage. This method is rarely used (e.g. Kingfisher is using it at Madras Airport). (Pilling, 2001)
4. A mixture of the above mentioned possibilities.
5. A new trend is for passengers without checked-in baggage:
   - web check-in: the boarding pass is issued through the web and the passenger has to print it at home
   - mobile check-in: the passenger can check in via his mobile and the boarding pass will be sent by SMS/MMS to the passenger's mobile phone
     Solutions are being prepared for this kind of check-in for passengers with checked-in luggage too.
6. Remote Check-in: in some cities (e.g. Las Vegas) it is possible to check-in in the hotel or in other cities (e.g. Hong Kong) at major interchanges and the airline will deliver the checked-in baggage to the airport.

The above mentioned check-in possibilities can use several tools too:

1. Boarding Passes:
   - Traditional Magnetic Strip
   - BarCoded Boarding Pass: using 2D barcode printed on a paper from the airport's check-in facility or outside the airport from the web or sent to mobile phones or

PDAs in SMS/MMS format. It should be used by all IATA member airlines by the end of 2010, and it should completely replace the magnetic strip

2. Baggage Tag:
   - Barcode: this is the commonly used solution
   - RFID tags embedded in the back of barcode paper: some airports and airlines have adopted it after some trials (e.g. Las Vegas, Hong Kong)

After the check-in the ways of the passenger and the baggage will separate, and unite again at the Baggage Claim of the final destination. The following graph (*Fig. 2)* shows the way and the steps a passenger and a luggage takes while travelling by an airplane:

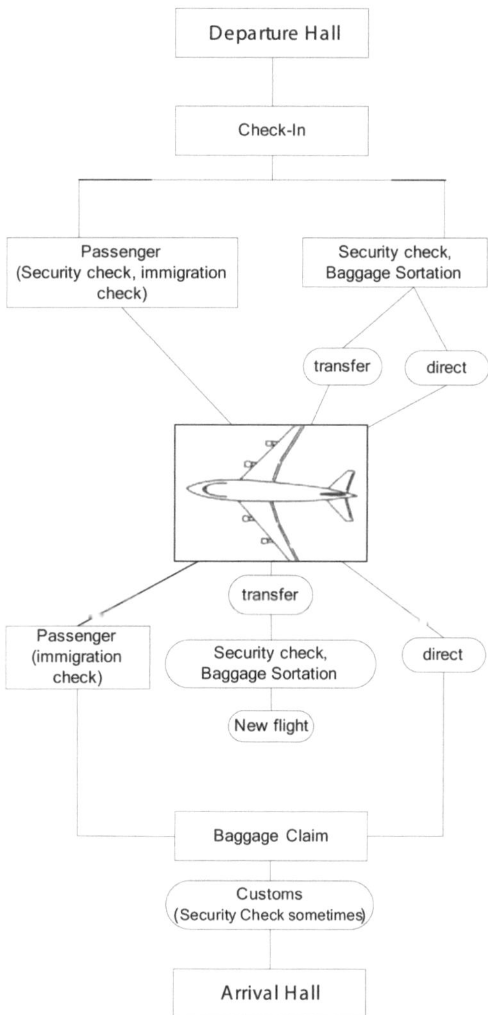

Fig. 2. Passenger and baggage flow during the flight procedure

The passenger is attending the security and immigration checks, the order depends on the airport and then at the time of boarding it will proceed to the plane. In the meantime, after the baggage check-in, the baggage passes through security check and baggage sorting. In the sorting room, with today's reconciliation technology, the stevedore scans the BagTag's barcode with a scanner that translates the encoded barcode and shows the stevedore to which baggage cart or container and airplane the luggage should be directed to. After arrival of the flight, in case of a direct passenger, the passenger continues to the immigration check and then to the baggage claim to collect his baggage and leaves the airport through the arrival hall. At the exit of the baggage claim nobody checks if the baggage was taken by its owner or another person. In case of a transfer passenger and baggage, the passenger stays in the transit of the terminal building after leaving the aircraft, and he passes through immigration check (depending on the destination) and a security check before re-boarding. The baggage passes through security check, re-sorting and then goes to the new aircraft. If the transfer time is one hour or less, the baggage is tagged with the ShoCon (Short Connection) sign. However, when the airplane is delayed, and the baggage would be needed to be transferred quickly to the next airplane, there is no special sign tagged on the luggage (originally it was supposed to arrive on time and supposed to have enough transfer time), the possibility of the non-arrival of the baggage is very high.

2.1 Problems with in-use barcode-scanner system for the checked-in baggage

Most of the world's airports use a scanner and a paper printed barcode for baggage identification.

The key problems with the barcode and scanner are as follows:

- The barcode needs optical sight, without the line of sight, it can not be read
- Concurrently the scanner is able to read only a single barcode, which is time consuming
- Barcode baggage tag read rates average 85%
- Barcode is printed on a paper that easily crumples, thus the scanner is not able to decode the information properly
- After printing the barcode it is not possible to overwrite the information (only by printing a new one)
- The paper of the barcode is long, full of information that comes off easily, thus making it impossible for the stevedore to identify where the luggage is supposed to be sent and the airline is unable to find it in the computer database. It will be regarded as the airline's mistake, and the airline has to compensate the passenger.
- *Fig.3* shows that the barcode is printed on a long-hanging paper, which is only attached at the middle or at a suitable part to the luggage. The most important part of the paper is just hanging down - without being fixed to the luggage- so it can easily come off or someone can tear it away.

Fig. 3. Today's Baggage barcode solution

2.2 Issues with the lost baggage

The causes of losing a luggage can be diverse: airline baggage system integration, the baggage process of an airport is overly complicated, new and tighter security regulations and more congestions at the airport, tagging error or mistake in the identification, sorting, loading or offloading of the baggage (it could simply fall of the trolley) at the departure or/and arrival /transfer airport, the transfer baggage could be directed to a false destination due to wrong identification or due to too short transfer times, due to human error at the check-in (e.g. wrong typing, passenger is checking in too late), weather or space-weight restriction, communication error between the agents (e.g. in case of rerouting) or the BagTag can fall off the baggage. In the last case the baggage is lost forever, the system for finding lost luggage can not find it as it is not possible to identify it, and according to data of IATA this is the case with 800,000 bags in the world every year.

The baggage can also get lost at the baggage claim without the error of the airline, airport or the operator: it can be taken by another passenger by accident (due to similarities) or it can be intentionally stolen. Irrespective of the reasons, it costs the airline and the airport a lot of money. The airline has to compensate the passenger in some form, depending on whether they find it and forward it to the owner within 24 hours, days or weeks or never and depending on whether the passenger was arriving at home or not. The compensation rules are standardised by IATA and the airlines.

The retrieval costs of a lost bag costs the airlines between US$100-150, excluding the eventual cost of an airplane being held up because of a mishandled bag (Ornellas, 2006).

Carriers	Passenger Checked-in baggage (million)	Mishandled
Airlines of the USA, 2005	440	2.93 million
US domestic airlines	Ca. 600	4.08
Southwest Airline	98	525.000
US Airways	49	420.000
Delta Air Lines	66	456.000

Table 1. Mishandled baggage in the USA in 2006 (Ornellas, 2007)

Year and Continent	Total passenger Checked-in baggage (million)	Lost luggage rate (bag/1000 passenger)
2006 Globally	N/A	6.73
2005 Europe	346	14.1
2006 Europe, airlines belonging to AEA	357	15.9

Table 2. Lost baggage rate (Ornellas, 2007)

Association of European Airlines (AEA) pointed out that this statistics do not differentiate between irrevocably lost baggage and bags later found and returned to their owners (Ornellas, 2007). It doesn't matter, if it is returned or not it costs the airline time and money and the passenger has hassle and is unsatisfied. Approximately 1% of the 1.7 billion bags that pass through the system every year are mishandled. Mishandled baggage is an annual US$3.8 billion problem for the aviation industry. It also affects about 42 million passengers annually and is the second most important factor in having a pleasant trip, according to the 2009 IATA CATS survey. Between 2005 and 2007, the passenger growth has increased by 9% in Europe, while the number of mishandled baggage has increased by 28% (IATA data source) In the USA the passenger numbers grew by 10.5% while the number of mishandled luggage has increased by 27% (IATA data source). According to IATA the key regions are: Europe, the USA and Latin America. In 2008 the rate of mishandling decreased around 20%, according to data of the Société Internationale de Télécommunications Aéronautiques (SITA) (Bondarenco & Price, 2009). The reason for this is the tremendous effort by airlines and airports and the fewer number of bags per passenger. In 2008 and 2009 the low cost airlines progressively introduced to pay extra for hold baggage and the global economy problems led to fewer passengers too.

2.3 Problems with passengers

The plane can only take off if the owner of the checked-in baggage is on board, if not, the agent has to find and remove the checked-in baggage. This takes a lot of time and can cause flight delay which can lead to further problems and delays costing the airline a significant amount of money and efforts.

There are several possibilities why a passenger is late at the boarding gate: the passenger can get lost, cannot find the way to the correct gate, it is stocked at the long queue of border control or security check, arrived simply too late at the airport, it was lost within the shops or any of the airport facilities, forgot the time and the flight, cannot understand or hear the loud speaker in case of gate change, or is simply to absent-minded, the signs of the airport are not clear enough, or even some medical problem or emergency occurred etc. Whatever the real reason, it costs money for the airline.

At Copenhagen Airport 4% of the flight delay for SAS (Scandinavian Airlines) are due to late passenger at the boarding gate (Ornellas, 2008).

3. Radio Frequency Identification as the improvement

Radio Frequency Identification (RFID) is a technology incorporated into a silicon chip that emits a radio signal which matches a user-defined serial number with an item. In this case the item is a piece of check-in baggage. This number can be read at a distance by an antenna. The following characteristics enable baggage to be sorted automatically and loaded faster than with barcode systems, while reducing the number of mishandled baggage and its associated costs at the same time.

The main differences between the RFID and the barcode-scanner technology are listed below:

- The tag is read by an antenna, it doesn't need optical sight
- Greater amount of baggage can be read simultaneously
- It is able to talk-write to a single tag allowing updating the status of the baggage as it is processed
- Barcode baggage tag read rates average 85% while RFID baggage read rates range between 95-99%

4. International approach to RFID Applications for airports

Current trends of the aviation industry, following the Simplifying the Business (StB) program of the International Air Transport Association (IATA), are: simple and seamless

Attributes	Bar-code	RFID
Optic view	Necessary	Antenna is reading from the distance
Reading possibility	Scanner points	Active tag: always, Passive tag: access points
Read rates accuracy	80-90 %	95-99%
Read - Write	Read only	Read-Write
Up-dating	1 time	Anytime
Real time bag matching	No	With the people
Data	Definite	Indefinite
Location	Top of bags	Everywhere
Removable, Vulnerability	Easily	Impossible
Reading after Vuln.	Mishandling	It can be identified correctly
Configuration	Long Paper	Can be embedded in everything
Technical equipment	Paper, Printer Scanner	Tag Read-Writer Antenna for the reading International Database possibility
Environments	Disposable	Re-usable
Speed	Slow	Fast
Automated (Manpowered)	Manually	Automated
Price	6-8 cent	20-42 cent
Cost	Cheap	The tags are expensive, + implementation costs
Maintenance	Has to be wiped daily	Little

Table 3. Comparing the technologies

travel experience with minimised hassle and more control by the passenger (e. g. less queuing, less time needed at the airport, more independency), meet the consumer friendly expectations, establish financially sustainable business environment, lower the costs of the airlines, environment friendly (paperless e-ticketing), faster and more efficient baggage handling, to create industry-wide standards (IATA b.)

IATA's StB Program had a part concerning RFID, but the related project was closed. On the website (www.iata.org) of the organisation it is written:"Because the value of RFID is subject to the individual merits of each business case, there is no mandate for the universal adoption of RFID from IATA." The project standardized the used RFID tag and frequency for the aviation industry, implementing it into the paper BagTag on the back of the barcode.

Within the StB Program, there was a goal to introduce the BarCoded Boarding Passes (BCBP) using 2D barcode. They can be accessed from anywhere, from mobile phones, PDAs, web, they don't need to be printed on expensive paper stock, and they facilitate off-airport check-in, they are cost saving and environmental friendly. Nowadays they are already in circulation or trials are undertaken.

Another program run by IATA related to passengers was Simplifying Passenger Travel (SPT). The goal was to facilitate the flight procedure for passengers, while emphasizing the simplified and secure passenger processing.

The aviation industry is already trialling and in some areas already applying the RFID technology. Tracking Ground Support Equipment (GSE), catering, cargo is becoming common.

Another useful application of the RFID technology is for the access control of vehicles to airport operational areas (Pilling, 2001). At London Heathrow airport American Airlines' access control system prevents unauthorized drivers from using American Airline equipment as the driver can only start the vehicle's engine by using Airport Security pass which is recognized by the use of RFID technology supplied by Vehicle Telematics Information System (VTIS) (Ornellas, 2007).

Airbus and Boeing cooperated in using RFID for the parts of the aircrafts. Airbus applied it to track tools and for inventory control on inbound shipping pallets (Mecham, 2005).

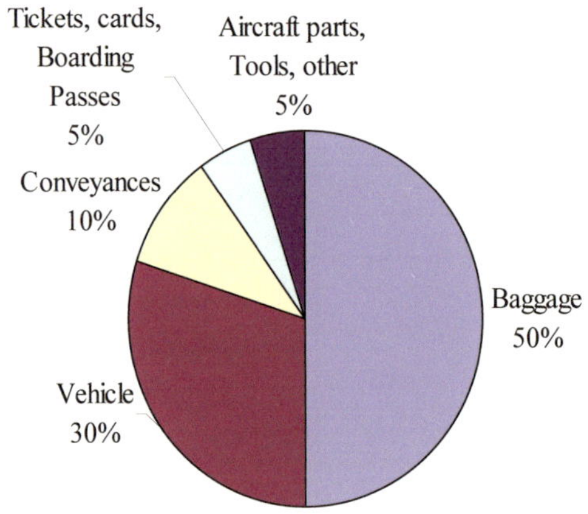

Fig. 4. Spent on RFID systems in the civil air industry in 2006, (Ornellas, 2006)

4.1 RFID for baggage handling

At the Passenger Services Conference in Geneva November 2005 the International Air Transport Association (IATA) has introduced a global standard for RFID baggage tags that paves the way for the use of RFID. In 2005, however, only two or three European countries have permitted site licenses for testing at the higher power level.

Various tests are conducted at major airports with RFID baggage tags. At Las Vegas McCarran International Airport, Hong Kong International Airport it is already operational. Paris, Amsterdam, Milan and San Francisco airport the trials are still going. Trials were completed in Vancouver, Philadelphia, New York, Honolulu, Nairobi, Frankfurt, London, Amsterdam, Rome, Kuala Lumpur, Beijing, Narita in Japan and some Korean airports (IATA a.).

Hong Kong Airport is using the technology for the checked-in baggage and is providing for the transfer baggage arriving to the airport without an RFID tag an extra tag. Currently, more than 70 airlines are involved in this project (Ornellas, 2009).

Heathrow started a six month trial of RFID technology for Emirates passengers (Ornellas, 2008).

For 2004 Delta Airlines has tagged its 40 thousandth passenger's baggage within a pilot program. While in 2004, the amount of RFID enabled baggage tags delivered was only a couple of thousands (and all in a trial setting), by 2005, however, this amount has increased to about 15 million, with an average price of 22 cents per tag.
The airports testing the RFID put the RFID tag into a paper and then attach the paper to the baggage. Even though the paper can come off, the identification is much easier. The other main problem is that if only some airport are applying RFID and the rest is using only barcode application it is not as efficient and still can get lost during the air travel. IATA calculates that airlines would save $768 million annually from reducing their mishandled baggage by only 1%. Even more can be saved by airports, about three times as much. The use of RFID in transfer processes was carefully analysed by IATA as part of the RFID transition plan. This analysis showed that only 80 airports needed to adopt RFID to deliver an annual benefit of over US $200 million to the industry (IATA c.).
Currently the trials and already applied systems are embedding the RFID tag into a BagTag. The baggage tag is a long paper and with embedded RFID tag it can still fall off the baggage.

4.2 RFID for passengers, and other improvements for passenger handling

In 2008, on the Amsterdam Passenger Terminal Expo and Conference international companies showed their latest developments, which provide the technologies necessary for printing and reading 2D Boarding Cards in any format. Different 2D boarding pass printers and technologies were on display, that could be built into Self Check-in kiosks or mobile printers. One of the latest possibilities was the boarding pass sent to mobile phones or PDA-s via SMS or MMS. The greatest advantage of this technology was that the passenger can be informed of the changes related to his trip, but this requires the user to have a modern mobile phone or PDA capable of receiving MMS, which is always on, and can communicate on all frequencies. These solutions are targeted for frequent flyers and business passengers, but they cannot be used for all passengers, because it cannot be expected that everyone has a mobile phone or PDA with such technology. Another problem with this innovation is that a mobile phone or PDA can run out of battery without possibility to recharge it. In the show, boarding pass readers and identification machines were also on display.
The automated boarding gate can read 2D barcode Boarding Passes from mobile phones, PDAs and paper. It is not only a boarding gate it has the facility of the basic process of automated passenger authentication. That means to have the biometric data of the passenger scanned while he is standing at the gate, to instantly compare this data to an existing biometric template of the same passenger and to check if both match. The second step is to check whether this passenger ID is included in the list of passengers who checked-in for the flight. The biometric data is sufficient to perform both steps and reading the

Fig. 5. Boarding Pass with bar code, and with RFID in a paper, 2D bar code in a mobile phone

Fig. 6. Automated Boarding Gate

boarding pass is not necessary (PTEC.) In such an automated boarding gate an RFID reader can be integrated easily.

Scandinavian Airlines tested at Copenhagen airport a more efficient passenger processing and reduced the flight delays due to passengers. The point was that each passenger having an RFID tag card were informed by SMS in case they were not appearing on time at the gate (Ornellas, 2008). The only problem with this is that the passenger has to have a boarding card and an RFID tag card as well.

Swissair conducted a trial at Zurich airport for checking passengers with RFID tags. Each TravelClub member was issued an e-pass, which is based on the membership number. Upon booking the flight this is entered into the reservation system of the airline. The trial involved only for members of the TravelClub without a checked-in baggage. They are automatically checked-in when they pass through the passport control and show their passport (the reservation system passes their data to the DCS), after the border control they had to proceed to the information desks and pick up the boarding pass (Pilling, 2000). The problem with this is that they still needed to print a boarding pass for the passenger and it was not the immigration who gave them the boarding pass, so they had to find the information desk and queue again, which was time consuming.

5. Using RFID tag implemented into a bracelet for passengers and baggage

The now common boarding pass with the magnetic strip or the new 2D barcode boarding pass has a very short lifetime, and is not used after the boarding, only for redeeming frequent flyer points. The recently introduced 2D barcode Boarding Pass, which should be adopted by all IATA members until the end of 2010, avoids printing a new boarding pass for each connecting flight, it can store the boarding pass data of all the connecting flights. But still, after the last boarding it will be thrown away and it is not possible to re-use it.

The airports testing RFID are only using the RFID tag as a baggage tag to minimize the costs for lost baggage. More can be profited from this new technology when it is applied for the baggage as for the passengers too. By giving the passenger an RFID chip implemented into a watch/bracelet, the passenger tracking at the flight procedure can be solved and it facilitates the orientation at the airports for the passengers. In case of a problem it is much quicker and easier to locate the passenger and its luggage. The passenger's way after the check-in could be tracked until the end of the flight procedure. This would make the followings possible:

- The RFID tag implemented into a watch/bracelet, etc would be machine fixed to the baggage at the check in. (It would be almost impossible to remove it.) No more barcode would be attached to the baggage. (It would be almost impossible to remove the bracelet from the baggage.) The possibility that the Bag RFID Tag comes off or someone tears it away is zero. The machine that fixes the bracelet to the baggage could be built into the check-in counter. At all the baggage checking points an RFID reader can be implemented instead of the now used scanner. An example is shown in the *Fig.7:*

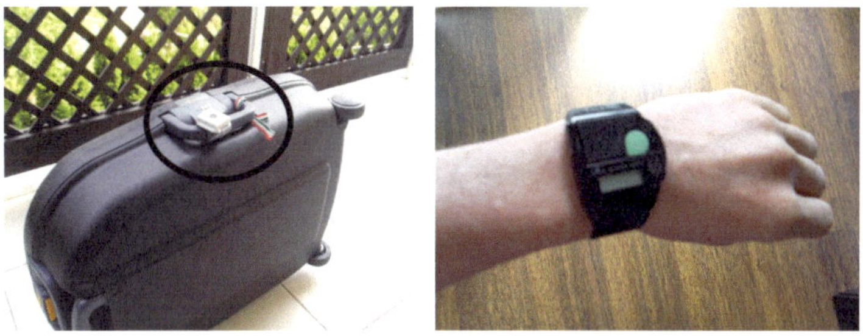

Fig. 7. RFID implemented into a bracelet

- The passenger's bracelet could have a small display to show the information of today's boarding pass, the flight information, and, at the arrival to the final destination, the details about the baggage claim. In case of transfer flight the actual details would be shown. This makes the passenger's orientation much easier.
- With the bracelet of the passenger having a speaker and vibration it would be possible to warn the passenger in advance so he would not be late at the gate. In case the passenger is late, it would be much simpler to find him. The flight coordinator could contact the passenger or simply find him in a second within the terminal and the time and costs of unloading its luggage would be minimized or totally avoided.
- In the transit hall right at the entry point or at several points, an information appliance could be installed to facilitate the passenger's orientation. The passenger has to hold his RFID tag against the machine and the machine shows automatically how much time the passenger needs to get to his gate and how far he is by illustrating the way with a printable map. It's also possible to offer the shop list and additional services on the way. In case of an arrival passenger it could show the number of the baggage belt and the way to it and whether his baggage is already circulating on the belt or not. This could be a good service provided to the passengers feeling lost at big, complicated airports.

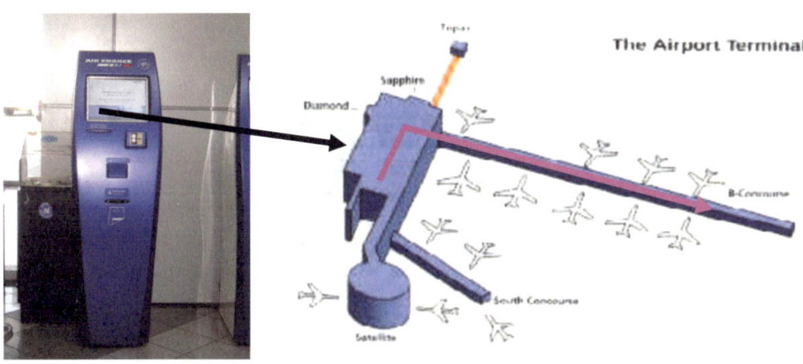

Fig. 8. Passenger Information kiosk

- At the boarding, where today the boarding passes are checked manually, the whole process could be automated with much less human intervention. The passenger would only have to hold his chip against the identifier and if he is at the wrong gate it would automatically alert with a sound or just simply say: 'Sorry you are at the wrong gate'. In case of automated boarding gate the door would not open if the passenger is at the wrong gate and would automatically display the gate number and time with a map where the passenger is supposed to board. While the passenger is crossing the boarding gate, the system is automatically checking if its baggage was already loaded or not, if not it knows where it is. At the end of the boarding procedure the system indicates to the flight agent whose baggage is already loaded but the owner hasn't passed yet the boarding gate and who boarded already but his baggage is still not loaded.
- If the passenger has an RFID tag with the same data as his baggage, the passenger-baggage check at the customs - before leaving the airport - could be effectively completed. The possibility that a baggage is taken away by another passenger by accident or intentionally stolen would be almost zero. It would be checked automatically, so it would be quick and it would not influence or slow down the passenger flow. Costs generated here for the airline could be reduced enormously.

Of course the passenger's privacy must be protected. The personal information of the passenger and his location should only be accessed by the flight coordinator if there is a problem. Until the system is not sending an alarm to the flight coordinator, the staff is only monitoring RFID tag numbers. To avoid mishandling of the personal data a one-way code must be applied and the tag of the passenger should be only activated when the passenger enters the transit hall of the airport at the security or passport check. In the exit doors of the airport an extra reader could be implemented: in case of leaving the airport after checking-in a baggage and not returning on time to reach the gate punctually, the system could send an alarm to the flight coordinator.

5.1 Reuniting passenger and their baggage at the baggage claim

Normally at the exit of the baggage claim nobody is checking if the baggage was taken by its owner or another person. Only at some American airports or at the Hanoi airport in Vietnam, dedicated service people are checking the bag tags manually right before travelers leave the customs area.

Nowadays on some airports the passenger-baggage check at the baggage claim is either a totally manual process or does not exist at all. But in case a baggage gets stolen, the airline is responsible for the compensation. It is not possible to manually check if the baggage belongs to the given passenger. It would simply cause enormous queues. There are not many security checks of the passengers leaving, so this problem is not dealt with.

If the passenger has an RFID tag containing baggage data, the passenger-baggage check at the customs - before leaving the airport - could be effectively completed. The possibility that a baggage gets taken away by accident or stolen by another passenger would be almost zero. It is not manually checked, so it would be quick and it wouldn't influence or slow down the passenger flow.

The realization of this check depends on whether the RFID tag is a one-time used tag or capable of reuse.

In case of a single use tag the check is very easy: In the exit door an antenna is placed which identifies the tags and in case of a problem a signal would be automatically forwarded to the security. Of course a couple of seconds are necessary to complete a check and the passenger

Fig. 9. Exit with reusable chips

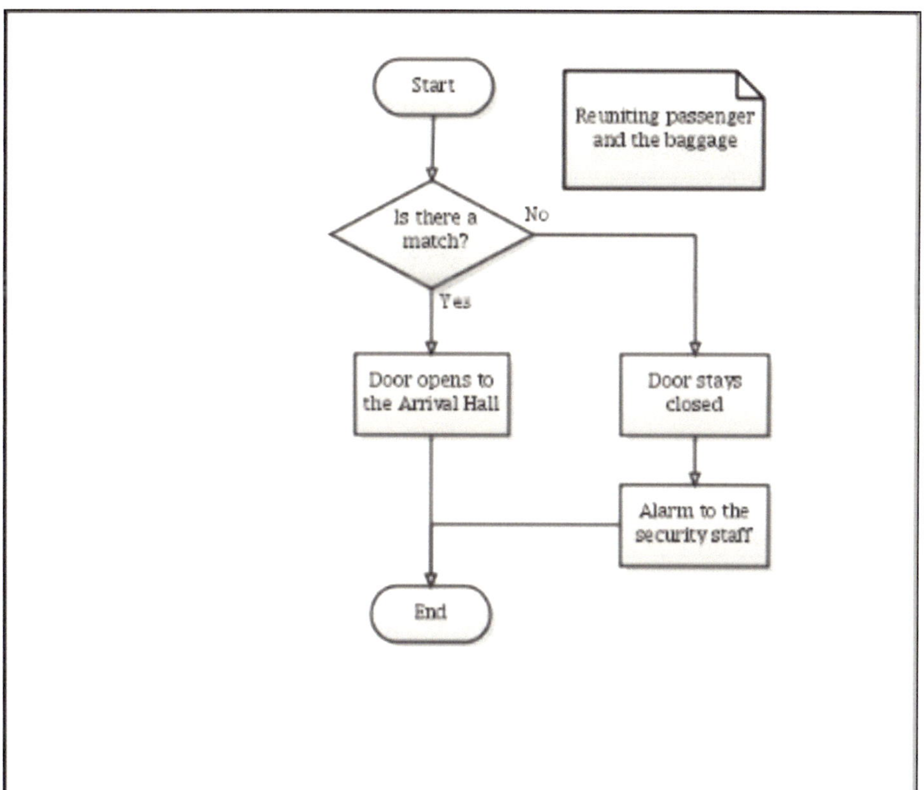

Fig. 10. Simplified flowchart diagram for Passenger and Baggage Identification machine at the Baggage claim

must walk into a narrow lane in order to do this. However the system to be implemented must be capable of handling all the ways passengers bring their luggage (besides, pulling behind etc.). It also has to be ensured that the reader is capable of identifying many passengers (and their luggage) exiting at the same time. After leaving this area the RFID chip should be de-activated to avoid mishandling of personal data and the chip itself.

In case of reusable chips, they must be returned at the airport. This can be processed with a box, where the passenger has to put the bracelet in a holder at the box, while pulling the luggage through. In the box the identification can be made and the chip removed. The exit door is opened if everything goes well. If there is a problem the door stays shut. To save time and space at the airport this identification procedure could be made at the customs.

If a passenger has more baggage it is encoded in the tags so the system knows that several baggage are coming through the box.

In case a baggage did not arrive on time, a tag writer could be used at the Lost and Found desk to overwrite the data so that exit problems are avoided.

As an additional service for the passenger, an information appliance at the arrival side could be used to inform the passenger where his baggage belt is located and other important information such as shops or money exchange, etc.

In case of RFID boarding pass integrated into the passport the passenger has to go through a door like at the boarding gate and has not even to take out his/her passport.

5.2 Combining passport with the boarding pass

The next step in the technology's innovation is the spread out of the biometrical passports with an RFID tag implemented storing the biometrical information. There are 2 possibilities for using the passports as a boarding pass too:

1. Passenger holding the new RFID tagged passports, with the passport having a small display, because then all the boarding information could be automatically stored in the travel document and no extra kind of boarding pass would be needed.
2. Following the trends of saving costs everywhere where it is possible, another option is that the passport's RFID has all the actual boarding information required, and the passenger simply memorizes or copies it to a small paper and the boarding pass would be absolutely paperless. It is only necessary to memorize the flight destination and the time because with these information any flight information display can provide further important information, or the passenger can proceed to a passenger kiosk proposed above.

At check-in desk the data of the flight would be written into the chip. With web-check-in becoming more common the passenger can check-in through the web and if he has no tag writer at the computer where he processed the check-in he just goes to the baggage drop (if he is having a baggage to check-in) and there he can write it to the passport's chip or simply at the border control (like the above mentioned solution of Swissair with the change that the officer is telling the information to the passenger and no pick-up is needed).

In case this is too complicated to apply for each person, then it would be definitely feasible to apply this method for diplomats, VIP person, frequent flyers, business persons etc. They would have their chips and their check-in would be much quicker. To implement a chip writer at home is just a question of time. As after a long period the CD/DVD and USB drivers and the paper printers spread out and now almost everyone has one, it is just question of time that the chips are getting more common and home tag writers can be used. And with the home chip writers the new problem (and the possibility of) printing the boarding passes at home can be solved too.

The Self-Check-in desks at the airport could have a chip writer and not a barcode printer. With this the tendency of Simplifying the Business wouldn't be stopped. No further boarding pass is needed; it reduces the cost and investigation for airlines and airports. There would be one unique technology, not like what the situation is now, each airline and airport having its own solutions. All nowadays applied technologies for passenger handling can be dealt with using the RFID tag as boarding pass whether implemented into a bracelet or in a passport. The next step is that the security check and the border control could be combined. As the security check technology can have RFID and biometrical identification integrated only by passing through the security check all necessary data can be checked. And only if there is a visa or "black list" kind of problem, then those persons would need to proceed to an extra check. This could speed-up the flow within the transit area.

RFID implemented into a passport with biometrical data as a boarding pass could be not only used within the aviation sector. A boarding pass can be lost easily but everyone is taking much care of the passport. It could be used for each transportation tool and no more ticket would be needed as everyone has only to refresh his own one. Furthermore, it could be used in hospitals and other applications too.

Where the passport check is not needed anymore, the passenger would receive his boarding pass as a bracelet and would return it at the baggage claim before exiting after the passenger-baggage identification.

In case the passport would contain the passenger-baggage information as well, the identification at the baggage claim could be combined with the passport check or simply the exit door would delete the fight data from the passport's chip after identifying. The before mentioned box wouldn't be necessary for the passenger's, there would be simply a reader, where the passenger has to show the chip. While the luggage's chip is a multiple used chip which remains at the airport flow and at the exit the before mentioned box the passenger has to pull his baggage through.

5.3 Advantages of the system

Integrating the RFID tool into a GIS system allows a very good visualization of the passenger and baggage flow, their connectivity, their way within the terminal, where they have spent much time. Many automatic recordings, reports, maps, statistical data information for further improvement can be generated from the GIS software emphasizing the strength and weak points of the infrastructure. Emergency alarm can be sent out and later analysed and visualised. In emergency situation a map is also helping the staff to find the place. Giving passengers and baggage RFID tag their tracking and tracing within the whole flight procedure can be visualized by map. The way of the baggage is seen: the weak points of the used infrastructure can be recognised (e.g. congestions) and corrected. Seeing the passengers flow within the transit hall the airport can see what type of shops, bars, services the passengers like, where they spend their time while waiting for the plane. All data is stored in a database, any type of queries can be made. The system can be connected with the border control and airport security services too.

Another advantage is for the passengers and for the airline and operator is that paper is minimally used. The passenger has only to take care of its passport (in case the passport is the boarding card) or only of the bracelet. As it is contact less, it is only important that the passenger has it, but doesn't really matter where it is. In the long term when everything will go with biometrical identification maybe not even the passport will be necessary, just an RFID tag.

The RFID technology is totally adoptable in currently industrial trends. Into any kind of check-in desk, boarding gate, security screening it can be integrated. It is fast, reducing the queuing times and congestions.

Airports and not airlines should be the owners of the new technology and issue the baggage tags in future. If only the airline was the owner they would only use it for their own baggage flow. Different airlines are applying different systems, and the main importance is the link between the airports and the standardization in this technology. It complicates the system if each airline signs a contract of cooperation with each other airline. If each airline applies its own solution it results in a very complicated baggage sorting that is much more time-and-space consuming.

The implantation costs and the chip prices are too high for the airlines. The airlines are using only a part of the airport and not the whole airport; this technology could be applied for the many airport operations. And it is unnecessary that each airline develops and implements its own technology. It is much more efficient, less complicated and cheaper if the airports develop and implement the applied technology and the airline is only renting it. The most efficient is to work out a strategy for the RFID bracelet for all the world's airports, and the airports are given a time period to implement the whole technology.

6. Conclusion

RFID technology can be used for identification, tracking, locating and monitoring both people and items. As the cost of the RFID technology has begun to fall, currently the baggage tracking is the field in the aviation sector where RFID has proved most useful, and is becoming widely adopted.

Using RFID for passenger and baggage handling makes the processes fully automated and minimises the manual tasks made by assistants, reduces the costs for airlines, airports and operators. As the system enhances the efficiency and service level, unnecessary costs of the paper-based technology can be minimised, and the return on investment will be payback in long term. The system is faster, needs less manpower, the maintenance is cheaper and the resources can be allocated elsewhere. It is following the current trends as it is environment friendly too as it is paperless, no printing and paper is needed which is a very important issue currently in the aviation industry. Another important issue is to make the air travel more customer friendly, less time consuming, hassle free, with less queuing, but the security of the passenger must be guaranteed. Airlines want to use less space and less infrastructure of the airport to save costs. The airports are tending to use more space for retails, bars & restaurants or other facilities for customers. The RFID printing can be integrated into check-in desks, self-check-in kiosks, it is not influencing the recent trend of using less space, and allow better use of resources for airports.

RFID can be used in the aviation industry not just for passenger and baggage tracing, furthermore for the better resource allocation of the Ground Support Equipment (GSE), inflight catering, couple of airlines (e.g. American Airlines) are using it to prevent unauthorised drivers accessing restricted areas of the airport, for access control and airport security, in the cargo handling (it is already commonly used). By giving passengers and baggage RFID tags, their identification and tracing can be combined with the security and immigration issues and together integrated into a GIS system for mapping (e.g. in real-time and non-real time, the points and flows depending on the RFID tag type) and reducing false security alerts, etc. RFID technology is a good method to increase the security, safety, and customer satisfaction. The technology could increase passenger throughput because baggage handling could be quicker, flight schedules could be kept more, and there would be fewer delays, all of which can be quantified as cost savings.

7. References

Lima, P.; Bonarini, A. & Mataric, M. (2004). *Name of Book in Italics,* Publisher, ISBN, Place of Publication

Bite, K. (2005). Effectively utilizing radio frequency identification at airport (In Hungarian: Biztonságos repülőtéri poggyászellenőrző rendszer fejlesztése), BUTE MSc. Final thesis, not published

Bondarenco, N. & Price, A. (2009). Baggage Improvement Programme Strategy Paper - 2009, *IATA Simplifying the Business Programme,* IATA, p.4

IATA a. RFID Trials for Baggage Tagging, IATA, http://iata.org/NR/rdonlyres/D319ADC0-ED5D-447E-9EEB-6CA1179C6BD9/0/RFIDtrialsforbaggagetagging.pdf (September 2009)

IATA b. Simplifying the Business Programme, www.iata.org/stbsupportportal (September 2009)

IATA c. RFID in Aviation, IATA, 2008

Mecham, M. (2005). Radio ID. *Aviation Week and Space Technology,* Vol. 162 No.18. (May 2, 2005) p. 42, ISSN 0005-2175

Ornellas, T. (editor) (2006). Bag Tag Reality: Growth Ahead. *Ground Handling International,* Vol 12, No 5 (October 2006) p. 14, ISSN 1364-8330

Ornellas, T. (editor) (2007). From bags to hitches. *Ground Handling International,* Vol 12, No 7 (April 2007) p. 52, ISSN 1364-8330

Ornellas, T. (editor) (2007). On the right track. *Ground Handling International,* Vol. 12, No 5 (October 2007) pp.38-42 ISSN 1364-8330

Ornellas, T. (editor) (2008). RFID: more trials ahead. *Ground Handling International,* Vol 13, No 2 (April 2008) p. 8, ISSN 1364-8330

Ornellas, T. (editor) (2008). A final end to late embarkation? *Ground Handling International,* Vol 13, No 4 (August 2008) p. 12, ISSN 1364-8330

Ornellas, T. (editor) (2009). More RFID tags for Hong Kong. *Ground Handling International,* Vol 14, No 3 (June 2009) p. 6, ISSN 1364-8330

Pilling, M. (2000). On track with E-track. *Airport World,* Vol .5, No.6, (December 2000/January2001), p.50, ISSN 1360-4341

Pilling, M. (2001). Security Spin-off. *Airport World,* Vol 6, No. 6, (December 2001/January 2002), pp.44-46, ISSN 1360-4341

Pilling, M. (2001). Queue busters. *Airport World,* Vol .6, No.6 , (December 2001/January2002), pp.48-49, ISSN 1360-4341

PTEC. (2008). Passenger Terminal Expo and Conference, Amsterdam, Netherlands

9

Fusion of Radio Frequency Identification (RFID) and Fingerprint in Boarding School Monitoring System (BoSs)

Herdawatie Abdul Kadir, Mohd. Helmy Abd. Wahab, Zarina Tukiran
Mohd Razali Mohd Tomari and Mohd Norzali Hj. Mohd
Universiti Tun Hussein Onn Malaysia
Malaysia

1. Introduction

This chapter discussed the implementation of RFID integrated with biometric sensor to improved Boarding School's (BS) management of student's and facilities. BS may offer a variety of extracurricular activities where conducive learning environment with good facilities such as computer room, gymnasium and entertainment room contribute to good way of mind relaxing and boost up brains development. Currently, most BS managements are having difficulties to monitor their students using the old-fashioned paper-based system; where the procedures are inefficient in monitoring the student's whereabouts. The application of RFID Metric Card system as a Boarding School Monitoring System is to improve the school management procedure, automatically monitor the interest group movements and increase their safety. By using RFID technology, it is easier and faster to detect students handling at that time and reduce assets losses. In this system the fingerprint recognition is also adopted to enable the process of identifying of BS student more reliable and secure for facilities management. Through enhancing security in BS access and facilities control, it help the school management to provide visibility of assets and effective users tracking.

This research work offers an important implication for monitoring the BS assets and eases the workload of the school management and save time for various student activities. In order to analyze and design the system, this application use student's metric card embedded with RFID tag for tracking their whereabouts and fingerprint registered to the access the facilities room. When the metric card passes through the RFID reader, it will trigger the system to read the data from the RFID tag to the database where the access data can be viewed online by BS management for monitoring purposes. For the facilities room access, RFID will be detected when someone with an RFID tag passes through the RFID reader and verify by matching fingerprint image. Upon verification with database is success, the system will enable the room to be access. Thus, the management people may know the students location and also view log record of room entry via internet. This will ease the management in monitoring the availability of boarding school, easy access to BS facilities and reduce unattended assets loses.

2. Related work

Recent advances in Information and Communication Technology leads to rapid growth of technology especially in monitoring system to improve any glitch and errors. Single-based authentication system seems unable to monitor and authenticate effectively due to easy to hack and break out. Since that, the introduction of Radio Frequency Identification (RFID) has taken place as an alternative to the previous authentication systems such as biometrics and bar code system as well as smart card technologies. However, the integration of those technologies is expected to provide an effective monitoring system and a very high secured system that can perform and monitoring to mitigate the ratio of unauthorized person access. Nowadays, the trend of RFID is clear that the technology is playing an essential role especially in various field such as in engineering, medical, business, construction (Corporation, 2005; EraBuilder, 2006), library (Shahid, 2005), farming management (Bakery, et al., 2007) and etc. Several methods and technologies have been used to develop monitoring systems.

i. Biometric Procedure
ii. Barcode
iii. Smart Card
iv. RFID

2.1 Biometric procedure

Biometric technology is one of the popular method that have been used to recognize a person identity based on the biological and behavioural characteristic, which reliably distinguishes one person from another, used to recognize the identity, or verify the claimed identity of an enrolee and enrolled into a template and store in a system database (Boatwright & Luo, 2007; Gil, et al., 2003). There are many type of biometrics recognition technology that exist today and the fingerprint is one of the widely used biometrics satisfying uniqueness and permanency (Boatwright & Luo, 2007; Gil, et al., 2003; Ratha, et al., 2007; Reillo, 2003), hand geometry, voice, retina, iris (Boatwright & Luo, 2007; Reillo, 2003) and face recognition (Boatwright & Luo, 2007; Ratha, et al., 2007).

Figure 1 illustrates biometric characteristic can be classified into two which is physiological and behaviour. Physiological are related with recognition of face, finger print, hand, iris and DNA meanwhile behaviour are recognition of sound, keystroke and signature (Simao, et al., 2008).

Wu stated that, the use of biometric depending on the application context, biometrics systems operate in either positive recognition or negative recognition. Generally, positive recognition must be performed in verification mode and negative recognition must be performed in identification mode. For these Biometrics-based network authentication systems, they have five subsystems: data collection, signal process, matcher, storage and transmission (Wu, 2008). Gil et al. developed an Access Control System that utilizing high level security fingerprint to verify user to gain access to property or service. It indicates data that use fingerprint can improve the security from token based security method (Gil, et al., 2003). The application of attendance system using biometric system has been demonstrated by Simao, Fonseca and Santos (Simao, et al., 2008) with integration with wireless communications. However, the weaknesses of the system has been discovered by Zhang et al. (Zhang, et al., 2003) by introducing palmprint which developed an attendance system to record the employee attendance including elderly people which may not provide clear

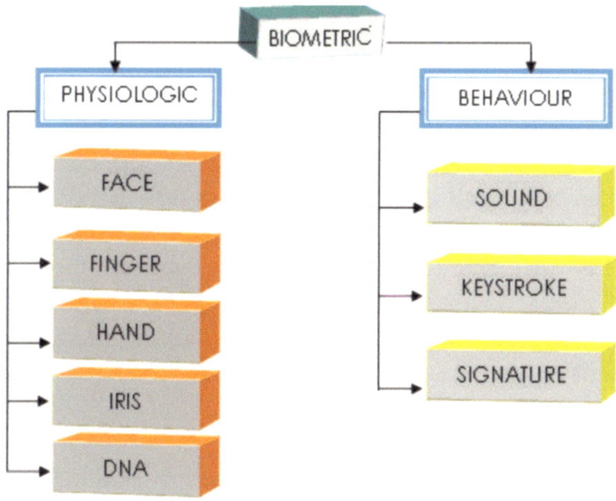

Fig. 1. Biometric Characteristic Classifications.

fingerprint due to problematic skins. An online system captures palmprint images using a palmprint capture sensor that is directly connected to a computer for real-time processing. An offline palmprint identification system usually processes previously captured palmprint images, which are often obtained from inked palmprints that were digitized by a digital scanner. Meanwhile, a similar project has been implemented by Kadry and Smaili (Kadry & Smaili, 2007) which applied wireless iris recognition attendance management system. This system is mainly used for employee identification. The biometric system is suitable for highly secured systems. The security of the system can be further enhanced by applying the multiple authentication system such as integrate with other techniques and technology.

2.2 Barcode

Barcode technology is a method of identification which is used to retrieve in a shape of symbol generally in bar, vertical, space, square and dots which have different width with each one (Gao, et al., 2007; Shepard, 2005). A reader or scanners are needed to identify the data that are represented by each barcode by using light beam and scan directly to barcode. During scanning process, a scanner measured the intensity of reflected light at black and white region. A black region will absorb the light meanwhile white region will reflected it.
There are several types of code bar scanner:

i. Pen Reader
ii. Laser Reader
iii. Charge Coupled Device (CCD) Reader
iv. Camera Reader

According to Gao, Prakash and Jagatesan (Gao, et al., 2007), the barcodes stored data in the form of parallel lines in different widths, and they are known as 1D barcodes, and could only encode numbers. The method of arranging the bars and spaces of barcodes are called symbology which is barcode symbology refers to the protocol that defines a standard for arranging the bars and spaces that comprise a particular type of barcode, such as UPC-A and EAN. It defines the technical details of a particular barcode type, including the width of bars, character set, method of encoding, checksum specifications and others Hebert et al. (Hebert, et al., 2004) designed a DNA barcode to identifying birds. Short DNA sequences from a standardized region of the genome provide a DNA barcode for identifying species. Compiling a public library of DNA barcodes linked to named specimens could provide a new master key for identifying species, one whose power will rise with increased taxon coverage and with faster and cheaper sequencing. Beside that, barcode technology also has been used with mobile phone (Kato & Tan, 2005)to carry more data that its ID counterpart. Previously, barcode has been used in attendance system which indicated by Susan, Mitchell, and Dudley (Susan, et al., 2002) which used handheld barcode readers and the scanner to keep track of students' attendance at LEAD event. Because of several disadvantage of barcode such as read range, data capacity and others, barcode technology is not suitable to implement for this project. This is because to implement an attendance record system for huge amount of student at wide area using barcode will take a long time.

2.3 Smart card

Smart card is built with variety of chip with a simple memory consisting of byte of information that may have range from 1K up to 64K of microcontroller or multi-application memory (Carr, 2002). Smart card are used as individual identification, building access and network access which are part of a multi-tiered program that is in the final stages of rolling out. The data in smart card can be read when there is a physical contact with a reader. Smart card has been used in a wide range of application such as to store operation history, medical record or telemetry (Hendry, 1995) as well as student identification in most organization with a multipurpose of usage (Omar & Djuhari, 2004).
Based on Halawani and Mohandes (Halawani & Mohandes, 2003), smart card has been developed at campus environment as identity cards for students and employees to grant access to certain data, equipment and departments according to their status. Meanwhile, a similar project has been implemented by Mustafa and Kyng (Mustafa & Kyng, 2007) which utilized MyKad Touch N Go features for student web-based attendance system. This system could be accessed by teachers, headmaster and parents via internet and intranet facilities. Because the smart card need physical contact to the reader before the data in the smart card can be transfer, the implemented attendance record system for huge amount at wide area using smart card will take a long time.

2.4 Radio Frequency Identification (RFID)

Nowadays, the RFID technology has been widely used in medical surgeries, animal identification, baggage handling, library services and real time location tracking (Sam, 2007). According to Yoon, Chung and Lee (Won-Ju, et al., 2008), RFID is an automatic identification method, whereby identification data are stored in electronic devices, called RFID tags (transponders), and these data are retrieved by RFID readers (interrogators) using radio frequencies. RFID systems can be classified into two categories according to the tags' power supply: active RFID systems or passive RFID systems. In active RFID systems, tags are equipped with their own batteries, whereas tags in passive RFID systems do not have an internal power supply. Therefore, compared with passive RFID tags, active RFID tags enable a greater communication range. Table 1 shown differentiation between active and passive RFID.
Sabri et. al. (Sabri., et al., 2007) used RFID as a method to record the web-based attendance application to overcome the problem in recording the student's attendance. The system only accepts five different levels of accesses which are the Administrator, Lecturer, Student,

	Active RFID	Passive RFID
Tag Power Source	Internal to tag	Energy transferred from the reader via RF
Tag Battery	Yes	No
Availability of Tag Power	Continuous	Only within field of reader
Required Signal Strength from Reader to Tag	Low	High (must power the tag)
Available Signal Strength from Tag to Reader	High	Low
Communication Range	Long range (100m or more)	Short (3m or less)
Multi-Tag Collection	• Collects 1000s of tags over a 7 acre region from a single reader • Collects 20 tags moving at 100 mph	• Collects hundreds of tags within 3 meters from a single reader • Collects 20 tags moving at 3 mph[2] or slower
Sensor Capability	Ability to continuously monitor and record sensor input; date/time stamp for sensor events.	Ability to read and transfer sensor values only when tag is powered by reader; no date/time stamp.
Data Storage	Large read/write data storage (128KB) with sophisticated data search and access capabilities available.	Small read/write data storage (e.g. 128 bytes)

Table 1. Comparisons between active RFID and passive RFID

University Administration and the Guest. Each user has its own limited access according to the user level. For example, the Administrator has all the security clearance. Meanwhile other users have limited access to the system. The disadvantage of this system is it unable to indentify the student who is late attending the class. Furthermore, Chen and Chang (Chen & Chang, 2008) also applied the RFID technology in developing the project. The project is developed using active RFID and wireless GSM message to construct an active student attendance system that sends the message to parents cellular phone informing whether their children has safely arrive in classroom in the morning. Meanwhile, the system is also used to relieve the traffic congestion around kindergartens especially while parents are driving to pick up their children after class at rush hours or on rainy days.
On the other hand, Qaiser and Khan (A. Qaiser & S. A. Khan, 2006) used RFID technology for the automation of time and attendance using RFID Systems. Students and faculty members are provided with RFID tags. When these tags pass through the reader generated interrogation field, they transmit information back to the reader, thereby identifying them. The RFID System makes it possible to monitor the movement of tagged users and record their real time data and pass it to processing system to maintain a system Log. A similar

project has been implemented by Herdawatie, Siti and Helmy (Kadir, et al., 2008) which is used to monitored boarding school students using RFID. The data will be sent online to the school management for monitoring purposes. However, Joseph and Yusuf have developed a similar student attendance system using RFID in which the information of student attendance will be sent to parents by using Short Message Service (SMS) (Joseph & Nakhoda, 2008).

Other than that, RFID technology also has been used to monitor a transportation system which is used to monitor the container tracking from Yokohama Port to Kobe Port using active RFID Systems. The results show that active RFID system has the capability to acquire a movement history and sensor data easily with low power consumption without GPS (Mizuno & Shimizu, 2007). According to Ogata et. al. (Ogata, et al., 2008), the paper proposed basic support for ubiquitous learning (BSUL) environments as an extension of e-learning systems. For the attendance-taking module, every student has a RFID tag. When the student enters the classroom, the system reads the RFID tag and sends a message to a web service based on the simple object access protocol (SOAP), asking to update the system database. There have four different statuses for the students: attendance, absence, delay, and a fourth one called remote attendance, which means that the student is viewing the class through the streaming video source. The criteria for deciding whether a student is late or not, can be configured by the teacher in charge of each course. The teacher can view the records of each student's attendance during the course using the environment website, but the students can only view their own records.

2.5 Fusion of RFID with other technologies

Multiple authentication system has been a trend due to the improvement of the security of either restricted area or to monitor the place which consider as private and confidential. The previous work presented on a transportation and logistics company which utilized RFID and cellular positioning automated tracking and tracing of general cargo in order to solve shipments which have all assumable sizes and shapes and, forwarders are very often organized in open logistics chains or networks (Hillbrand & Robert, 2007). The integration of RFID and Accelerometer Sensing were used to monitor an effective and obstructive activity recognition based on 10 housekeeping performed by 12 subjects. The results indicates that recognition accuracy can be significantly improved by fusing the two different types of sensors (Stikic, et al., 2008). However, there are fusion of RFID and wireless sensor network to build a wireless localization system in a children's theme park. The main purpose of the project is to localized system to track and locate children within a certain range near some landmarks in the park. The design experience in this project can be exported to other applications such as object tracking and surveillance (Chao Chen, 2006). The work presented by Won-Suk Jang, Mirosław J. Skibniewski (2008) utilizing RFID and GPS to produce a prototype framework for automated tracking and monitoring system for construction materials on Project Sites (Jang & Skibniewski, 2008).

3. BoSs architecture and design

Boarding School Monitoring System (BoSs) is an integrated monitoring system that enable the school management to fully centrallized the data of each registered student movement and permit access within the define area to improve the security, reliability and tacking efficiency of student's. It's combines the RFID technology to track the student movement and fingerprint reader to enable the management to give authorized to selected student's to access facilities room thus enables the school management to monitor individual location and also view the evidence of room entry via internet. An overall view of BoSs is shown in Figure 2.

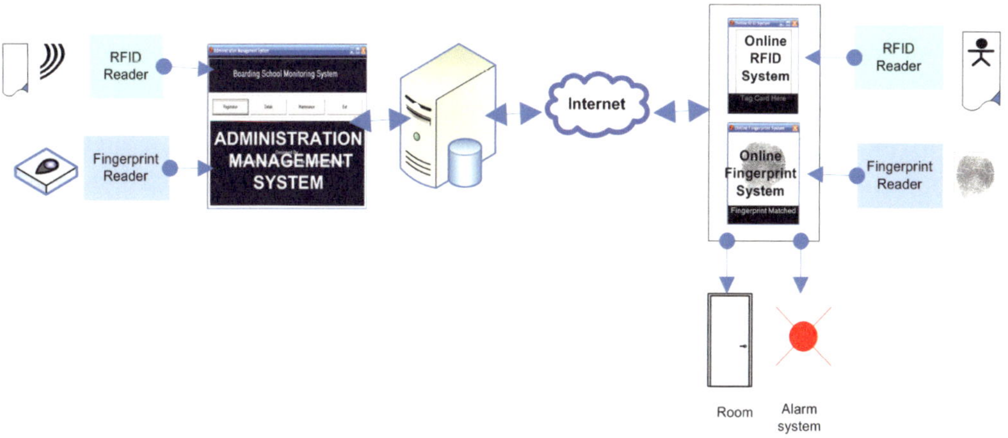

Fig. 2. BoSs Architecture

Generally RFID is an auto-id device technology that used to store and extraction data within range (up to few metres depends on the reader) by using radio waves The use of RFID system can reduce operating costs and simplify the data storage and retrieval (Nement et al., 2008). In a security perspective, biometric recognition was used, and more specific fingerprint was chosen to provide correct identity of the card holder due to its low cost for data authentication (Bazakos, et al., 2005).RFID technology is composed of three main components; a RFID tag, which contains the identification number, a RFID reader, which activates the tag to broadcast its identification number, and a RFID Middleware, which integrates the information from the reader to the backend database system. Nowadays, RFID is used in many areas and every where possible. However at present, the popularity of RFID technology opens several issues within the security umbrella. Although the combination RFID and fingerprint is not an answer to unauthorized access however it is the key to reduce forgery and counterfeiting of passive RFID tags (Smith & Coetzee, 2008) and solve the problems in the traditional system thus increase the efficiency of management process (Jiang, et al., 2005).

In BoSs, the RFID tags enable the school management to track the student around the boarding school area and access in and out of the facilities rooms. An individual without the RFID card or without the correct fingerprint image will disable the entry to the facilities room. The system will also trigger the security signal and this will notify the school management upon the unauthorized access using an online monitoring system. This system used the main component of passive RFID system and fingerprint sensor, database management system and networking i.e. wireless. When the RFID tag pass through the RFID reader in a range zone and match fingerprint image for the scan tag, the system will record the data from the RFID tag to the database system. Then the data will be sent online to the management for record and supervision of rooms' access. This could ease the management to observe access of selected area and facilities rooms.

4. Application scenarios

For easy monitoring the area were divided into four areas: zone one, zone two, zone three and zone four. RFID reader will be located in appropriate area in each zone to enable the student movement monitoring. In zone one, the centre point is the area that are consider as restricted area; it provide the facilities such as computer room, society room, television room and mini library for student activities. The RFID reader will be located at the in/out door of the building and fingerprint reader will be located at each facilities room to enable easy access of usage by the authorized individual. Thus, restricted the access for each room in order to enhance the security of the assets. The boarding school hostel, canteen, study room, pantry and residential advisors are located in zone two and zone four and last zone is located in sport centre area. To illustrate the BoSs concept a sample of layout of boarding school map were provided in Figure 3.

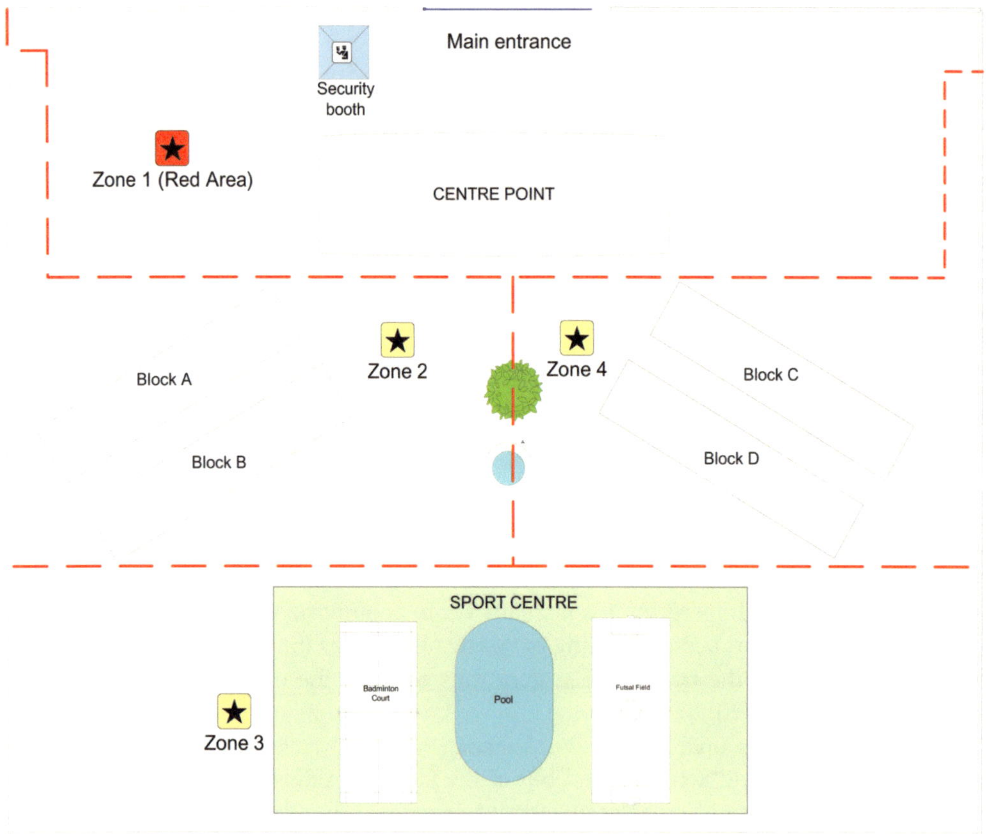

Fig. 3. Boarding School map layout

By implementing BoSs, the school administration people will be able to search within the database for the student data and able to scan the last location of the student whereabouts, if the student is found missing or a disaster happens. By defining the selected area the whereabouts of the students are easier to pinpoint whereas the student location and time were recorded by the BoSs system and provided appropriate emergency information for tracking and life saving.

5. System implementation

BoS system records students' presence at Zone 1 - Zone 4 area and privileges on accessing hostel facilities at Zone 1. In order to records student's presence at particular zone area, RFID technology is used. The overall system flow is shown in Figure 4.

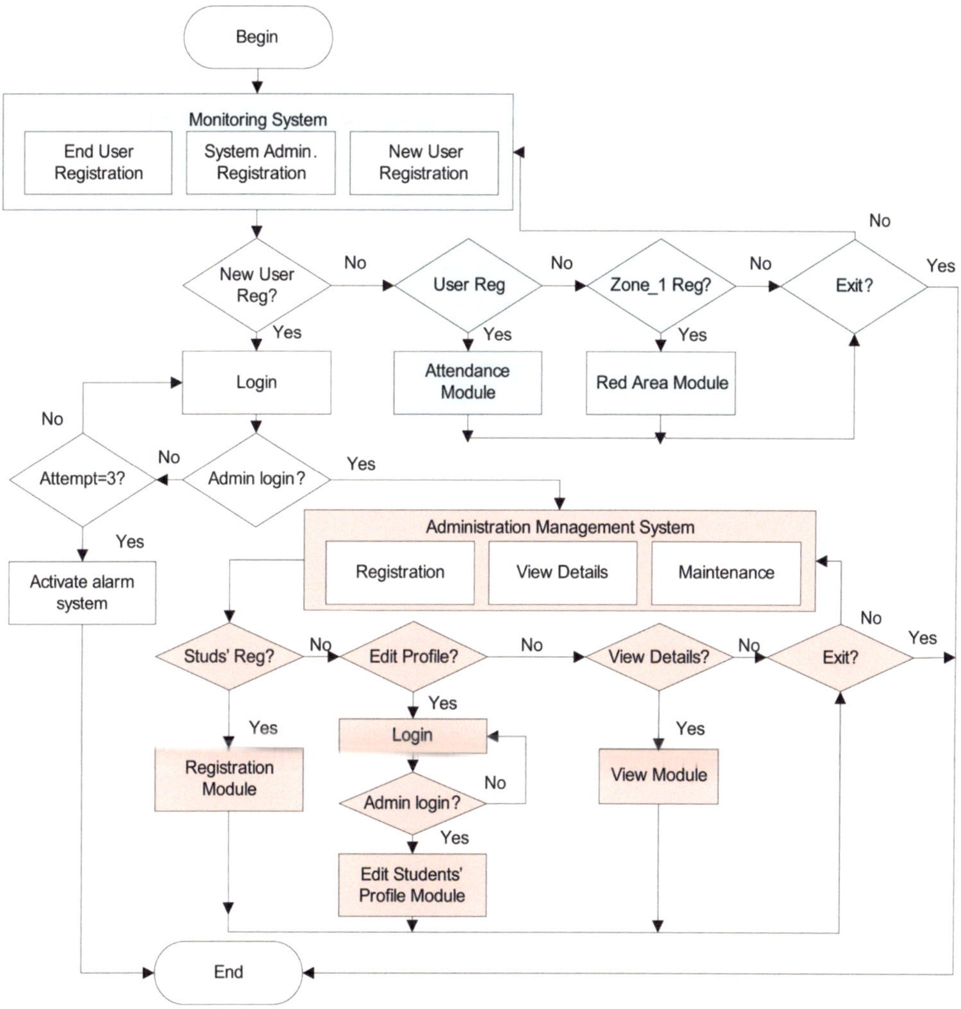

Fig. 4. Boarding school monitoring system flow

BoSs consists of two main subsystem; monitoring system and student management system. Monitoring system is used to keep track of student's movement around the school area. New users' registration or modification on users' profile should be done by system administrator through Student Management System. To be part of the BoSs's entities, new user must be registered by system administrator. In order to make sure an authorised personel login to the system, the correct password of system administrator should be entered. Any unsuccessful login will be stored into BoSs's database with a maximum of three times login attempt before activating an alarm system. This type of security mechanism is also applied to protect the data inside BoSs's database. A successful login instantly activates the RFID reader to read a RFID tag. Once the reader reads a unique number of RFID tag, it will be stored into the database. Each tag that has been stored in the database is assigned to a student that indicates the tag and the user have been registered into BoSs. Therefore, if registered user is in RFID range, BoSs's automatically store the user's presence into the database. Other than holding information on student's profile, the tag also carrying information on student's permission level of accessing a certain restricted school area (mentioned afterwards as red area). Because of the characteristic of the tag; easy to change from one hand to another, an additional security mechanism is created to act as a second level of security system and to ensure an authorised person entered to the system. The second level of security system employed a biometric system using fingerprint. The detail on flow on registration module is shown in Figure 5.

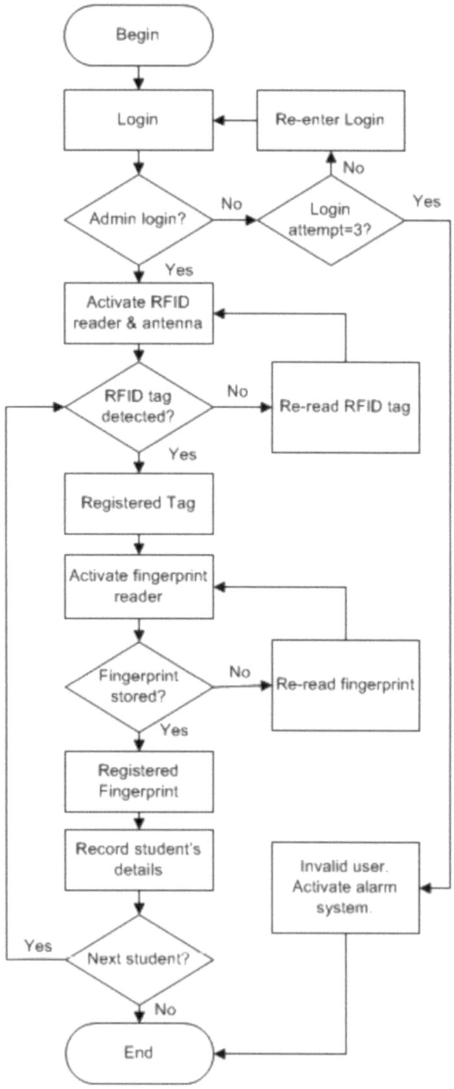

Fig. 5. System flow on registration module

To access Zone 1 (red area), user must be registered to the database to enable RFID and biometric reader to give entry permission; these two technologies were combined to produce two levels of security system. Besides that, the permission level of each user at Zone 1 is also determined where the fingerprint verification consists of two steps: enrollment and verification as shown in figure 6.

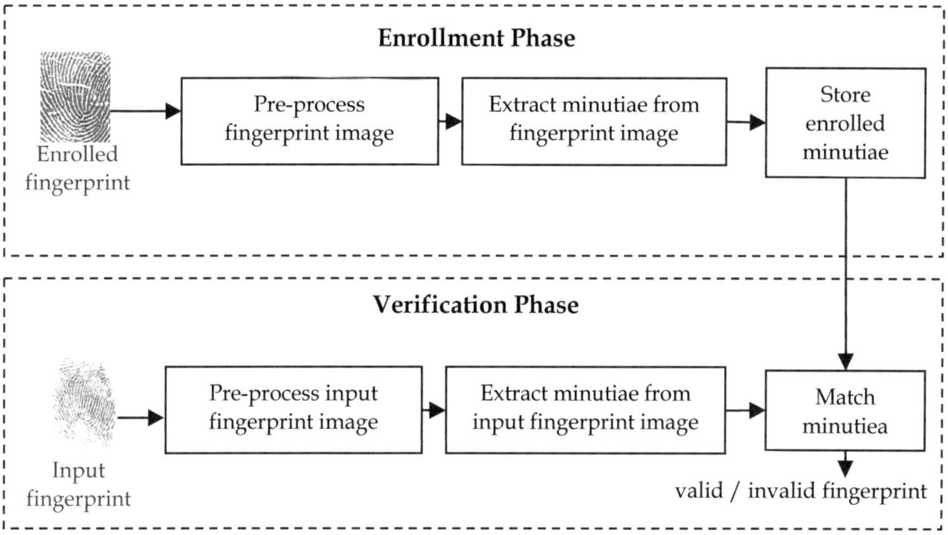

Fig. 6. User authentication system using fingerprint

The red area determines the type of permission level that user had before the system allowing any user to entered the area; for instance a computer room. The permission level for each user is determined by the system administrator during user's registration. In accessing the red area, the user must provide a valid RFID tag and a valid fingerprint. At this area, there are two types of readers; RFID reader and fingerprint reader, which will be in a stand-by mode. These two readers will indicate two security parameters are needed and should be correctly inserted to enter the red area.

The intended user must present his/her RFID tag within RFID range. BoSs also needs user's fingerprint by placing his/her registered finger onto fingerprint reader. The reader reads the tag and the fingerprint and sent signals to BoSs. The entered tag and the entered fingerprint are compared to the same parameters in the database to check parameter validity and accessing level. If these two security parameters matched, meaning that the user belong to the system and this will create a successful attempt. A successful attempt will makes a secured door opened and starts recording user's activities such as date, time in, time out, room no. etc. Any unsuccessful attempt from any of these two security parameters will allow the user to re-enter with a maximum three times attempt before BoSs's activates the alarm system and logged such attempted. The details of system flow and system interaction on red area are shown in Figure 7 respectively.

6. Boarding school monitoring system interfacing

The development of BoS system is divided into three access parts; users' access, system administrators' access and online access.

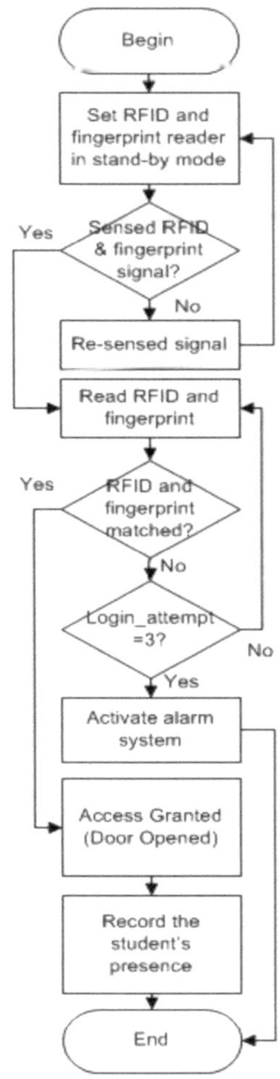

Fig. 7. System flow on red area module

6.1 Users' access

There are two interfacing technique: for user and system aministrator. For user interface, the system, its will give the user permission to prompt the access facilities from zone 1 until zone 4. There are two reasons for this interfacing (a) to record the user movement in all defined zone (b) to show and determine the user's permission level on accessing zone 1. The recording of user's presence in each zone is done using RFID technology. User need to scan their RFID tag into RFID reader as prompted by Online RFID System (refer Figure 8). If the tag was registered into BoS system, the system will displays the user's name and opened the gate. If not, the system will displays a message that indicates unsuccessful login. For such case, the system will allow the user to re-tag up to three times before activating the alarm system. The activation of alarm system will alert security staff to take an appropriate action.

Each successful login indicates that the user is allowed to enter the system. Therefore, the user is allowed to be at any zone area and their presence will be recorded into the BoS system. This can be accomplished by placing a pair of RFID device at each zone entrance.

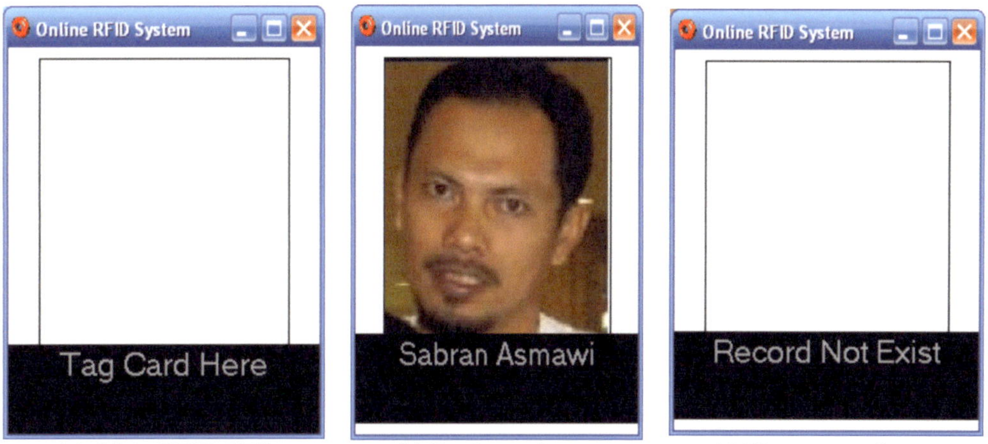

(a) Main RFID GUI (b) Registered user (c) Unregistered user

Fig. 8. RFID Tag status

Even though user can access all zone area, extra identification system that uses biometric technology is needed each time the user wishes to enter a facility rooms in Zone 1. The identification system needs user to present their fingerprint (Figure 9) onto fingerprint reader.

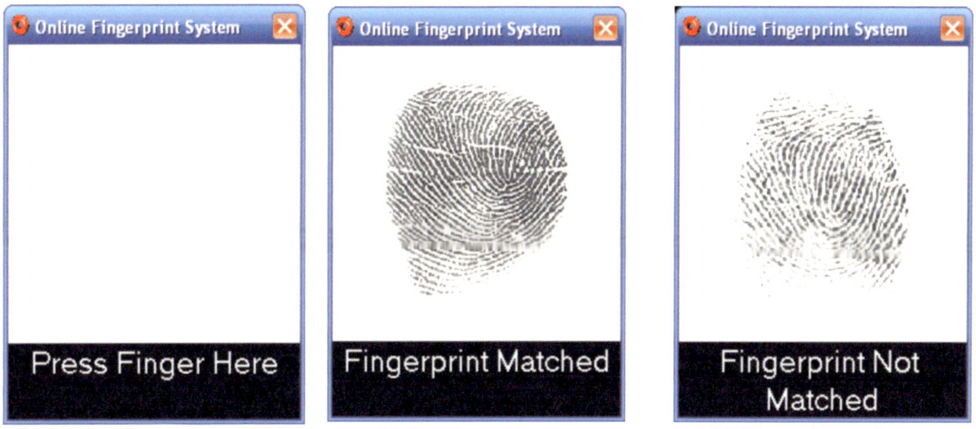

(a) Main Fingerprint GUI (b) Registered user (c) Unregistered user

Fig. 9. Fingerprint status

BoS system will matches the fingerprint with the one stored in the database system. If the fingerprint does not match, the system will displays the message and allows the user to re-enter up to three times before activating the alarm system. If it matched, the door will be unlocked and the system begins to record the room uses.

6.2 System administrators' access

User interface for system administrator is needed to allow the administrator to maintain the BoS system. There are three modules; Registration, Details, and Maintenance. Before they proceed to each module, the system administrator needs to enter administrative password (Figure 10). Any successful login will allow the administrator to manipulate the data.

Otherwise, the administrator needs to re-enter the security password up to three times before activating an alarm system.

Fig. 10. System Security for Administrative Staff

The main GUI of maintaining the BoS system is shown in Figure 11. For the purpose of this chapter, we will focus on two modules that use RFID and biometric technology; Registration module and Details Module.

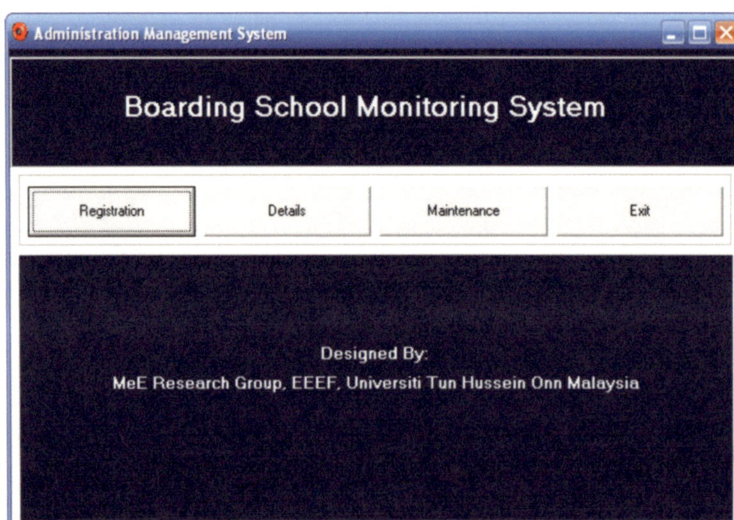

Fig. 11. Main GUI of BoS system

a. Registration Module

Registration module can be access by pressing on *Registration* button. This will allow the administration to add new user. The following are two purposes of registration module.

1. Register a new user to be part of BoS entities.
2. To access all facilities in Zone 1.

The user can be part of BoS entities by inserting user's information as shown in Figure 12. In order to enter hostel area, the user should have a RFID tag. The tag is registered into BoS system by clicking the *Register RFID Tag* button. By doing so, the RFID reader reads the tag ID and sends it to the system before displaying it to the user and save the information into the database.

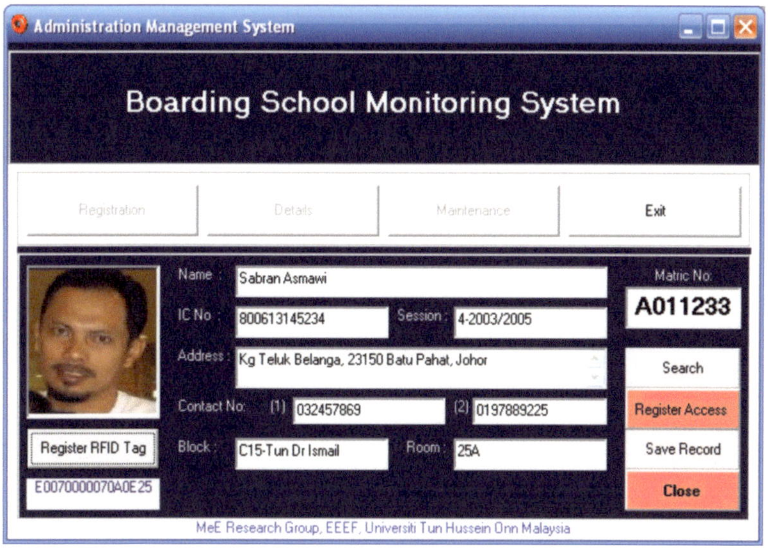

Fig. 12. Registration Module

Each new or registered user could access all facilities in Zone 1 by clicking the *Register Access* button. For registered user, the *Search* button in Figure 12 is used to gain user's information before proceeding to zone 1 registration (Figure 13).

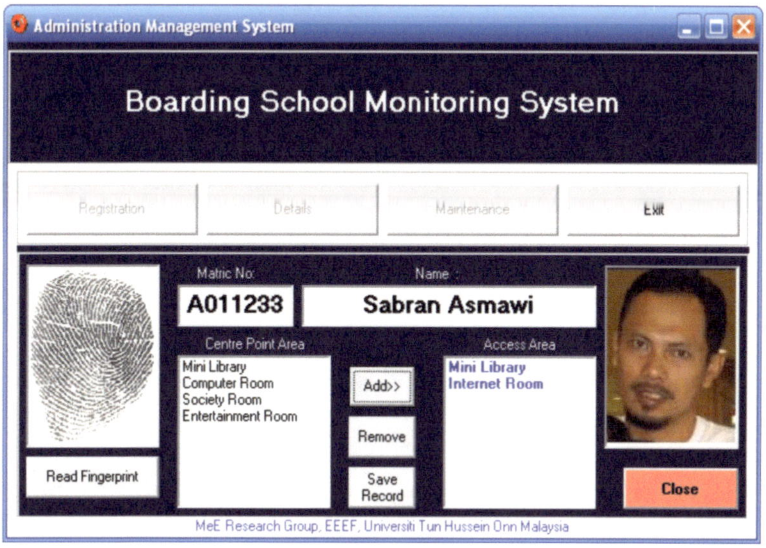

Fig. 13. Access Registration of Facility at Zone 1

Accessing Zone 1 acquire extra identification system. In this application, the identification used user's fingerprint. Therefore, each user needs to register one of their fingerprints into the system. The fingerprint is registered into BoS system by clicking the *Register Fingerprint* button. By doing so, the fingerprint reader reads the fingerprint and sends it to the system before displaying it to the user and save it into the database. The overall flow of registration process is shown in Figure 14.

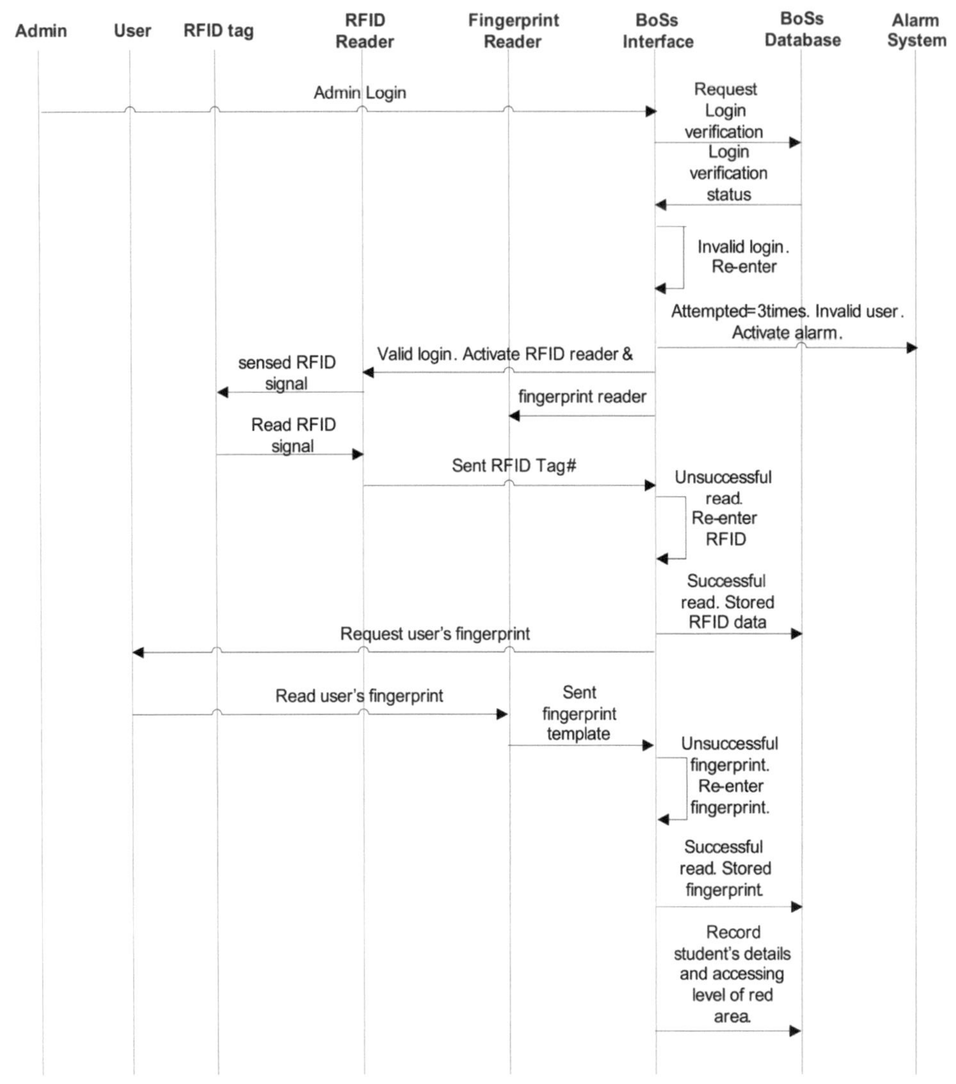

Fig. 14. Overall flow of Students' Registration and Access Registration on Facility Room at Zone 1

b. Details Module

System administrator can access Details module by pressing on *Details* button. This will allow the system administrator to view information on user's presence at hostel area (Figure 15). This detail will give information such as time entered and time out of the zone, date entered and out of the zone and etc.

c. View access information at all zone

The number of records of user presence at particular zone is depending on the number of user presence at the zone and it may have significant number of records. Therefore, in order for the system administration to views any particular record, the filtering is provided. The filtering process can be done by entering the zone code and the date before clicking on the *Search* button as illustrated in Fig. 16.

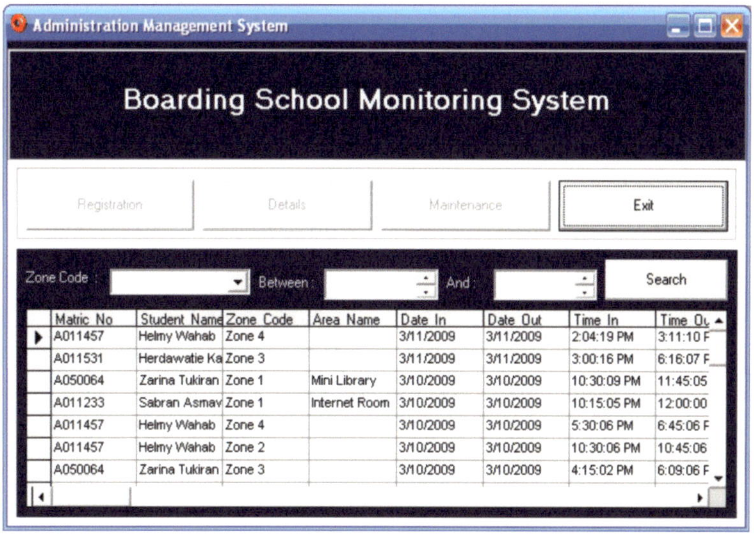

Fig. 15. View record on user's presence at hostel area

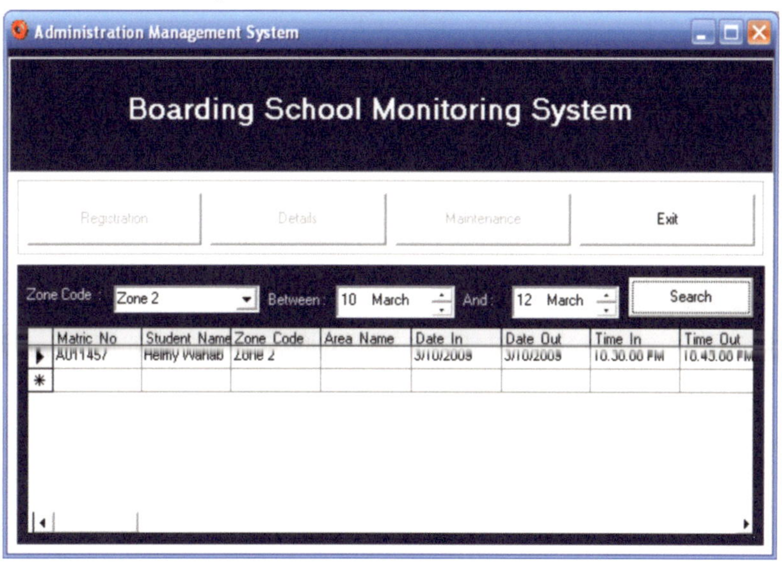

Fig. 16. Filtering information based on zone code and date

6.3 Online access

An online user interface has been developed to ease the management system to monitor the student movement and to search the student data easily. Some snapshots of the online system are provided in Figure 17.

7. Conclusion

In this chapter, we discussed the integration of RFID with boimetric technology to enchanced the security level in the boarding school managment system. First we discussed

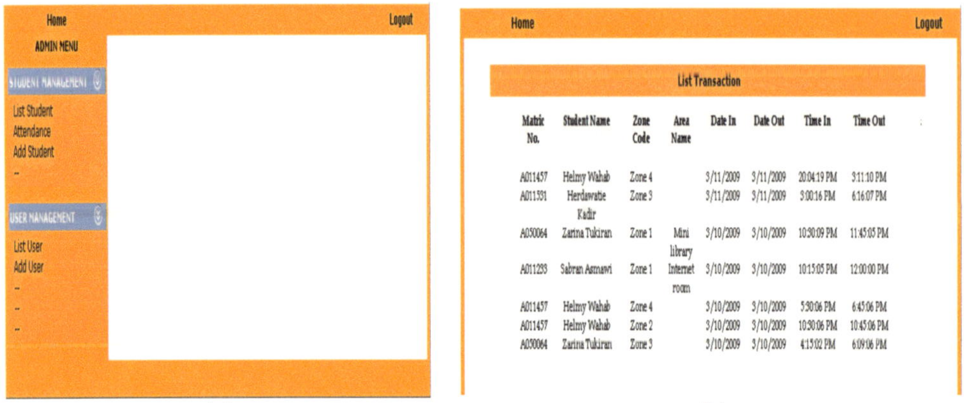

(a) (b)

Fig. 17. Online system sample

the technologies used to develop monitoring systems. We found that by adding the second layer of authentication to the system enchanced the security level. Finally, based on the application scenario proposed we develop a monitoring system and discussed the flow of the overall system. The application of BoSs will improve school management procedure, monitor the interest group movements automatically and increase assets security thus offers an important implication for monitoring the BS assets and eases the workload of the school management and save time for various student activities.

8. References

A. Qaiser, & S. A. Khan (2006). Automation of Time and Attendance using RFID Systems. Second International Conference on Emerging Technologies, November 13-14, 2008 Peshawar, Pakistan.

Bakery, N. S., Johari, A., Wahab, M. H. A., & Danial, M. N. (2007). RFID Application in Farming Management System. Paper presented at the 3rd International Conference on Robotics, Vision, Information and Signal Processing.

Bazakos, M. E., Ma, Y., & Johnson, A. H. (2005). Fast access control technology solutions (FACTS). IEEE Conference on Advanced Video and Signal Based Surveillance, 15-16 Sept. 2005 Como, Italy.

Boatwright, M., & Luo, X. (2007). What Do We Know About Biometrics Authentication. Fourth Annual Conference on Information Security Curriculum Development.

Carr, M. R. (2002). Smart Card Technology with Case Studies. 36th Annual 2002 International Carnahan Conference on Security Technology, 20-24 October 2002 Atlantic City NJ.

Chao Chen, C. A., Justin Ebaugh, and Christina Hong (2006). Development Of A Child Localization System On RFID and Sensor Networks in An Undergraduate Capstone Senior Design Project. 2006 Illinois-Indiana and North Central Joint Section Conference, March 31-April 1, 2006

Chen, W. D., & Chang, H. P. (2008). Using RFID Technology to Develop an Attendance System and Avoid Traffic Congestion around Kindergartens. First IEEE International Conference on Ubi-Media Computing 2008, 31 July-1 Aug 2008 Lanzhou University, China.

Corporation, V. (2005). RFID and the construction industry: What you need to know Contractor Magazines.

EraBuilder (2006). A Review of current state of Radio Frequency Identification Technology, and its use and potential future use in construction.

Gao, J. Z., Prakash, L., & Jagatesan, R. (2007). Understanding 2D-BarCode Technology and Applications in M-Commerce- Design and Implementation of a 2D Barcode Processing Solution. 31st Annual International Computer Software and Applications Conference 23-27 July 2007 Beijing.

Gil, Y., Ahn, D., Pan, S., & Chung, Y. (2003). Access Control System with High Level Security Using Fingerprints. 32th Applied Imagery Pattern Recognition Workshop (AIPR'03), 15-17 October 2003 Washington DC, USA.

Halawani, T., & Mohandes, M. (2003). Smart Card for Smart Campus KFUPM Case Study. 10th IEEE International Conference on Electronics, Circuits and Systems, 14-17 Dec 2003.

Hebert, P. D. N., Stoeckle, M. Y., Zemlak, T. S., & Francis, C. M. (2004). Identification of Birds through DNA Barcode. Public Library of Science, 2(10), 1657-1663.

Hendry, M. (1995). Smart Card Security and Applications (Second Edition ed.). London: Artech House Publisher.

Hillbrand, C., & Robert, S. (2007). Shipment Localization Kit: An Automated Approach for Tracking and Tracing General Cargo

IEEE Sixth International Conference on the Management of Mobile Business (ICMB 2007).

Jang, W.-S., & Skibniewski, M. J. (2008). A Wireless Network System for Automated Tracking of Construction Materials Journal of Civil Engineering and Management 14(1), 11 - 19.

Jiang, M., Fu, P., Chen, H., Chen, M., Xing, B., Sun, Z., et al. (2005). A Dynamic Blood Information Management System Based on RFID. 27th Annual International Conference of the Engineering in Medicine and Biology Society, 1-4 September 2005 Shanghai, China.

Joseph, D. I., & Nakhoda, Y. I. (2008). Students Attendance by Using RFID Informed Through SMS. Fourth International Conference on Information and Communication Technology and System, 5 August 2008 Surabaya, Indonesia.

Kadir, H. A., Kanafiah, S. N. A. M., & Wahab, M. H. A. (2008). Boarding School Students Monitoring System (E-ID) Using RFID. Fourth International Conference on Information and Communication Technology and System, 5 August 2008 Surabaya, Indonesia.

Kadry, S., & Smaili, K. (2007). A Design and Implementation of a Wireless Iris Recognition Attendance Management System. Information Technology and Control, 36(3), 323 - 329.

Kato, H., & Tan, K. T. (2005). 2D Barcode for Mobile Phone. Second International Conference on Mobile Technology, Application and System, 15-17 November 2005 Japan.

Mizuno, K., & Shimizu, M. (2007). Transportation Quality Monitor Using Sensor Active RFID. International Symposium on Applications and the Internet Workshops, 15-19 Jan. 2007 Hiroshima, Japan.

Mustafa, M., & Kyng, L. Y. (2007). TITO: Utilizing MyKad Touch N Go Features for Student Attendance System. First International Malaysian Educational Technology Convention 2 - 5 November 2007 Johor Bahru, Malaysia.

Ogata, H., Saito, N. A., Paredes, R. G. J., San Martin, G. A., & Yano, Y. (2008). Supporting Classroom Activities with the BSUL. Educational Technology & Society, 11(1), 1-16

Omar, S., & Djuhari, H. (2004). Multi-Purpose Student Card System Using Smart Card Technology. Fifth International Conference on Information Technology Based Higher Education and Training, 31 May-2 June 2004 Australia.

Ratha, N. K., Chikkerur, S., Connell, J. H., & Bolle, R. M. (2007). Generating Cancelable Fingerprint Templates. IEEE Transactions on Pattern Analysis and Machine Intelligence, 29(4), 561-572.

Reillo, R. S. (2003). Smart Card Information and Operations Using Biometrics. IEEE Aerospace and Electronic Systems Society, 16(4), 3 - 6.

Sabri., M. K. Y., Aziz., M. Z. A. A., Shah, M. S. R. M., & Kadir, M. F. A. (2007). Smart Attendance System by Using RFID. Asia-Pacific Conference on Applied Electromagnetics, 4-6 December 2007 Melaka, Malaysia.

Sam, P. (2007). The RFID Case Study Book-RFID Application Stories From Around the Globe: Abhisam software.

Shahid, S. M. (2005). Use of RFID Technology in Libraries: a New Approach to Circulation, Tracking, Inventorying, and Security of Library Materials. Library Philosophy and Practice, 8(1), 1-9.

Shepard, S. (2005). RFID Radio Frequency Identification. United State of America: McGraw-Hill

Simao, P., Fonseca, J., & Santos, V. (2008). Time Attendance System With Multistation and Wireless Communications. IEEE International Symposium on Consumer Electronics 2008, 14-16 April 2008.

Smith, G. S., & Coetzee, M. (2008). Analogue Fingerprinting for Passive RFID Tags, Third International Conference on Broadband Communications, Information Technology & Biomedical Applications (pp. 156 - 163). Pretoria, Gauteng, South Africa.

Stikic, M., Hu`ynh, T. a., Laerhoven, K. V., & Schiele, B. (2008). ADL Recognition Based on the Combination of RFID and Accelerometer Sensing

Second International Conference on Pervasive Computing Technologies for Healthcare, Jan. 30 2008-Feb. 1 2008 Tampere, Finland.

Susan, M. S., Mitchell, V., & Dudley, E. (2002). Using the Campus Web Site to Track Student Attendance. 30th Annual ACM SIGUCCS Conference on User services, 20-23 November 2002 Rhode Island, USA.

Won-Ju, Y., Sang-Hwa, C., & Seong-Joon, L. (2008). Implementation and Performance Evaluation of an Active RFID System for Fast Tag Collection. Computer Communications, 31(17), 4107-4116.

Wu, Z. (2008). Authentication System on Open Network and Security Analysis. International Symposium on Electronic Commerce and Security 2008, 3-5 August 2008 Guangzhou, China.

Zhang, D., Kong, W. K., You, J., & Wong, M. (2003). Online Palmprint Identification. IEEE Transactions on Pattern Analysis and Machine Intelligence, 25(9), 1041-1050.

10

Accelerating Time to Value for RFID Solutions with Reusable Assets

Han Chen, Paul Chou, Sastry S. Dury and Jim A. Laredo
IBM T.J. Watson Research Center
USA

1. Introduction

Applications of RFID technologies have progressed from isolated proof-of-concept projects to integrated business enablers that provide up-to-date visibility of physical objects in the context of business processes within an enterprise and across partners. The increased scale of the projects makes the drive for Return on Investment (ROI) more important than ever; reducing the RFID solution lifecycle costs and accelerating the time-to-value have become the key challenges to broad-based RFID adoption and deployment. This calls for that solutions be built using proven architectures based on well-understood patterns and making use of the lessons learned from one project to the next. If new solutions can be built based on existing ones with reusable parts to reduce the cost of development and integration, the ROI case becomes a lesser hurdle, and situations where once RFID would have been considered for being too costly to implement now become affordable to deliver business returns.

Like any technology, the continuous innovation and identification of new applications promote asset reuse. It reaches levels of maturity where components are better understood from many aspects of their lifecycle, such as features/functions, dependencies, performance, and ease of integration. It is common to see these components evolve bottom-up from the perspective of technology maturity. Early RFID applications focused on components that process RFID read events generated from the readers. As the reader technology improved and tags became more common due to reduced price points, components were hardened and broadened to support larger event loads and additional usage patterns. Standards started to emerge, such as Application Level Events (ALE) and Low Level Reader Protocol (LLRP), developed under the auspices of EPCglobal (EPCglobal). They provide the basis for building more reusable components regardless of the choice of specific RFID reader technology. The evolution of RFID-enabled goods shipping/receiving portal serves as a good example of component maturity. Higher-level standards, such as the EPC Information Service (EPCIS) (EPCgloba), further address the needs for component interoperability, enabling supply chain level integration supported by standards-based components.

These more robust components opened the door for more business oriented solutions, built around business processes that deliver value as measured by specific key performance indicators. These solutions can handle a greater inflow and variety of events or integrate to other components in the supply chain. Some of these RFID-enabled business processes, as they matured and were better understood, became common business building blocks that are applicable for a larger set of use cases, in some cases across industries. Their realization into the supporting infrastructure becomes possible as long as it is possible to choreograph those processes across the components, in some cases requiring some level of customization, as a function of the industry where they were being used. As the components get reused, it is easier to identify its applicability and integration requirements with others. It is then possible to identify the integration points and configuration parameters, also known as Points of Variability, and create a Template that captures the expertise of those that have performed the activity many times. Follow-on integrations only require the instantiation of the template and the configuration of the points of variability to take advantage of the expert experience packaged in the template, a much simpler and faster activity compared to earlier implementations that often requires more time and skills to complete. Creating templates that facilitate the use of underlying components and help realize a high-level business process ultimately accelerates the time to value of a solution.

In this chapter, we closely examine the evolution of technology, architectures, and approaches to solution building in the RFID field. We examine those components that have benefited from standards that now have become the building blocks for solutions that when combined with business processes unleash a greater business value. Furthermore, we look at the options to templatize parts of the process to accelerate the integration and reuse of these components and to ultimately deliver a greater ROI in a shorter period of time. We conclude looking forward to what is coming up next and our possible choices to solve the open problems that remain.

2. Pharmaceutical products track and trace – a running example

In this section we describe one of the most important applications of RFID for the pharmaceutical industry - the track and trace of pharmaceutical products throughout their supply chains. We use this example to illustrate the common processes across the supply chain for enabling tracking and tracing of goods, and how these processes are supported by reusable components built around open industry standards.

In addition to the common benefits associated with RFID-enabled supply chain management, such as better inventory visibility, reduced out-of-stock, and so on, pharmaceutical track and trace addresses two even more pressing issues in this industry—drug counterfeit and diversion. Many governments including European Union, Japan and several states in United States have proposed or adopted pedigree laws or regulations to address the counterfeit problem. For example, the California law requires that each prescription drug must be serialized by the manufacture at its smallest packing unit. Every transaction starting from the manufacturer must be appended to an electronic pedigree; and wholesalers and retail pharmacies cannot take possession of a product or provide it to others without validating the chain of custody. The problem of drug diversion is a result of the multi-tiered pricing structure employed by the Pharmaceutical industry. For the exact same product, lower price is used for less affluent markets while much higher price is targeted for communities which can afford to pay more.

The RFID-enabled Pharmaceutical Track and Trace application addresses both problems. It allows serialized, RFID tagged product items to be quickly identified at each transaction.

The serial identification is used to look up and update an electronic pedigree system that is built on a standard-based information service, namely EPC Information Service (EPCIS). It provides product traceability for diversion detection and also enables product tracking for recalls.

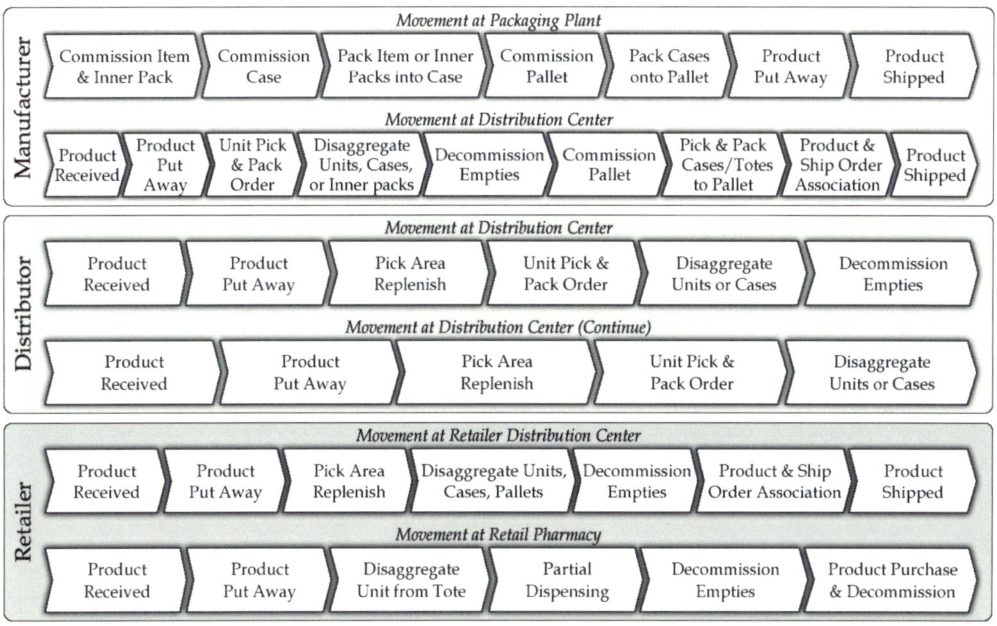

Fig. 1. Pharmaceutical Product Process Flow

Fig. 1 shows at a very high-level various steps in the pharmaceutical supply chain as drugs move from manufacturer to retail pharmacy. Many of the business process steps, such as product packing (aggregating), shipping, receiving, put-away, unpacking (disaggregating) are common, albeit with some variations, across different part of the supply chain. For example, one could define an abstract packing process to consist of the following steps: 1) determine the number of smaller units per large container, 2) fill large container, 3) verify large container contains expected number of smaller containers, and 4) create containment association in an information service.

Fig. 2 shows a simplified operational setup for an example case packing process ("Pack Items or Inner Packs into Case" in Fig. 1). The example process involves physical processing of goods wherein the pack line operator responds to sensory feedback, and initiates appropriate actions to pack case containers with a predetermined number of product units (in this case bottles).

In a typical scenario, some drug is manufactured during a production run and is filled into unit containers. The pack line operator authenticates to the system and using drug's National Drug Code, batch and lot numbers to allocate a set of unique Electronic Product Code (EPC) numbers for both unit and case containers. The pack line operator feeds this information to a label generating application to commission unit and case tags. These activities correspond to the steps "Commissioning Item and Inner Pack" and "Commissioning Case" in Fig. 1. The next step, "Pack Item or Inner Packs into Case," is described in detail below.

In the example packing process, unit containers are affixed with high-frequency (HF) tags, and cases are affixed with ultra-high-frequency (UHF) tags. Therefore, packing process needs two different RFID readers: one for reading HF tags and another for reading UHF tags. Starting from the left the bottling machine fills unit containers with specified drug. A tag slapping machine applies HF tags to bottles. A predetermined number of bottles are packed into cases. The tunnel reader reads HF tags. The UHF reader to the right of the tunnel reader reads the case tag. The light stack to the right of UHF reader shows the status of the read, specifically indicating whether the number of tags read match the number of unit containers expected in the case and whether the case tag was read successfully. If all counts match expected values, then the system creates containment association between case tag and unit tags, and saves this information in a repository.

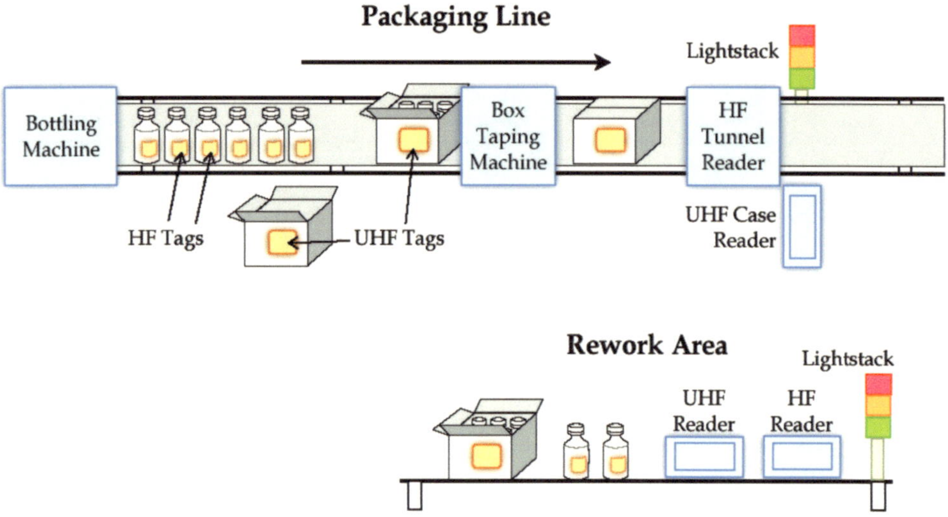

Fig. 2. Packaging Line Operational Setup

Otherwise, the conveyer is stopped and the case diverted for manual inspection later. After removing the defective case from the conveyer, the operator starts the conveyer and continues with the remaining cases. The rework area is used for the operator to manually examine/repack the cases with tag read problems as indicated by the light stack.

3. Evolution and current technologies

While RFID is still being gradually rolled out in many industries. It has reached a certain level of maturity to allow us to reflect on the evolution of its adoption and the associated software architecture. Using the pharmaceutical track and trace application as an example, we examine the trends that we have identified.

3.1 RFID technology is better understood and integrated into business process

In the early days of RFID adoption, companies have focused mainly on the technology itself. A lot of demonstration projects were conducted as proof-of-concept or technology validation. The overriding goal of these early projects is to see if and how well the RFID technology works in specific operating environments and business processes. Read accuracy was a key focus. Due to the complex physical nature of radio transmission, many factors may adversely affect the read accuracy, for example, the operating frequency of a tag, the dielectric property of the material being tagged, the radiation power of a reader, and the relative positions of nearby readers. In the exemplar pharmaceutical packing use case described earlier, the content inside the bottle has a direct impact on the choice of tags. For non-conductive material such as pills, UHF tags, which are electrically coupled, may be used to tag the bottle. However, if the bottle contains medicine in liquid form, which is electrically conductive, a magnetically coupled tag, such as HF tag, must be used.

Some early adopters viewed RFID simply as a high-speed technology that replaces barcode. Many had the unrealistic expectation of achieving 100% accurate tag reading and hoped that RFID would be the magic bullet that can increase the operation efficiency without affecting the existing business processes. As these early technology validation projects have shown, missed reads and duplicate reads can still happen even in a carefully designed, engineered, and deployed RFID data collection system. With this realization, most users have abandoned the assumption or requirement that tags be read at 100% accuracy. Instead, these conditions are taken into account in the involved business processes during later-phase pilots and rollouts. Various form of temporal- and spatial-based filtering techniques are used to deal with duplicate reads. For missed reads, the solutions are more finessed. In situations where absolute accuracy is required, exceptions are raised in a business process which triggers a rework subflow so that the error can be properly remedied. For example, in the aforementioned pharmaceutical packing application, every tag must be read correctly in order to comply with the pedigree regulations. The packing process must be designed so that when fewer than expected bottle tags are read in a box, a manual rework subprocess is initiated. In some applications, a business may elect to tolerate a small number of missed reads by inferring the presence of the tagged items using heuristics from other sources, for example, an Advanced Shipping Notice (ASN). An example would be a receiving dock of a consumer goods retailer; the serialized IDs of the merchandise within a shipping container can be inferred by looking up the ASN through their container's ID. When compared with IDs reported by the reader, any missed reads can be identified and rectified.

3.2 Software architecture becomes more encompassing; reuse happens at higher level

As RFID technology matures and gets better understood and integrated into business processes, a similar trend also happens in the software architecture that supports RFID solutions.

During the technology-centric early phase of RFID adoption, the software solution architecture for RFID was correspondingly device-centric. Software and solution providers were mostly concerned with creating device-level middleware to simplify the operation of RFID devices. Some of the major issues were device drivers for readers and printers on different platforms, parsing and converting tag formats (for example, EPC), mapping between physical and logical devices, and simple aggregation and filtering. Solution-level software reuse was very limited if happened at all. Because most of these middleware and software components were vendor specific, interoperability was also hard to achieve. Using the pharmaceutical packing use case as an example, early solution providers would have a very limited repertoire of low-level software components. For every new deployment of the same use case, whether it is for a different customer or for a different location of an existing customer, a lot of resources are required to re-engineer a solution, sometimes almost completely from scratch, due to the inevitable variations present in the new environment. Clearly, this is not cost-effective and would become a major impediment to the wide adoption of RFID.

Many of these issues are addressed by open standardizations such as EPCglobal. Fig. 3 shows various EPCglobal standards that are applicable to components in the packing process. In the figure, the simplified packing process flow is spread across three layers: at the bottom is the *device layer* where the focus is on controlling the device infrastructure to facilitate reliable data collection. In the example packing process this layer is responsible for turning on the reader for data collection when a case approaches, detecting when the case has left read zone and aggregating tags collected and sending them to next layer. Components in this layer also interact with other types of sensors and actuators in addition to RFID readers. For example, motion detectors, various types of displays, or controls that start/stop conveyer. For example, if tag counts do not match the conveyer is stopped so that the offending case can be subjected to manual inspection later. The tag data from the device layer flow into components in the *filtering and collection* layer. Filtering and collection capabilities are essential features of every RFID based supply chain use case. The *Application Level Events* (ALE) specification which provides for a standard way to declaratively specify filtering and collection requirements greatly enhances the ability to develop reusable components. At the next layer, the *EPC Information Capturing layer,* the aggregated tag information is augmented with business step information and stored in a repository. The *EPC Information Service* specification standardizes formats and interfaces to combine business activity information with RFID tag data and to capture relationships between RFID tagged entities, for example, the unit bottles and the containing case.

The emergence of standards enabled middleware and solution vendors to invest more in the upper levels of the architecture stack and consider RFID and its surrounding business processes in a holistic way. Many vendors have promoted their own specific RFID solution architectures. Although details may vary in these architectures, they share many common design features. In such architecture, RFID is but one component of the overall solution which may consists of multiple inter-connected layers. Thanks to standardization, RFID can now be abstracted away properly as a type of device, alongside with other types of devices such as barcode reader. A network service layer deals with the information exchange within a business and across partners in the framework of EPC network. A process layer consists of the necessary tools and runtimes for orchestrating RFID enabled business processes. An integration layer provides connectors and adapters to external information sources, e.g., trading partners. Finally, the business layer allows RFID enabled processes to be monitored and controller using tools such as dashboard.

With this level of thinking, technologies, such as RFID, are put into their right place as an enabler for business process improvements and innovations that result in the real return on investments. Now, solution providers can start treating higher-level solution artifacts such as process as reusable software assets and leverage their accumulated expertise in process engineering, integration, and optimization to deliver RFID solutions more efficiently.

3.3 New software programming models are gradually adopted

Appropriate software programming environment is needed to process the events generated by the RFID devices. Such an environment includes programming models, tools supporting the programming models, and runtime platforms for executing the artifacts created by the tools. As we have discussed in the preceding subsection, with the maturing of technology and advent of standards, lower-level components that process RFID events have become more reusable. Increasingly, solution providers are looking into reusing higher-level solution artifacts, such as, use cases and processes.

Enterprise application integration methodology and design patterns, such as, Service Oriented Architecture (SOA), have been successfully used in various business process integrations scenarios. In SOA, a reusable piece of software code is packaged as a component, with clearly defined interfaces, which specify the services the component is capable of providing, and references, which indicate the services that this component requires in order to carry out its designed functions. In addition to traditional programming language, such as, Java, C/C++, higher-level business-oriented methods, such as, business rules, decision tables, or process flows, are also used to implement SOA components. Applications or solutions can then be created by connecting individual components via their interfaces and references to form a composite.

Because of the availability of expertise in SOA-related technologies, many early RFID solutions have been built using these existing tools and runtimes from device-level components all the way to processes. As more and more RFID solutions are developed and deployed, the distinction between solution composition at process level and RFID event processing level becomes clear. At process level, reusable assets are abstracted as services, which can be choreographed as procedural steps to form a process. Control flow languages such as BPEL (Andrews et al., 2003) are well suited for solution composition at this level. On the other hand, at the event processing level, reusable assets are more suitably abstracted as agents, which generate and react to events asynchronously. Data flows instead of control flows connect the components to form composites. Although it is possible to implement this data flow semantics using traditional SOA middleware, it is often cumbersome at design time and inefficient at runtime.

For this reason, researchers have promoted event-driven architecture (EDA) as an alternative and complement to SOA to better support the event processing portion of an RFID solution. Event-driven architecture shares many of the same design principles of SOA. A piece of reusable software code is packaged as an agent with an interface, which defines the types of events that it consumes (input ports) and the types of events that it generates (output ports). Hierarchical composition is used to create complex applications or solutions by connecting individual agents to form a data flow diagram. Because this programming model offers a closer mapping to the physical data flow in an RFID event-processing system, it allows solutions to be created more easily, thus lowering the development cost. Because the agents in a data flow exist in their own rights and are not instantiated for each invocation, an application can be executed very efficiently by an EDA runtime to cope with the large volume of RFID event data, further reducing the operating cost of a solution. The reader is referred to (Chen et al, 2008) for an example of an event-driven RFID programming model and run-time system.

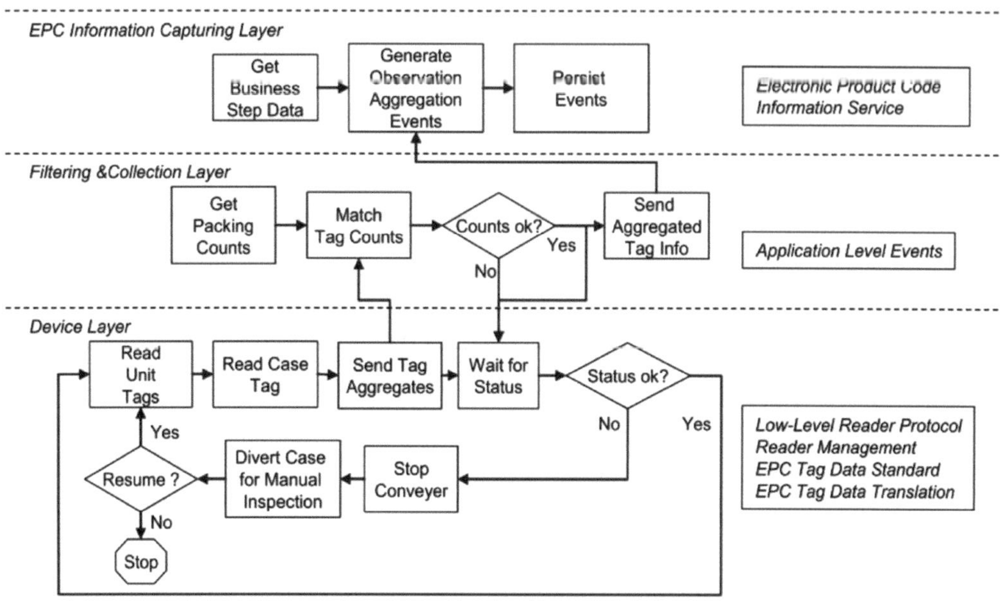

Fig. 3. Mapping of packing process across EPC Network Layers

4. Templatization approaches

There are some factors that could influence how the packing process is carried out in a given setting. The kind of material filled into the unit containers influences the type of tags used to track them. If the unit containers contain liquids, typically, high-frequency (HF) RFID tags are used. Cases, typically, are tagged with ultra-high frequency (UHF) tags. In this setup, shown earlier in Figure 2, two RFID readers are needed. If the unit containers are filled with dry goods, UHF tags can be used and only one reader is needed, and all tags are read at the same time. These are variations that will determine what to set up for the RFID infrastructure. At validation step other factors come into play depending on what type of accuracy we need to attain; in the case of electronic pedigree we must ensure that all individual tags are counted, while in other scenarios we may allow for some degree of error and reconcile with other sources. The process logic should explicitly distinguish unit tags from case tag. Depending on the industry we are addressing we may choose one type of validation over the other. In another variation, when the process is used for packing cases into pallets unit tags are usually filtered out. Furthermore, as this process involves physical movement of goods the underlying device infrastructure may differ: conveyers are normally used for items/units and fork lifts are used for moving pallets. This means the sensing and reporting infrastructure may need to be configured accordingly to support variations at the device layer in Fig. 3.

Let us now observe how a System Integrator that has a team working with a customer to build a packing and verification solution and is using assets from different Independent Software Vendors (ISV) would approach the problem. One of those ISVs has built a verification component that could be configured to integrate with the packing portion of the process. This component is now in its third release and the ISV understands better how clients are approaching the validation process and integrating it with other components to build solutions. Han the architect of the component has directed his development team to implement a new requirement to support electronic pedigree that originated from the use of the previous versions. Han is planning to generalize the component and expose points of variability or configuration points to parameterize the degree of accuracy that a given component requires before triggering an exception. Back at the System Integrator, Paul, the Business Analyst that has mastered the Packing process and its variations is working with his Solution Architect, Dalia. Dalia understands the RFID domain and the components and processes that could realize that packing process. Based on standards such EPCIS and ALE, Dalia chooses components that are able to interoperate with each other. Dalia assembles an RFID solution that is EPCIS compliant and thus can interoperate with other components they have built in house. Dalia sets the Points of Variability, both pre-deployment and runtime parameters, according to their agreed values and integrates the validation component. A second scenario helps Dalia gain more insight on the component, this time she sets the Point of variability differently but still according to the specification to accommodate the new use case. Dalia has observed that some Points of Variability do not change from use case to use case and could be preset, furthermore she has established reasonable ranges for some parameters, and created some relationships with parameters from other components that conform the solution. Next time the composition of components is deployed fewer parameters will need to be preset and the guidelines and constraints will accelerate the instantiation process. At this point Dalia's experience is captured into what is called a template, a composition of components and possibly other templates that others can use, with the benefit of the encapsulation of default parameters and simplified settings of those that still need to be addressed.

Lily a practitioner at the System Integrator, working with another customer, a manufacturer of shampoos and conditioners, finds the template applicable to her customer's problem and decides to use and apply the template into her solution, the template advises on the RFID infrastructure needed such as the use of two RFID readers, and allows to configure the degree of accuracy required by removing the need of 100% accuracy and lowering the corresponding threshold. This approach requires a much smaller learning curve to use the assets and apply the lessons learned during Dalia's iterations and maturing process while building the template. Sastry, the customer, on the role of solution administrator, is the person that at on-boarding time and during steady state configures a deployed solution to the specific customer needs. He has another opportunity to further change the behavior of the solution; for example, he may select specific external services that fulfil a required function, or other available services to manage the policies that govern the thresholds of scan retries on a per product basis.

4.1 User roles

In the above description we identified the user roles involved in the templatization process following those described in (Fu et al., 2008). We also extend their classification with the Solution Administrator as one that has another opportunity to configure the deployed instance. The Template Creator represented by Dalia, Solution Creator carried on by Lily, the Component Builder impersonated by Han, and the Solution Administrator realized by Sastry.

The Template Creator is usually the Subject Matter Expert, she has enough technical skill to abstract a solution and create related templates such that they can be used multiple times. The Template Creator determines the points of variability and expresses them into the template.
The Solution Creator is able to identify the right template candidate and may be able to perform the configuration tasks needed to instantiate the template. She may perform additional compositions to complete the solution.
A Component Builder has the more in-depth knowledge about each component used in a template. He understands the underlying technology and is able to create new reusable components based on requirements. He creates versatile components with its own customization points that later can be exposed as Points of Variability in the template.
The Solution Administrator belongs to the end-user group; he implements decisions at runtime to customize the solution based on the business needs.

4.2 Template properties

According to the Object Management Group, Inc., Reusable Asset Specification (OMG, 2005) you can look at an asset in 3 dimensions: granularity, variability, and articulation. Larsen (Larsen, 2006) synthesizes them as follow:
Granularity is the spectrum of asset size and shape. Asset may range from fine-grained, meaning they are small in size and purpose, to coarse-grained, providing larger size and purpose and often containing or referring to fine-grained assets.
Variability refers to the asset's degree of customization. This, too, is a spectrum. On one end, the asset is fixed, and on the other, it is widely visible and changeable.
Articulation is the level of completeness of artifacts in an asset that can provide a solution. Some assets have very few artifacts that assist its consumer; whereas others are fully articulated, providing artifacts that include, for instance, requirements and testing validation.
We look at templates through this same asset view. We will explore granularity by looking at the template continuum that we believe are relevant to RFID solutions. Variability will show us the options to compose a solution; it will let us determine the amount of configurability, flexibility on setting parameters, and even potential for late bindings of relevant pieces of the solution. Articulation will show us the degree of realization of a solution.

4.3 Template granularity

Assets range through all levels of the solution building process. At the highest level we can have a solution description that comprises a set of business processes that captures the subject matter expertise around the problem being solved. Our use case in Section 2 around the electronic pedigree is such solution description. The business process may express the possible variations and constraints that allow the process to be valid and yet fulfill distinct solution recipients. An asset may be purely descriptive, documentation only or be a model that captures actors, interactions, and data artifacts on which they operate. An asset may also be accompanied by a set of components, data definitions, business process orchestrations, and user interfaces that could be deployed customized and connected to a given customer scenario. Each of these elements may have a realization in one or more target environments. For example, a UHF or HF physical configuration depending on the nature of the items (liquid or dry) may fulfill the implementation of the packing infrastructure or at a lower level from a user interface a simple HTML User Interface or a Rich Client Ajax based interface could implement the same interactions between actors for the solution. Each of these elements may be considered an asset on its own, often the distinction between an asset and just an element depends on the value it holds on its own, usually described by its variability, which we will describe later.

Templates as assets behave similarly. At the higher level, a process description may be templatized, suggesting a methodology on how to approach the solution, for example the Track and Trace process could be a template that starts with a process model with its associated artifacts. Next, each particular portion of the process such as packing, or shipping could be a template, the template may compose lower level components into what could translate to a fully functional part of the solution once properly configured and integrated with runnable assets.

If we think how a template comes about, it is the result of multiple iterations of applying an asset or group of assets to a solution that is repeated within some range of change. As we understand better that range we can parameterize those values that affect the behavior that is relevant to the solution. The smaller size assets lend themselves better to incremental modifications, and we will only incorporate other assets at the point that we understand how those interact with the ones that already are part of the template. Figure 4 illustrates this point; the lower layer shows the candidate components and how they can be assembled into a template that supports part of a process.

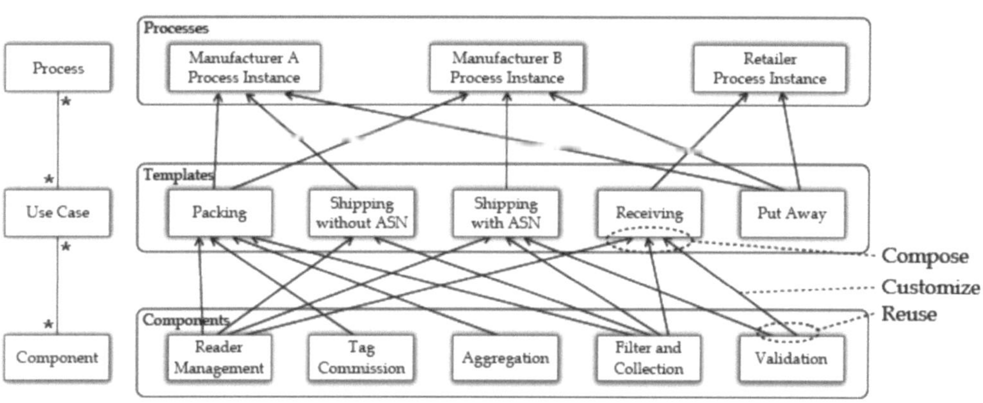

Fig. 4. Solution building using templates

It is often the case that a composition will grow to a point that represents a very particular case, possibly of little interest to anyone else, and with the lack of opportunity to iterate to make any judgment on the amount of combinations that are available it often leads to a set of assets that provides just an example of the integrated components, usually for others a proof point that the technologies employed work and can be used in similar scenarios yet requiring an integration of their own. In a way we could argue that such end to end solutions represent a template yet lacking proper articulation.

4.4 Template variability

An asset is as valuable as its capability to integrate with other assets in the solution. Lack of integration points will limit its useful life. In the past, vendor technologies for COTS (common off the shelf) packages have provided their own mechanisms to integrate; however with the advent of SOA it has become much easier to integrate assets from different sources. As long as the asset exposes a WSDL or REST interface the integration does not require special libraries and just needs the standardized definition of the interface or the resource and the documentation of its use.

There are several implementations of such notions, for example, Apache iy Project (Tuscany) implements the Service Component Architecture or SCA (Besicgel et al., 2007). SCA provides a model for creating composite applications by defining the services in and their relationships with one another. The services can be implemented in any technology. Protocols are pushed out of business logic and are handled through pluggable bindings. Applications can easily adapt to infrastructure changes without recoding since protocols are handled via pluggable bindings and quality of services (transaction, security) are handled declaratively.

As the components are composed into templates the Template Creator builds into them a degree of variability to support the experience he has gained. He then expects the Solution Creator to make use of them through a valid configuration to realize a working solution. Let's consider the more common components in a solution composition: process flows and adapters. A process flow will encompass all related business process constructs such as packing or its associated validation process as described in our example while adapters will comprehend all necessary logic to interact with an external system, an adapter to RFID reader for example. The next subsection examines how variability is built into a template.

4.4.1 Points of variability

To address the customization aspect of a template, the notion of points of variability is introduced. A given artifact may offer customization points or points of variability that identify the options to modify its behavior. We see Variability in two dimensions at Design or Solution Composition time, and post-deployment at runtime, often at customer on boarding time, when the solution is first configured and later tuned up during steady state. The former may require making decisions earlier, usually at pre-deployment time and set parameters one time with little or no opportunity to modify them again, while the latter presents the options at onboarding time to reflect the decisions of the end-user.

For a process flow variability may be introduced at design time on the interfaces that are being offered, some interfaces may be enabled or disabled based on the choice of parameters that may be configured, for example. At runtime configuration parameters that complete the setup of the process flow may take place. (Fu et al., 2008) and their solution template tool focused on design time templates offered the following points of variability for process flows:

- Interface selection - Depending on the features being enabled in the process, interfaces could be conditioned based on property values. The tool supported three types of conditions to govern the point of variability: mandatory, optional and conditional. Some interfaces of the process flow are mandatory and need to be present when the template is configured. Some interfaces are optional and can be enabled or disabled at configuration time at the discretion of the Solution Creator. Some interfaces will be automatically enabled if the condition associated with the point of variability is met; otherwise, it will be disabled. For more sophisticated conditions rules can be used to evaluate the condition that determines the enablement of the point of variability.

- Properties - An artifact may offer some configuration parameters in the form of name and value pairs, a type may further constrain and provide some level of validation on the possible values. More importantly this value can be extrapolated at the template level to create a global notion that once configured its value propagates to the individual components.

Similarly, for an adapter, points of variability can be defined for the inbound and outbound interfaces, use of properties to set a behavior, etc. One type of variability is the implementation choice, in the case of an adapter, it may be determined by the third party software or service or device we choose to connect to. Depending on the setting of the point of variability a different underlying adapter will be instantiated.

We expect tooling to detect if a variability point is available and present the Solution Creator with options to choose, further more would a mismatch occur, tools offer options to resolve the incompatibility. The idea is to minimize the second guessing by the user and drive the configuration process through completion into a valid configuration where the artifact may operate.

Note that we have used the terms configuration and customization interchangeably throughout this chapter. In practice we refer to configuration when we have a set of parameters that needs to be set and they are strictly defined usually with proper ranges and consistency rules. We refer to customization when a more complex and likely longer cycle takes place to vary the behavior of a component. For example, a validation function may have to be implemented to support a user requirement, or an adapter for a non-standard peripheral may need to be incorporated to complete the solution. Although configuration is preferred over customization the nature of these complex solutions often require a degree of more sophisticated flows, managed by a varieties of policies and constraints that may require additional implementation. One approach that has been used to bridge the gap effectively is with the use of Business Rules as described by Linehan (Linehan, 2005). Rules as another artifact in the SOA composition can be referenced from a component to dictate a point of variability in the behavior of such component.

4.5 Template articulation

In essence, the use of SOA allows for a modeling capability that could be platform independent, as long as the component builders are able to express the components interfaces as service interfaces and provide its configurability through a service. A given component could be realized with various implementations. Templates themselves could also be specified as an SOA assets allowing the notion of sub-templates. Once the template is configured per the user requirements and a platform selected, it can be provisioned as a solution instance in that target platform. The template itself does not pose any tie-in to the underlying technology; the resulting realization of the solution should be standalone and work under the design and runtime tools of that platform. This is also a common use of templates, where the template acts as a starter and the native design tools of the target

platform continue to enrich the solution. Integrated template tooling such as SIMPLE (Laredo et al., 2009) further provide the traceability to the design tools allowing to capture any refinements made and expose them again in a new version of the template for future use.
Other technologies such as Model Driven Business Transformation (Bhattacharya et al., 2007) allow to capture business process models including the business artifacts relevant to the process and the roles that are involved. It provides transformation capabilities to create platform independent state machines for each business artifact and user view, and easily translate to a target platform and map into existing components. Customization opportunities exist along the way; from modifying the base Business Process to extending the compositions at each step of the process. This approach provides for a better articulation of a template as the transformations and customizations ultimately end in an executable solution.

5. Conclusions

RFID has been shown to deliver business benefits in supply chain management such as reducing stock-outs and improving on-shelf availability (Corsten, 2005) (Hargrave, 2006). Such benefits can only be realized by judiciously integrating the technology into related business processes. In this chapter, by using a pharmaceutical product track and trace application as an example and drilling down the manufacturer's product packing process, we examined the evolution of RFID projects and the emergence of solution templates. In essence, templates capture the knowledge and assets applicable to a problem domain and allow for rapid solution realization and deployment. We discussed the attributes of templates to be considered for addressing points of variability and asset reusability that are crucial for increasing ROI and accelerating time to value.
The exploitation of SOA and its integration with EDA allow developers to capture the invariant process logic and defer the decisions about the specific service implementations and device choices to deployment time. The result is that the logical design of a process can be preserved while the underlying components and devices can continue to evolve over time. With solution templates, in most cases, changes in business process or requirements can be easily addressed by 1) changing parameter value, 2) changing the composition flow, 3) adding, deleting or choosing a different implementation of components., and 4) applying business rules None of these requires low-level coding.
The pharmaceutical industry is a good example of an industry segment that has standards applicable to all layers of EPC network application stacks. Such a standard based stack promotes the development of components that cover functional requirements while leaving many non-functional aspects for vendors to compete in implementations and allow for later customizations, thus laying the foundation for template building. We expect the development of industry oriented standards to accelerate and cover many more industries and applications.
The decision whether to reuse an existing asset is a complex one and a number of factors have to come together to tip the balance markedly in favour of reuse. To better facilitate asset reuse decision and process, additional technology enhancements are required in the following areas. One of them is easy-to-use asset repositories for discovering and depositing

assets. Another is an environment to quickly learn various aspects of an asset in question such as usage examples, runtime simulation of asset compositions etc. Yet another factor is the existence of standards reducing asset learning curve and increasing its understanding.

6. References

Andrews, T., Curbera, F., Dholakia, H., Goland, Y., Klein, J., Leymann, F., Liu, K., Roller, D., Smith, D., Thatte, S., Trickovick, I., Weerawarana, S., Business Process Execution Language for Web Services, *Version 1.1, OASIS,* Needham, MA, 2003.

Bhattacharya, K., Caswell, N. S., Kumaran, S., Nigam, A., Wu, F. Y., Artifact-centered operational modeling: lessons from customer engagements, *IBM Systems Journal, v.46 n.4, p.703-721,* October 2007

Beisiegel, M., Blohm, H., Booz, D., Edwards, M., Hurley, O., Ielceanu, S., Miller, A., Karmarkar, A., Malhotra, A., Marino, J., Nally, M., Newcomer, E., Patil, S., Pavlik, G., Raepple, M., Rowley, M., Tam, K., Vorthmann, S., Walker, P., Waterman. L., SCA Service Component Architecture - AssemblyModel Specification, version 1.0. *Technical report, Open Service Oriented Architecture collaboration (OSOA),* March 2007.

Fu, S., Yang, J., Laredo, J., Huang, Y., Chang, H., Kumaran, S., Chung, JY., (2008) Solution Templates Tool for Enterprise Business Applications Integration, *Proceedings of IEEE International Conference on Sensor Networks, Ubiquitous, and Trustworthy Computing (SUTC 2008) 2008,* pages 314-319, June 11-13, 2008, Taipei, Taiwan.

EPCglobal, www.epcglobalinc.org.

Laredo, J., Yang, J., Jeng, JJ., Perez, G., (2009) SIMPLE: Template Based Service Integration, *Proceedings of IEEE International Conference in Web Services, Industry Summit*

Larsen, C. (2006) Model-driven development: Assets and reuse, IBM Systems Journal, Volume 45, Number 3, pp 541-553

Linehan, M. (2005) SOA programming model for implementing Web services, Part 9: Integrating rules with SOA, *IBM DeveloperWorks,* http://www.ibm.com/ developerworks/webservices/library/ws-soa-progmodel9/

Object Management Group, Inc., (OMG), 2005, Reusable Asset Specification, Version 2.2, http://www.omg.org/technology/documents/formal/ras.htm.

H. Chen, P. Chou, N. Cohen, S. Duri, C. Jung, "DRIVE: A tool for developing, deploying, and managing distributed sensor and actuator applications," *IBM Systems Journal, Vol 47, No 2,* 2008.

Corsten, D., and Gruen, T. (2005) On shelf availability: an examination of the extent, the causes, and the efforts to address retail out-of-stocks. In Doukidis, G. J. and Vrechopoulos, A. P. (eds.) Consumer Driven Electronic Transformation: Applying New Technologies to Enthuse Consumers and Transform the Supply Chain, pp. 131–150, Springer, Berlin

Hardgrave, B., Waller, M., and Miller, R. (2006) RFID's impact on out of stocks: a sales velocity analysis. Research article ITRI-WP068-0606, Information Technology

Research Institute, Sam M. Walton College of Business, University of Arkansas, June 2006.

Tuscany home page, 2008. http://tuscany.apache.org/.

11

RFID Applications for Sanitary Environments

Giuliano Benelli, Stefano Parrino and Alessandro Pozzebon
University of Siena, Department of Information Engineering, Siena, Italy

1. Introduction

Healthcare represents one of the most significant sectors where the diffusion of RFID technology is growing day by day. Many different applications have already been studied and developed, with both active and passive devices working at all the available operative frequencies.

Sanitary environments are nowadays extremely complex structures employing several thousands of people with very strict safety requirements: in emergency situations for example 5 minutes can make the difference for a patient between survive and die. RFID is especially indicated to be employed in these scenarios for two main reasons: first of all because it's a particularly reliable technology, with good performances, few errors and fast interaction, and secondly because, due to the presence of many different technological systems, ad-hoc solutions can be designed on the specific requirements of the application to be realized.

At present the most common RFID applications in healthcare can be divided into two main categories: the items tracking and the tracking and identification of people, patients and sanitary operators. The items tracking is performed in order to avoid the loss of expensive devices and to reduce wasting of time during assistance operations: systems studied and realized for this purpose cover all the range of RFID systems and provide different services according to their different performances. Passive systems are mainly thought for the tracking of large amounts of items, providing identification also to low value items, and are mainly based on gate structures. Active systems are designed for the tracking of big and expensive items and use long range antennas located at specific key points.

RFID systems for the identification of people probably represent the most interesting sector, due to the variety of different applications that can be studied and realized. The most common systems foresee the use of RFID for the tracking of sanitary operators or patients during their assistance operations: these solutions are very similar to the one described for the tracking of items and are already working in many structures around the world. Along with this many other applications have been implemented, including systems operating the unambiguous matching between the patient and his treatments (for example the medicine or the blood sack) or between the mother and the child in the paediatrics departments: these applications mainly foresee the use of passive technology, due to the lower costs, and ad-hoc items like electronic bracelets have already been realized and can be commonly found off the shelf. Finally the availability of a memory on the transponders fostered to the use of RFID technology also as a mean to promptly store and retrieve patient related information: for example electronic case history or electronic medical prescription applications have been studied and developed.

The chapter is divided into five sections: in the first section we will describe the main characteristics of sanitary environments, focusing on scenarios where the use of RFID technology is mainly required.

In the second part we will describe some standard applications for the "items tracking" scenario. The applications described will include systems covering all the operative range, including both passive and active devices.

The third part will be focused on the applications for the people. All the scenarios described before will be analysed and different systems and solutions will be described.

In the fourth section we will discuss potential problems deriving from the introduction of a radio technology in an environment with sharp safety requirements, focusing on the possible interferences with other devices and on the law recommendations.

Finally, in the conclusions we will analyse the possible future applications and the potential improvements deriving from the refinement of the technology.

In Table 1 is shown a summary of RFID technology and its possible applications in healthcare scenarios (Supply Insight, 2006).

	APPLICATIONS	HEALTHCARE
Low Frequency (125-148kHz)	Mainly passive systems, Animal tracking, Access control	Applications with organic materials, Tracking of cadavers, Implantable Under-The-Skin tags
High Frequency (13.56MHz)	Mainly passive systems, Smart cards, Payment system, Tracking, Access Control	Access Control, Tracking of assets and patients, Smart Cabinets, Electronic Case Histories
Ultra High Frequency (300-1000MHz)	Passive and Active systems, Items and people tracking, Integration with sensors	Tracking applications, Monitoring of patients, Localization, Sensors systems
Microwave (2.4-5.8GHz)	Mainly active systems, applications also based on Wi-Fi, ZigBee, Localization and Tracking on large environments	Long Range Localization, Integration with Sensor Networks
Ultra Wide Band (3-10GHz)	New Technology, Emerging Tracking Applications	Pilot applications for the items and patient tracking

Table 1. RFID frequency bands and their applications

2. Sanitary environments and RFID

When we speak about sanitary environments we cover a wide range of different structures, with specific features deriving from their different functions and kinds of patients treated. Along with hospitals and Emergency Rooms under the term "Sanitary environments" have also to be included retirement houses, pharmacies and every other kind of structure where the patients assistance and care is the main goal.

All these kinds of structures present specific requirements and limitations according to the operations performed. While hospitals have many difficulties in the management of items and people, Emergency Rooms have to perform fast and safe assistance operations, retirement houses have to ensure the safety and the health of elderly people and pharmacies have to guarantee the proper medicine.

All these operations can be performed more efficiently introducing RFID systems, which can speed up all the operations enhancing the quality of the assistance operations. Moreover, RFID can improve the management of items reducing the losses and optimizing the operating expenses.

Currently killer RFID applications in the healthcare domain are (Datamonitor, 2004):

- Medication Administration: the right association between patient and medication achieved with RFID tags can reduce errors and speed up all the assistance operations;
- Medication authentication and restocking: medications have to be stocked and their expiration dates and quantities kept under control;
- Hospital equipment tracking: RFID can improve the management of inventories, keep trace of lent and lost items and ensure proper medical instruments handling, for example in operating rooms;
- Medical supplies tracking: medical supplies like surgical parts or pacemakers need not only to be tracked but also to be monitored. This means that RFID can become also a sensing structure, controlling parameters like temperature, localization or expiration date;
- Asset and substance tracking: sensible substances, like medical supplies, have not only to be tracked but in many cases also preserved in ideal humidity and temperature conditions, and RFID can become also a sensing platform;
- Medical waste tracking: medical waste is not the same business as common rubbish. RFID is used to ensure its correct storage and disposing;
- Patient tracking: RFID tags and wristbands are used not only for the tracking of patients inside the sanitary structures but also as devices to store vital information such as allergies, blood type or medications to be undertaken;
- Blood banking: one of the first applicative fields of RFID in healthcare has been the correct association between blood type and patient, in order to avoid errors and to optimize the managing of blood reserves;
- Lab and pathology sample tracking: next to blood, also other organic materials like serum or tissue can be tracked, performing also in this case monitoring operations
- Self-medication for seniors: RFID can improve the life conditions of elderly people helping them in common operations like medicine administration or first aid callings.

All these applications can be roughly divided in two main groups according to the nature of what has to be identified: in fact on one side we can find the tracking of items, from blood sacks, drugs, assets or medical waste while on the other side we find all the applications dealing with patients, from the simple tracking to most complex systems involving electronic medical prescription, electronic case history or applications focusing on t patient safety.

In any case, every time that a technological system has to be introduced in public environments many limitations appear, mainly due to electromagnetic compatibility requirements and space constrains. This fact is even more critical in sanitary environments for the following factors:

- The obvious limitations in electromagnetic emissions are even more severe due to the presence of children, elderly and infirm people;
- The presence of other electronic devices requires ad-hoc studies to verify the compatibility of the new systems with the existing technological infrastructure. This problem also involves wearable and inner devices from which in many cases the life of patients is directly dependant;
- Space constrains are even more strict and new devices introduced must be the less cumbersome possible avoiding to get in the way of stretchers or other sanitary equipments moved along corridors and rooms.

3. The tracking applications

3.1 Applications based on passive technology

While the common passive RFID systems work at Low (125-135kHz) or High (13.56MHz) Frequencies some examples of applications can be found also in the UHF (Ultra High Frequency - 860-930MHz) band and even in Ultra Wide Band systems.

Before moving to the description of the applications some consideration on the devices have to be made (Finkenzeller, 2003). First of all, the identification devices in passive tracking systems are obviously the common smart labels: passive systems present lower performances than active ones, but their cost is notably lower. This means that they are suitable for applications where huge quantities of items have to be identified. In this sense, the simpler is the transponder, the lower is its price. Anyway, some ad-hoc devices have been studied, mainly to be directly embedded at the moment of the fabrication in the item to be identified. The second consideration refers to the reading device. In most cases two different categories of devices can be employed: handheld devices or fixed structures. While handheld devices provide a short read range and require the operator to manually detect the transponder before reading it, with fixed structures transponders are automatically located and the information is read without the direct intervention of an operator. Finally, while handheld devices present quite the same shape even at different operative frequencies, fixed structures can vary a lot according to their operative frequency. In particular, Low Frequency and High Frequency gates are big devices whose positioning can be very difficult while at higher frequencies the different physical protocol forces to use smaller antennas with a lower impact on the surrounding environment.

Starting from the lower RFID band, 125kHz passive transponders have been used in applications where the item to be identified is made of organic materials. In fact this kind of materials present a high percentage of water, which is a partial insulator. In particular, pure water is an insulator, but the presence of other materials turns natural water into a partial conductor. Attenuation depends in inverse proportion from the frequency of the electromagnetic wave and transmissions in presence of water is possible only at low frequencies: for example all under-the-skin RFID tags work at Low Frequencies, from 125kHz to 135kHz.

While implantable tags present many legal limitations due to privacy violation risks and to possible health hazards deriving from electromagnetic emissions, thus limiting their use for patient identification purposes, some applications have been developed with low frequency used to track organic materials like human organs or body parts.

An interesting example comes from the United States, where The University of California's Anatomical Services Department uses this technology to track human cadavers (O'Connor (a), 2009): an ISO 11784 or 11785 RFID transponder storing an alphanumeric identification code is sutured to every body and then read with handheld devices. When a specific part is removed form the body, for example a section or an organ, an additional transponder is issued, and then sutured to the body part or, when not possible, attached to the box where it is stored. With an ad-hoc software is then possible to create an electronic inventory with the real-time association of all the body parts with the original cadaver, with a substantial reduction in the operating times. A last consideration has to be made on what happens at the end of the usefulness of the specimen: due to the very low costs of the transponders it's more convenient to leave the tags on the cadavers when they are cremated than recollect them.

Moving to the High Frequency band many examples can be found of systems operating at 13.56MHz: this is in fact the most common operative frequency for passive systems, and many devices can be found on the market, with prices pretty low and a good level of reliability. Due to its vast diffusion several solutions have been studied, mainly for what concerns the readers:

- from the common handheld readers ad-hoc solutions for hospitals have been studied, with mobile computing platforms designed to be employed by doctors or sanitary operators, embedding HF readers;
- from fixed readers many different solutions have been studied, including smart cabinets and ad-hoc analysis devices, and new technical solutions have been tested to improve the performances for what concerns the accuracy of the readings and to reduce the environmental impact of bigger devices.

Examples of High Frequency applications can be found worldwide. Many hospitals use this technology to manage the storing of materials and assets in order to reduce the losses and the number of unused items. Transponders can be affixed on the items to be identified using adhesives, tie-wraps or mechanical hardware. They can be incorporated directly into the equipments and they can be designed to be used in presence of metallic materials. Once positioned, their identification number is stored inside an informative system. Then through ad-hoc devices like gate antennas, smart cabinets or simply handheld readers, every time that the item is moved through a structure, its new location is recorded and updated inside the database, speeding up the retrieval of a specific item and reducing the expenses for the materials. Obviously, next to the simple tracking many other operations can be performed, getting advantage of the memorization capabilities of the transponders.

An interesting system, covering different technical solution, has been set up at the Memorial Hospital, Chattanooga, Tennessee, where passive HF RFID is used to track high value items (Swedberg (a), 2009). In this structure every new item received is tagged, and the identification number is read with a desktop reader and then stored into a database with many other related information, like product type or expiration date in case of perishable materials. All these materials are then stored inside cabinets equipped with one RFID reader per shelf which, periodically, perform a reading of all the items stored in the cabinet. When an item is removed the reader notifies the missing item to a management system. Then, if the removed item is brought for example in an operating room the employees use a handheld reader to read the transponder and notify to the system the new location of the item. If an item remains in the "missing" state for a predetermined span of time an alert is generated, pointing out the lost item.

With a similar structure the University of Michigan Health System manages the storage of all the organic tissues (Swedberg (b), 2009). Each tissue is tagged with a Texas Instrument

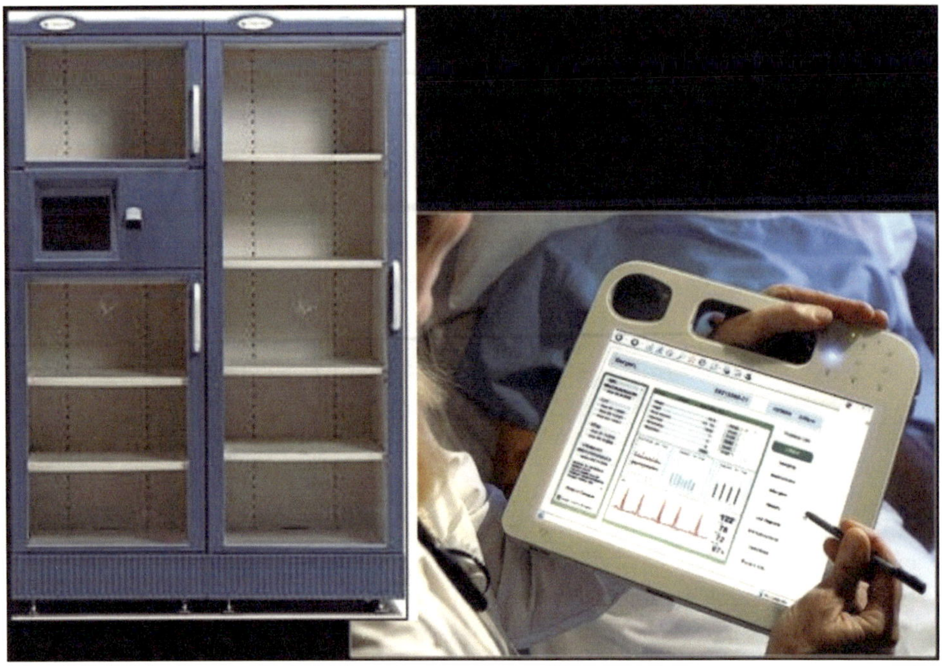

Fig. 1. ad-hoc RFID devices (Smart cabinet and computing platform)

ISO 15693 tag, attached to the bottom of the box or to the plastic bag containing the tissue, then all the items are constantly checked by the readers embedded in the cabinets in which they are located. The cabinets are also equipped with a Low Frequency (125kHz) reader used to control the opening by the operators. The choice of a different operative frequency is mainly justified by the necessity to avoid interferences between the two systems.

The two systems described before foresee the use of short range readers: this means that both in the case of the handheld reader and in the case of the smart cabinet the identification of an item takes place only when it is positioned in a specific place (the cabinet) or when the operator brings the mobile reader close to the transponder. To perform effective tracking operations the localization and the identification of an item should be automatically made, and the only way to perform such an operation is to build a fixed wireless infrastructure covering all the environment where the items should be tracked or to provide all the doors between two adjacent rooms with gate structures identifying all the items crossing them. The short ranges achievable with passive HF systems don't allow the creation of a wireless infrastructure. Moreover, also the creation of efficient gate is extremely difficult: in fact common commercial HF gates provide reading ranges up to 1.60m which is evidently a range too short to cover common hospital doors which are usually up to 2m wide.

An alternative gate structure has nevertheless been realized providing a 2m reading range and, most important, reading transponders in all the orientation (Benelli et al., 2009). In fact it can be easy to enlarge the reading range of a system when transponders are put in the best coupling direction, but an efficient reading in all the orientation is very difficult to be achieved. The structure is made of 4 partially overlapped antennas and it uses, to widen the reading range, the mutual coupling phenomenon between all the antennas. Anyway, due to its dimensions, its introduction in crowded environments like hospitals can be quite difficult and new solutions are being studied in order to reduce the dimensions keeping the same performance levels.

The HF band is the most common for passive systems, but some examples of applications can be also found at higher frequencies: in particular some systems operating in the UHF have been implemented, and applications working with the new Ultra Wide Band technology are also emerging.

For what concerns the UHF, smart cabinets similar to the ones described before have been realized, working with the same protocol: both items and operators are provided with a transponder, which is used to manage the supplies and to control the identity of the operators accessing to them. This ensures a cost effective optimization of the storages and provides higher efficiency for what concerns the times of assistance.

UHF technology is used also for the tracking operations and while by one side it has the big advantage of the small dimensions of the antennas, on the other side it presents some problems of electromagnetic compatibility. First of all at this band the interference created by the presence of different materials makes more difficult the reading of the tags and their positioning on different items. Moreover specific studies indicate that problems of interference with other biomedical devices may occur, making the introduction of such a kind of system in sanitary environments quite complex, requiring specific studies customized on the features of each environment.

Finally few words have to be spent on the Ultra Wide Band systems: passive tags have been developed to track the location of blood specimens in medical laboratories, with transmission frequencies of 5.8GHz and 6.7GHz (Swedberg (c), 2009). Next to good localization performances these devices present extremely low dimensions (around 2mm), making them suitable for a large range of other applications.

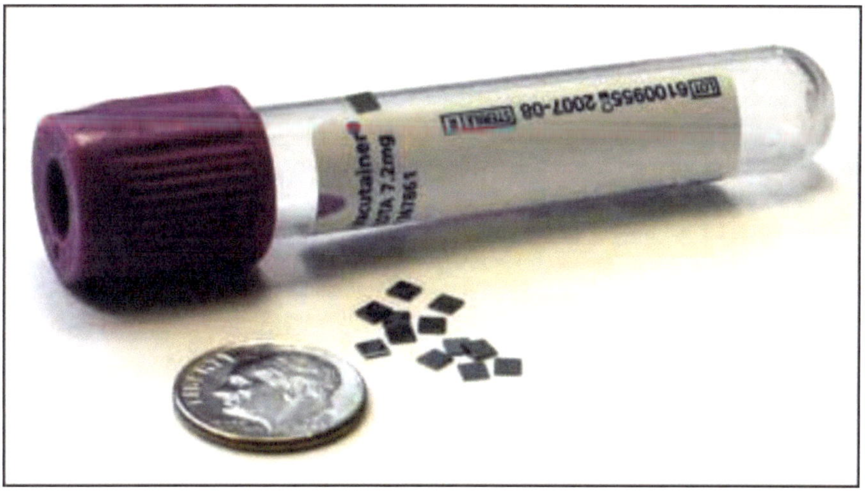

Fig. 2. UWB tags next to a tube for blood samples

3.2 Applications based on active technology

As in the case of passive RFID, active systems present many different operative frequencies, with different features and subsequent different possible applications. While the lower frequency bands (LF and HF) are used only by passive systems active healthcare applications operate at 433MHz, in the UHF band and up to 2.45 and 2.48GHz. Moreover we can list as active RFID also systems realized with technologies like ZigBee or Wi-Fi performing however automatic identification.

Active technology is obviously more expensive than passive, and it's then more suitable for the tracking of high value items, whose amount is not too high. On the other side while costs are higher performances are significantly better, with reading ranges up to hundreds of meters and dimensions of the readers extremely reduced: the positioning of the devices doesn't require then specific interventions.

Active asset tracking systems exist at all the frequencies: active tags are located on valuable items and their position is constantly monitored with an adequate number of antennas covering all the surface of the structure. While most of this kind of systems is not of particular interest, due to its standard functioning, some words have to be spent of 2.45GHz Wi-Fi systems: the biggest advantage in using this technology derives from the fact that the readers can be the common Wi-Fi access points used for the wireless infrastructures that in many structures are already existing. Even if sometimes the number of access point to perform the localization of an item with a good approximation has to be increased (this kind of systems is able to localize an item with a precision around 4-5 meters), the chance to use an existing system can notably reduce the final realization expense.

Together with the tracking systems many other solutions, performing more complex operations, have been studied, in some cases also in combination with other technologies like Infrared (IR) or sensors.

Starting from the lower frequency (433MHz) an interesting example on how a common tracking system can be turned in something more useful comes from the El Paso County's 911, Texas, where 433MHz active tags are used with temperature-tracking sensors to control the temperature of the equipment rooms and the offices. In particular, a tag with a built-in sensor is positioned in all the most important rooms, while two readers are located on the two sides of the building, thus covering all the structure (the devices used provide a maximum 50m reading range depending on the type of antenna). Tags transmit the temperature associated with their ID number every minute to the readers, which turn them into an informative system analyzing the eventual overtaking of the thresholds. This kind of devices can be found also with other sensors: in particular the same producer also sells tags equipped with humidity sensors.

UHF is probably the most common solution for active RFID: many kinds of different applications can be found, even though some limitations in its use have already been underlined in the previous section. Sensor equipped tags have been studied to track and monitor blood sacks (Fraunhofer Institute for Integrated Circuits, Germany) (Wessel, 2009), joining this function with the right association between patient and blood, to track the temperatures of refrigerator and freezers (Wake Forest University Baptist Medical Center, USA) (Swedberg (d), 2009), joining it with tracking of assets. To improve the efficiency of the tracking systems RFID has also been integrated with the infrared technology: combining the two localization techniques the performances increase in a considerable way reducing the risk of errors.

At higher frequencies RFID tends to merge with other technologies: in particular 2.45GHz systems are, as described above, basically Wi-Fi systems, while 2.48GHz systems operate with the ISO 802.15.4 ZigBee protocol. Applications using Wi-Fi RFID for the tracking of assets are emerging day-by-day all around the world, with more complex systems adding sensing capabilities, due mainly to the presence of many companies offering efficient solutions for what concerns both the hardware and the software applications.

Fig. 3. Wi-Fi RFID reader and tag

ZigBee RFID systems are less common, but the features of the ZigBee protocol allow solutions not possible with the other technologies. First of all ZigBee allows mesh networking: ever node of the network (in this case every RFID tag) can receive and forward the data sent by other nodes. This means that the number of gateways (the RFID readers) can be significantly reduced, with a reduction of the total cost of the system (even though ZigBee nodes have higher costs than other active tags). Another important feature of ZigBee derives from the fact that this technology has been developed mainly for Wireless Sensor Networks. As a result, the introduction of additional sensors on the nodes is very simple and new functionalities can be easily added. Two examples of ZigBee systems can be found in the USA where the Advocate Good Samaritan Hospital, Downers Grove Illinois uses this technology to track surgical trays (Bachelor (a), 2009) and the Jackson Memorial Hospital, Miami, Florida uses it for the tracking of medical equipment (Bachelor (b), 2009). In the first case ad-hoc transponders have been designed able to withstand steam autoclave cycles and liquid sterilization methods: this means that, once localized, the trays can be sterilized without removing the tags and then being transferred to the surgical room. In the second case temperature sensors are also associated with some tags, monitoring not only the items position but also the environmental conditions of the rooms in which they are kept.

4. Patient centred applications

4.1 Applications based on passive technology

While the standard technical features of passive RFID don't change moving to applications focusing on the safety of patients, some new interesting devices have to be introduced before describing some applicative solutions. Some words have already been spent in describing the under-the-skin implantable tags: currently the use of this kind of devices is strongly discouraged mainly due to privacy concerns and electromagnetic compatibility hazards. Applications have been studied at prototypical level and functioning systems don't seem to exist in any area of the world. Moreover the number of producers of this kind of devices is extremely limited (only one company is offering such a commercial solution) and some local authorities around the world are presenting drafts of a law totally banning their use.

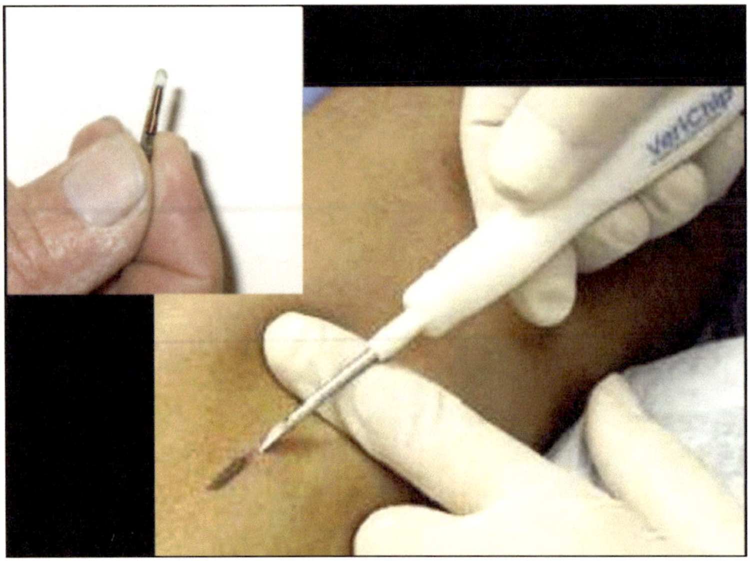

Fig. 4. Implantable Under-The-Skin RFID tag

Another interesting device especially realized for patient applications and with a vast diffusion is the RFID wristband: this is simply a plastic bracelet with an embedded passive tag (typically of small dimensions) to be read with a handheld reader. Several producers sell these devices and many applications have already been developed worldwide.

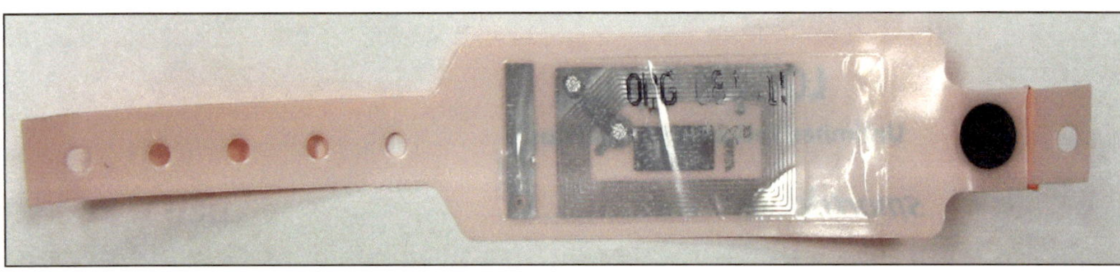

Fig. 5. Passive RFID wristband

While LF RFID is used mainly for access control, several interesting applications have been studied with HF and UHF technologies, with complex systems operating to improve the safety and the life conditions of patients, with a specific attention to patients affected by chronic diseases like for example Alzheimer disease.

Before moving to higher frequencies only an application using Low Frequency has to be cited, mainly for it peculiar use. Cambridge Temperature Concepts, a UK company, has realized a system using passive 125kHz transponders to help women to predict their ovulation cycles (O'Connor (b), 2009). The RFID transponder is used in combination with a basal temperature sensor: due to the fact that the body temperature of a woman increases of one-half to one degree Fahrenheit during the ovulation, this change can be used to predict the fertility periods. The module integrating the sensor and the tag is attached to the skin of the woman under the armpit through an adhesive module. Then through a handheld reader the temperature information is downloaded periodically and analyzed with an ad-hoc software predicting exactly the fertility periods.

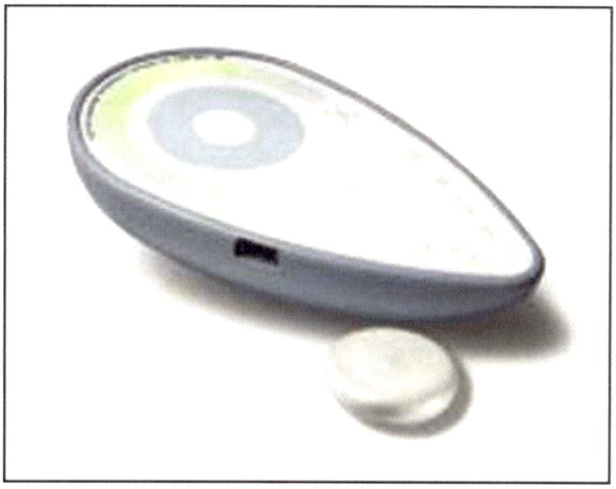

Fig. 6. Tag and Reader of the ovulation prediction system

High Frequency wristbands are used in many applications. The Children Hospital "Vittore Buzzi" in Milano, Italy uses this technology to ensure the right association between mothers, newborn children and their case histories. Both mother and child are provided with a wristband and every time that mother and child meet their identity is checked comparing the identification codes of the tags. The same happens every time that a medicine is administered to the child. Similar applications are also used to ensure the right association between patients and blood sacks and between patient and medication.

The presence of a memory on transponders has also encouraged their use as electronic case histories or electronic prescriptions. Common passive transponders provide up to 4kBytes of memory: this space can be used to store vital information on the patient in order to avoid possible errors during the assistance operations. Through an ad-hoc codification of the information a big amount of data can be stored on a single transponder: presence of allergies and vaccines, blood types or personal data can be stored inside a smart card or a wristband and then retrieved and decoded with a handheld device like a PDA or an ad-hoc laptop like the ones already mentioned in section 3.1. Finally, an interesting application operating at 13.56MHz has to be described before moving to higher frequencies: RFID has been used by the New York Public Schools District 75 in Queens, USA, to help non verbal children to communicate through an ad-hoc reading device incorporating 5 readers put side by side (Swedberg (e), 2009). The child is then provided with a big quantity of ISO15693 smart cards with images corresponding to words printed on the front: when the child wants to "speak" he simply places the cards on the antennas and the required word is read by the device.

Finally, some applications operating in the UHF band also exist: in particular an interesting system has been set up in Shady Palms, Florida, USA. It's basically an assisted-living facility where some patients suffering from dementia are housed: in order to avoid their attempts to leave the structure a tracking system has been realized. While initial studies focused on the introduction of 125kHz transponders inside the soles of the shoes to be read with antennas buried underground, the final solution foresees the use of waterproof UHF EPC Gen2 tags sewn into the clothing. Readers have been mounted near the entrances and the exits of the structure and every time that a monitored patient crosses one of these doors an e-mail containing an alert is sent to an appropriate staff member.

4.2 Applications based on active technology

As in the case of items, active systems mainly operate at higher frequencies. Applications operating in the UHF band are very common while applications compliant with Wi-Fi technology perform the tracking of patients exactly in the same way as they track items. Anyway, an interesting application operates at the two unregulated frequencies of 262kHz and 318MHz (Bachelor (c), 2009. Swedberg (f), 2009): in this system, installed in two different structures in the USA, RFID bracelets are used to monitor the position of children, and to detect whether a tag is tampered of a child is taken away from the structure. The two frequencies are used to communicate with interrogators, located in doorways (262kHz) and with readers, deployed in hallways and other locations (318MHz). The receivers are mainly used to check the right functioning of the tags, while the interrogators generate an alarm every time that a tag comes close to a doorway.

Moving to higher frequencies some applications exist operating at the new ISO 18000-7 433MHz frequency, in combination with IR (Infrared): six hospitals in Alaska, Washington and Oregon (USA) are using this technology to track patient location and treatments, while the University of Miami-Jackson Health System Center for Patient Safety is studying a system to check the washing of hands by their operators. In the first application RFID badges are issued to the emergency patients to track their position throughout the facility (Swedberg (g), 2009). The combination of RFID and Infrared provides a long read range (RFID) and the location precision (IR). Every tag transmits a signal every 3 seconds but, if its position remains unchanged for a long time, it switches to dormant mode conserving the battery power. With a precise tracking of patient the assistance operations can be optimized, with a significant reduction of the times and with an increase in the quality of service. The second application is a little bit complex, with a checking software developed to control the performing of the hand washing (Swedberg (h), 2009): every staff member in the structure is provided with a hybrid RFID-IR tag. When the operator presses the hand-sanitizing dispenser containing soap a sensor embedded in the dispenser reads the tag number through IR signal. Then, through the RFID tag the badge number is transmitted to a reader linked to a PC, confirming the washing of the hands. The system measures the time elapsed from the moment of the washing: a reader positioned above a patient bed captures then the ID of the tag and if too much time has passed or if the washing has not been performed an alarm is generated.

As already underlined many patient tracking applications exist in the UHF band: in this kind of systems readers are usually located in strategic positions in the rooms of the structure, allowing a quite accurate localization of the patients (usually with a precision of 1-5m). These data are used by the operators of the structure to optimize the assistance operations but can also be shown, in the case of Emergency Rooms, to relatives in the waiting room through monitors following all the movements throughout the structure. An interesting application enhancing the services provided is the system developed at the Tan Tock Seng Hospital in Singapore (Friedlos, 2009), where RFID is not used only for the tracking of patients but also to monitor their health conditions. In particular, ad-hoc studied transponders operating at 868.4MHz can constantly read the body temperature of the patients, also monitoring their vital signs.

The last application described is the system Developed by the University of South Florida to diagnose early dementia through Ultra Wide Band RFID (O'Connor (c), 2009): analyzing the movement patterns of patients it is possible to find clues indicating early stages of dementia. The tags used in this system emit radio pulses over multiple bands (from 6GHz to 8GHz) simultaneously. These signals can be transmitted for a much shorter duration of time, improving the reading rate and the final accuracy of the survey.

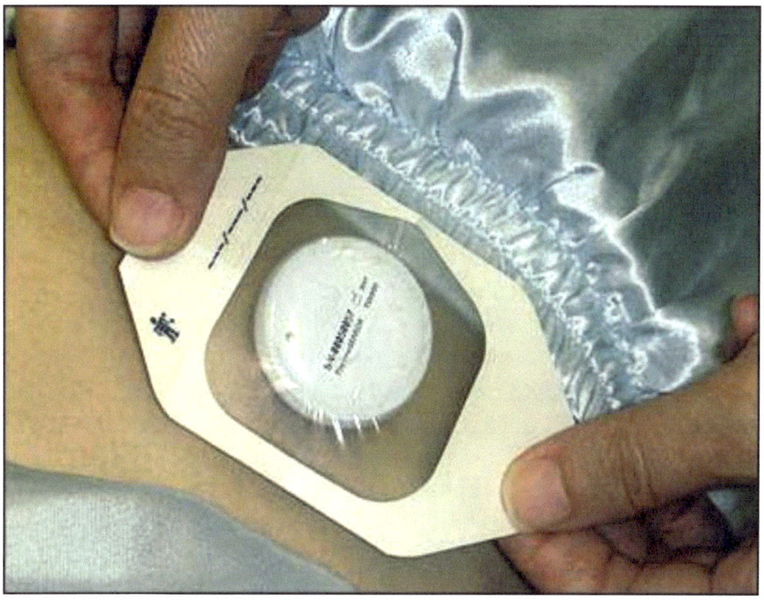

Fig. 7. RFID tag integrating a temperature sensor

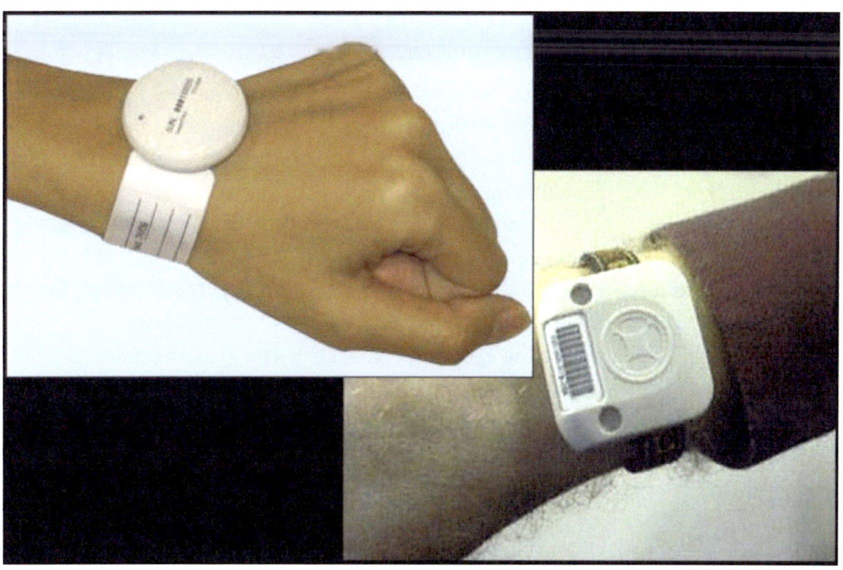

Fig. 8. Active wristbands

5. Laws and regulations

RFID technologies must be compliant with European legislation, which is actually divided in two different levels:

- CEPT recommendations (the European regional organization dealing with postal and telecommunications issues);
- Directives of the European Commission, whose decisions are binding on E.U. member countries and their failure is subject to economic sanctions and legal procedures.

Each user has to verify that products are compliant with the laws. At low frequencies the density of the induced current in tissues must be monitored, while at high frequencies we have to refer to the specific absorption rate (Specific Absorption Rate - SAR), defined as the power density absorbed by human body per mass unit. These quantities depend on the intensity of electromagnetic field emitted from the device and on its position compared to the human body. There are different thresholds for general population and for hospital staff exposed to continuous emission of electromagnetic field.

Usually the compliance of RFID systems is certified following the EN50357-2001 standard procedures. This standard provides three distinct stages:

- Direct measures: they provide the measure of the amplitudes of the electromagnetic field near to the under-test device, without the presence of human bodies, at a distance indicated by the standard for different configurations. Only if averaged values on the volume occupied by the human body are needed, the standard indicates the geometry and position of measuring grids, that usually refer to the human head and torso.
- Measurements of compliance with the Basic Restrictions: it's required to take into account the spatial variation of the electromagnetic field and whether the exposure occurs in the near or far field zone. The electromagnetic field produced by the device can be measured on a grid or with electromagnetic simulation softwares. The induced density current (low frequency) or SAR (high frequency) must be evaluated and compared with the Basic Restrinctions.
- Numerical dosimetry: it's made on models of the human body. These models are obtained from three-dimensional images from Nuclear Magnetic Resonance or from pictures of human dissections. Then the incident electromagnetic field can be obtained as above, while the interaction with the model of the human body is finally calculated using the appropriate electromagnetic simulators. The parameters confronted with the Basic restrictions are still the induced current density (low frequency), and the SAR (for devices operating at high frequency).

For frequencies up to 13.56 MHz, the coupling between the reader and the tag is an inductive coupling, and is said the system works in near field conditions.

The frequency or the band (as in the case of the UHF), the transmission power and the maximum time for communication between tags and readers are the regulated parameters; in particular in this case the transmission power is represented by the maximum field strength (H-Field), which is expressed in dBμA/m (In the Fig. 9 we can see an example of the following ISM frequencies: 6.78MHz and 13.56MHz).

The recommendation CEPT ERC/REC 70-03 Annex 9 establishes the technical requirements and regulations for the use of harmonized of Short Range Devices (SRD) among the countries belonging to the CEPT.

For HF RFID CEPT ERC/REC 70-03 a.9 establishes:

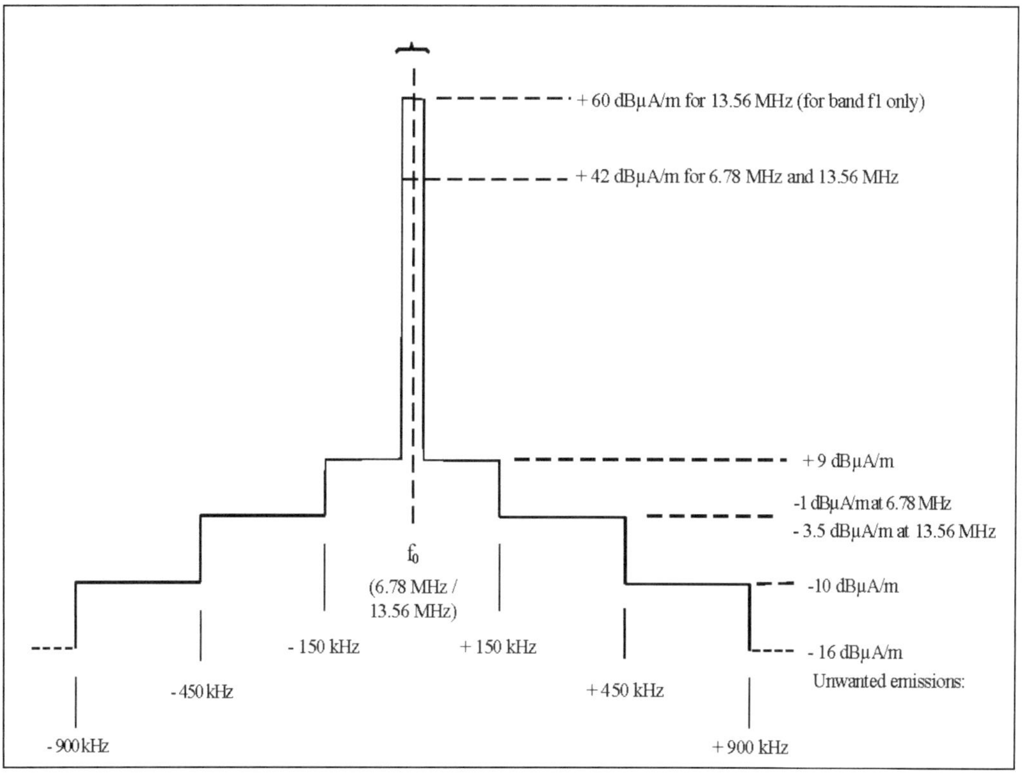

Fig. 9. ISM frequencies for 6.78MHz and 13.56MHz

- Frequency: 13.553-13.567MHz for RFID and EAS (Electronic Article Surveillanc only;
- Intensity of magnetic field: 60dBµA/m at 10m;
- Duty cycle: No Restriction;
- Channel spacing: No spacing.

RFID technology in the Healthcare is understudied yet, but in the June 2008 an issue of the Journal of the American Medical Association (JAMA), (Van Der Togt et al., 2008) reported results from a nonclinical study, completed in the Netherlands in May of 2006, demonstrating that active and passive radio frequency identification (RFID) systems can be manipulated to produce electromagnetic interference (EMI) with medical devices commonly used in hospital critical care units (e.g., infusion pumps, external pacemakers, defibrillators, monitors). The study was a non-clinical study, in fact no patients were involved, and the manner in which the tests were performed was not analogous to the way RFID systems are conventionally used in a modern hospital. Without a patient being connected, EMI by 2 RFID systems (active 125 kHz and passive 868 MHz) was assessed under controlled conditions during May 2006, in the proximity of 41 medical devices (in 17 categories, 22 different manufacturers) at the Academic Medical Centre, University of Amsterdam, Amsterdam, The Netherlands. In 123 EMI tests (3 per medical device), RFID induced 34 EMI incidents: 22 were classified as hazardous, 2 as significant, and 10 as light. The passive 868-MHz RFID signal induced a higher number of incidents (26 incidents in 41 EMI tests; 63%)

compared with the active 125-kHz RFID signal (8 incidents in 41 EMI tests; 20%); difference 44% (95% confidence interval, 27%-53%; P < .001). The passive 868-MHz RFID signal induced EMI in 26 medical devices, including 8 that were also affected by the active 125-kHz RFID signal (26 in 41 devices; 63%). The median distance between the RFID reader and the medical device in all EMI incidents was 30 cm (range, 0.1-600 cm). In a controlled nonclinical setting, RFID induced potentially hazardous incidents in medical devices.

Another study, in contrasts with the University of Amsterdam study, instead showed that UHF systems created no EMI when used with antenna positions and power settings that would be seen in a typical hospital setting. Implementation of RFID in the critical care environment should require on-site EMI tests and updates of international standards.

A key factor contributing to a wireless medical device's safety is the limited amount of RF spectrum available and potential competition among wireless technologies for the same spectrum. This is managed in different ways for different RF wireless communication technologies that may be available for use in healthcare communication and health informatics exchange.

6. Conclusion

The worldwide diffusion of RFID technology in the healthcare scenario is evidently an unstoppable process: new applications are emerging day-by-day, and new technical solutions provide the means to realize systems performing operations once unthinkable.

Among the technologies whose diffusion in the next years will probably modify many of the actual habits of common people one of the most important is without doubt Near Field Communication (Benelli et al. (b), 2009). NFC is a new short range communication system based on RFID technology. NFC systems can work like traditional RFID systems, where a master device reads some information from a slave device, but they can also set up a two-way communication between two items. In particular, NFC devices can be integrated on mobile phones, widely enhancing the intercommunication capabilities of the users. NFC phones integrate the functionalities of RFID tags and readers: this means that a phone can act as a smart card, with information stored inside its internal memory card and readable with an external reader even if the telephone is turned off. But this also means that a telephone can read the data stored in a common smart card, acting then as an handheld reader.

Evidently many of the applications described before could be transferred on such a device, turning it alternatively into an electronic case history or into personal assistant for doctors, able to read and write smart cards.

This and many other technologies will probably change in a considerable way all the technological infrastructure in sanitary environments, but estimations are very difficult to be made, because the evolution rate of technologies is so fast that probably tomorrow some new devices will appear, whose existence was just a dream only the day before.

7. References

Benelli, G.; Parrino, S. & Pozzebon, A. (2009) (a). Possible configurations and geometries of long range HF RFID antenna gates. *Proceedings of The Sixth International*

Symposium on Wireless Communication Systems 2009 (ISWCS'09), Siena, Italy, September 2009

Benelli, G. & Pozzebon, A. (2009) (b). NFCARE, Possible Applications of NFC Technology in Sanitary Environments. *Proceedings of The Second International Conference on Health Informatics (HEALTHINF'09),* Porto, Portugal, January 2009

Bachelor, B. (2009) (a). Advocate Good Samaritan Hospital Tracks Trays Packed With Surgical Instruments. *RFID Journal,* http://www.rfidjournal.com

Bachelor, B. (2009) (b). Jackson Memorial Enlists Thousands of RFID Tags to Track Assets. *RFID Journal,* http://www.rfidjournal.com

Bachelor, B. (2009) (c). St. John's Children Hospital Deploys RFID to Protect Children. *RFID Journal,* http://www.rfidjournal.com

Datamonitor (2004). RFID in healthcare. *Datamonitor,* http://www.datamonitor.com

Finkenzeller, K. (2003). *RFID Handbook: Fundamentals and Applications in Contactless Smart Cards and Identification,* John Wiley and Sons

Friedlos, D. (2009). Tan Tock Hospital Uses RFID to Take Patient Temperatures. *RFID Journal,* http://www.rfidjournal.com

O'Connor, M. C. (2009) (a). California Researchers Tag Cadavers, Body Parts. *RFID Journal,* http://www.rfidjournal.com

O'Connor, M. C. (2009) (b). U.K. Startup Sees Pregnant Opportunity. *RFID Journal,* http://www.rfidjournal.com

O'Connor, M. C. (2009) (c). RFID Helps Diagnose Early Dementia. *RFID Journal,* http://www.rfidjournal.com

Supply Insight (2006) (e). RFID Applications in Healthcare. *Supply Insight Inc.,* http://www.supplyinsight.com

Swedberg, C. (2009) (a). Tennessee Hospital Tracks High-Value Items. *RFID Journal,* http://www.rfidjournal.com

Swedberg, C. (2009) (b). University of Michigan Health System Tag Surgical Tissue. *RFID Journal,* http://www.rfidjournal.com

Swedberg, C. (2009) (c). Chip-size Passive RFID Tag Promises Long Range. *RFID Journal,* http://www.rfidjournal.com

Swedberg, C. (2009) (d). Wake Forest Med Center Launches Vaccine-Tracking RTLS. *RFID Journal,* http://www.rfidjournal.com

Swedberg, C. (2009) (e). RFID Gives Voice to Nonverbal Children. *RFID Journal,* http://www.rfidjournal.com

Swedberg, C. (2008) (f). Tamper-Resistant RFID Infant-Tracking System Improves Security. *RFID Journal,* http://www.rfidjournal.com

Swedberg, C. (2009) (g). New Oregon Hospital Adopts IR-RFID Hybrid System. *RFID Journal,* http://www.rfidjournal.com

Swedberg, C. (2009) (h). Patient-Safety Center Tests RFID-enabled Hand Sanitizers. *RFID Journal,* http://www.rfidjournal.com

Van der Togt, R.; Van Lieshout, E. J. ; Hensbroek, R. Beinat, E. & Binnekade, J. M. Radiofrequency Identification Devices Might Cause Electromagnetic Interference. *Journal of the American Medical Association (JAMA),* June 2008

Wessel, R. (2009). German Researchers to Test Networking Tags for Assets, Blood. *RFID Journal*, http://www.rfidjournal.com

12

RFID-based Information System for Patients and Medical Staff Identification and Tracking

Tudor Ioan CERLINCA, Cristina TURCU,
Cornel TURCU and Marius CERLINCA
Stefan cel Mare University of Suceava
Romania

1. Introduction

Radio frequency identification (RFID) technology is having a major impact on the health care industry. By attaching radio frequency tags to different entities (people and objects), RFID technology can provide identification, tracking, location, security and other capabilities.

The goal of this chapter is to show how RFID technology can be used to reduce medical mistakes, improve patient safety and enhance the quality of medical service in hospitals.

After briefly introducing the eHealth domain and some of the healthcare issues, this chapter describes how the RFID technology can be used in healthcare. Thus the third section describes some hospital use cases that could benefit from RFID technology. Also it briefly presents some of the existing projects that successfully implement this emerging technology in healthcare.

The next section shows how to use a medical staff and patients tracking application, called the RFIDHospitalTracker, to improve the quality of the hospital services. We have developed the RFIDHospitalTracker to support the high requirements for scalability, reliability and security. An overview of its distributed software architecture is given. Also, this section enumerates some open problems that still have to be solved before RFID technology will be fully embraced by the healthcare community.

The last part presents some of the future developments proposed by our research team.

The conclusion summarizes the main achievements of this chapter.

2. Healthcare issues

The U.S. Institute of Medicine says that at least 98,000 people in the United States die every year because of medical mistakes. According to an investigative report called "Dead By Mistake" released by the Hearst Corporation, an estimated 200,000 Americans will lose their lives in 2009 alone due to preventable medical errors and hospital infections (Dyess, 2009).

Some of these mistakes involve errors related only to prescription medicine, such as (Lebowitz & Mzhen, 2007):

- prescribing the wrong pharmaceutical drug;
- not finding out a patient's medical history;
- not finding out whether a patient is allergic to a certain drug;

- prescribing the wrong series of treatment;
- filling a prescription incorrectly with the wrong medicine or the wrong dosage;
- giving a patient someone else's prescription.

These kinds of medical errors can lead to serious injuries and death.

3. RFID in healthcare

RFID is now generating significant interest in the marketplace because of its robust application capabilities. Thus, RFID systems are becoming popular in logistics operations, inventory and materials management, and industrial automation, replacing optical identification technologies such as barcodes. But RFID technology can provide a number of benefits to the healthcare industry, improving overall safety and operational efficiency because it operates without line-of-sight while providing read/write capabilities. In fact, RFID can contribute to create the hospital of the future by improving patient care and safety, optimizing the workflows, reducing the operating costs, and reducing costly thefts. There are a number of ongoing trials and studies at hospitals and healthcare centers around the world utilizing and integrating RFID into their hospital information systems. One study estimates that the market for RFID tags and systems in healthcare will rise rapidly from $90 million in 2006 to $2.1 billion in 2016. Primarily, this will be because of item level tagging of drugs and Real Time Locating Systems (RTLS) for staff, patients and assets to improve efficiency, safety and availability and to reduce losses (Harrop & Crotch-Harvey, 2008).

By attaching RFID tags to different entities in healthcare industry (people and objects), RFID technology can provide some capabilities, such as: identification, tracking, location and security. These capabilities directly affect the major issues currently experienced by healthcare organizations while helping to drive down costs (BlueBean, 2007).

Next we present a short "state of the art" of the RFID-based healthcare applications.

Patient Identification

Patient identification error may lead to improper dosage of medication to patient, as well as having invasive procedure done. Other related patient identification errors could lead to inaccurate lab work and results reported for the wrong person, having effects such as misdiagnoses and serious medication errors (Sirivattha, 2003).

RFID technology is used in order to cut these clinical errors, to improve patient care and security.

Anti-Counterfeiting

Drug counterfeiting is an increasing global problem:

- patients lose because they are paying good money for inferior products, which not only is a bad deal economically but also can be a significant health threat as they can contain dangerous substances;
- pharmaceutical companies lose money because of counterfeit drug trade;
- government also loses due to significant loss in taxes and spending considerable resources to combat counterfeiting.

According to the Center for Medicine in the Public Interest, worldwide counterfeit sales are increasing at about 13 percent annually - nearly twice the pace of legitimate pharmaceuticals - and could become a $75 billion industry by 2010 (*, 2009).

RFID technology is more and more used to help pharmaceutical companies, distributors, and hospitals to combat and deter drug counterfeiting.

Inventory Management

By increasing the efficiency of the supply chain and inventory visibility through RFID technology, hospitals can gain some benefits, such as (BlueBean, 2007):

- reducing out-of-stocks;
- procuring less inventory;
- decreasing lost and stolen supplies;
- decreasing the rate of non-charged consumables;
- reducing the time staff spends searching for inventory;
- optimizing the utilization of current inventory;
- improving inventory replenishment within hospitals;
- identifying inventory to be ordered.

Tracking Entities

An RFID system can be used to track people (patients, doctors, nurses), expensive and critical equipments in hospitals in real time:

- tracking patients – meet the need of patient identification and location to ensure patient safety when urgent medical attention is needed and to protect vulnerable patients (infants, Alzheimer's patients, etc.);
- tracking medical staff - identifying the location of caregivers in hospitals to ensure the most efficient response to emergencies;
- tracking pharmaceuticals from the manufacturer, distributor, and pharmacy to the point of administering medication to the patient;
- tracking movable equipment, furniture, medical devices, and other high-value items (surgical equipment, wheel chairs, etc.) - to provide ready access when needed and to reduce losses. For example, a bed could be moved through the hospital and as it passes through the rooms of the hospital it is detected by the RFID readers that send the information to the hospital server;
- tracking blood – RFID-based systems are used to track bags of blood to record transfusions and to minimize the risk of patients receiving the wrong type of blood;
- tracking documents - by attaching an electronic label on the cover, files containing important documents like prescriptions, clinical reports, medical bills, etc. can be easily traceable in hospitals.

Other Applications

- ensuring the proper identification of laboratory specimens, including biopsy samples and containers of blood or urine to reduce medical errors;
- managing controlled substances, pathogens, and other materials that pose a public health risk;
- restrict access to drugs, pediatrics and other high-threat areas to authorized staff.

Next we present some RFID-based systems that are already successfully implemented within worldwide hospitals. Case studies are presented from the following countries (alphabetically): Finland, Germany, Italy, Taiwan, United Kingdom and USA.

Finland: *RFID and medication compliance*

A Finnish company developed a system using RFID and mobile phones to make sure Alzheimer's patients take their medication.

Italy: *RFID and blood transfusion*

RFID technology is used to increase efficiency and safety in the management of the transfusion process in Italy's National Cancer Institute in Milan and Ospedale Maggiore hospital in Bologna. Thus, RFID tags are placed on blood bags and patient wristbands.
Staffs are provided with RFID ID cards and personal digital assistants (PDAs or handheld computers) in order to:

- register patients upon arrival;
- verify patient-blood group;
- recognize patients and transfusion units at any time.

Germany and USA: *RFID and medication*

At Jacobi Medical Center in New York, nurses use a Tablet PC to match the RFID tags on patients' wrists with bar-coded information on packets of medication. The match ensures that each patient receives exactly the right dose and only the medication that has been prescribed to him or her. It also automatically creates an electronic record of the nurse's visit, allowing the nurse more time to provide direct patient care (Crounse, 2005).
Germany's Jena University Hospital implements a system using RFID to track medication from the point of dispensing in the hospital's pharmacy to the drug's administration up to 24 patients in intensive care, in order to avoid drug errors (Vilamovska et al., 2008).

Taiwan: *RFID and SARS*

A number of RFID-related software applications were developed in Taiwan. Some of these applications focused on the use of radio frequency identification (RFID) technology to prevent spread of the severe acute respiratory syndrome (SARS) disease. These applications include computerized systems for monitoring the body temperature of healthcare personnel and patients in the hospital, track potential virus carriers and, when necessary, map their movements throughout hospitals and keeping track of people under quarantine in facilities separate from hospitals.

UK: *RFID and hospital equipment library*

The Royal Alexandra Hospital uses a hospital-wide RFID asset tracking virtual asset library in order to (Vilamovska et al., 2008):

- improve the use of its assets;
- ensure the availability of medical devices at the point of need;
- streamline routine scheduled maintenance;
- reduce health and safety risks resulting from failure to meet scheduled inspection plans.

USA: *Tracking and tracing assets and equipment*

Southern Ohio Medical Center has deployed the Radianse Reveal Asset Tracking platform to increase its efficiency of asset and equipment tracking.
Since 2004, Bon Secours Richmond Health System has deployed one of the largest RFID-enabled mobile asset management programs in US healthcare industry. The system is providing tracking and management services for critical mobile medical equipment for three Bon Secours Richmond hospitals (Vilamovska et al., 2008).

USA: *Emergency Department workflow improvement*

To improve the operation of its Emergency Department (ED), including cutting down patient waiting time, Memorial Medical Center in Long Beach (CA) and Shelby County

Regional Medical Center relied on a RFID-based people/asset tracking software. The new system provided unprecedented data on ED use and patient trajectories, and became the key tool in the quality improvement policies the hospital implemented (e.g. new triage procedures, staffing). These led to a decrease in the waiting time for the first triage nurse from 1hr 20 min to 9 min for incoming patients in Memorial Medical Center.
This leads to (Vilamovska et al., 2008):
- increased patient safety;
- better use of staff time;
- increased facility capacity.

For the future

Continuing advances in nanotechnology and robust wireless infrastructures are now making it possible to envision an environment in which RFID devices ingested or implanted in patients could actually (Crounse, 2005):
- provide real-time information on health indicators and vital signs;
- regulate the release of medications;
- monitor and report on the results of surgeries;
- communicate with other devices in medication labels to alert caregivers to potential allergies, errors in dosage, or drug interactions.

The researchers estimate that with the help of RFID technology, a single nurse could more effectively and accurately monitor the status of an entire ward of patients, with care automatically dispatched in response to changes in a patient's condition. And outpatients could be remotely monitored, receiving nearly the same level of attention as those within the walls of the hospital (Crounse, 2005).

The main benefits of adoption RFID technology in healthcare include:
- patient's safety improvement by preventing errors;
- resource use improvement;
- costs saving;
- security enhancement.

Although the adoption of RFID technology by healthcare community is a relatively slow process, its usage is expected to accelerate when RFID technology prices drop down.

4. RFID-based information system for patients and medical staff identification and tracking

Our team proposed an integrated RFID-based system for patients and medical staff identification and tracking, called RFIDHospitalTracker.
The different components of this scalable and robust distributed system are depicted on figure 1.
Relating to this architecture we can note that the integrated RFIDHospitalTracker system includes the following main components:
- RFID 13.56 MHz passive tags (so that both patients and doctors can be tagged with unique codes);
- fixed RFID readers placed at strategic points in specific areas of the hospital: offices, laboratories, emergency rooms, operating suites, etc.;
- Pocket PC devices or mobile phones with RFID readers attached;
- a database server that will store patients' tags, tag templates and tracking information;

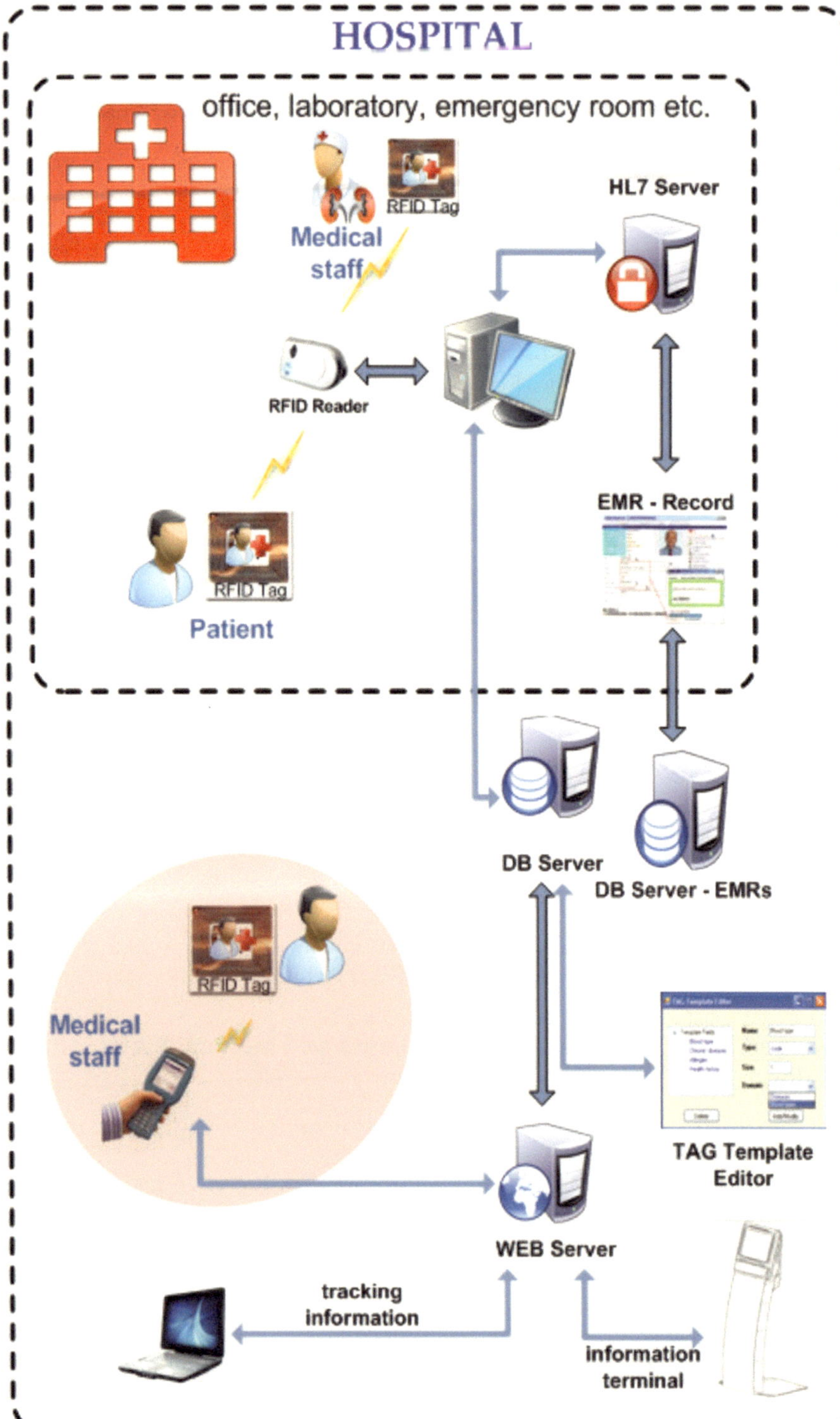

Fig. 1. RFIDHospitalTracker system architecture

- a web server that will allow database querying (to determine the location or the trajectory of a patient or doctor, etc);
- an EMR (Electronic Medical Record) server that will store the electronic medical records of patients in the hospital;
- information terminals that will allow patients to check their own medical records by scanning their wristbands;
- a HL7 server for interoperability with other clinical software already developed by other companies or organizations.

Next we present the way that RFID tags will be used for both patients and medical staff. The doctors, nurses, caregivers and other staff members wear an RFID card storing their employee ID number. These cards are used only for tracking purposes. Thus, they require a very small memory capacity. Every time a card is read by a fixed RFID reader, the actual location of the holder and current date/time are recorded into the central database.

On arrival, each patient receives a wristband with an embedded RFID tag storing a unique identifier, and some information about him (e.g. blood type, medical allergies, or other health history). In addition, all the patients' paper medical histories and other important documents are tagged with self-adhesive RFID labels containing patient identifier. With the help of this additional information, the risk of administrating wrong medication in case of an emergency will be highly reduced. Moreover, the hospital staff will be able to take the best medical decisions according to the actual health state of the patient.

These wristbands and RFID tags are waterproof and heat-resistant and can be used in the darkish environments. Moreover, RFID tags memory can be erased and written more than 100,000 times. Thus, to reduce the cost of the RFID healthcare system these wristbands and RFID tags can be reusable after high-temperature sterilization.

The data format that will be used for writing additional information into tags can be defined through an advanced template editor (figures 2 and 3) which allows user to establish all the necessary fields (e.g. blood type, chronic diseases, allergies, etc.), their type (string, code, codes list). The template is created at the PC level and it can be transferred to Pocket PC devices or mobile phones through specialized web services. A typical tag template can have the following fields:

Field	**Size**	**Type**	**Description**
Blood type	1	Code	code for blood type – e.g. **3 (A Positive)**
Chronic diseases	3	Code	a list of codes for chronic diseases – e.g. **1 (Asthma)** **4 (Diabetes Insipidus)** **5 (Bronchiectasis)**
Allergies	100	String	a list of allergies – e.g. **Penicillin** **Sulfa drugs** **Dilantin**
Health history	150	String	the most recent health history – e.g. **Appendix Surgery in 2002** **3 heart attacks in last month** **coronary artery bypass surgery in 2009**

Table 1. Typical tag template

Note that all types in column 3 of Table 1 are not built-in but application specific types. The use of codes has some important advantages, such as:

- less storage space is required for storing the same data: RFID tags with smaller memory capacity reduce the overall cost of the RFID healthcare system;
- increased security: a tag that stores codes means nothing for people outside the hospital;
- one definition, multiple uses.

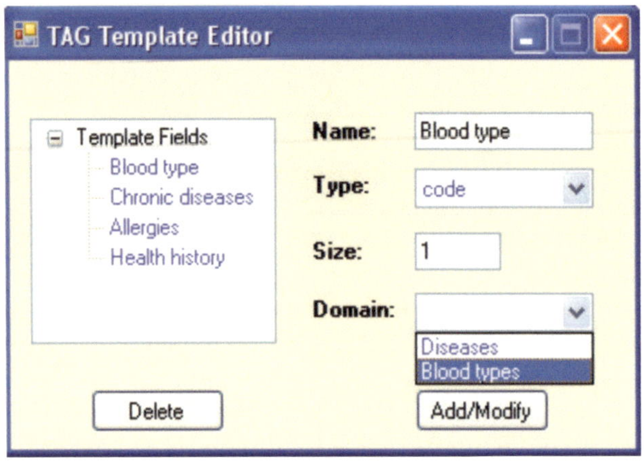

Fig. 2. Tag template editor

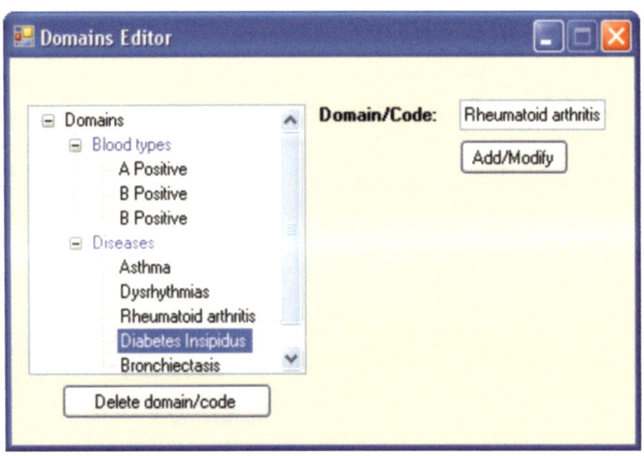

Fig. 3. Domains/codes editor

By default, an RFID tag does not store the name of the patient. To enable the tracking feature, the RFID tag is first linked with the patient's name. This is done by reading the RFID tag's unique ID and adding a new record ***(TAG_UniqueID, PatientNAME)*** into the central database. After that, every time the tag is read by a fixed RFID reader, the actual location of the patient and current date/time are recorded into the central database.
When a new patient arrives in the hospital, the following actions must be performed (figure 4):

- the RFID tag has to be linked with the patient's name;
- the RFID tag has to be linked with the patient's EMR, if any. This operation is done by adding a new record ***(TAG_UniqueID, SocialSecurityNumber_EMRID)*** into the central database. This way the hospital staff can access for review and update the desired

patient's information from hospital EMR system, only by reading the patient's RFID tag embedded into the wristband;
- the additional information has to be written into the tag's memory so that the information will be available for reading/writing at any time both at PC and Pocket PC levels.

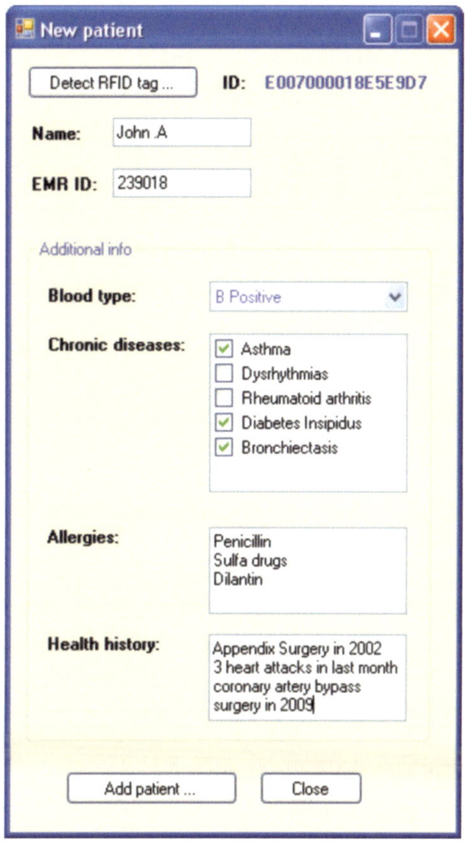

Fig. 4. Registering a new patient

The application located at the PC level can be used for the following purposes:
- to design tag templates;
- to register new patients;
- to remove an existing patient from the database;
- to track patients and hospital staff members;
- to read/write information stored on RFID tags;
- to display patient's EMR.

Perhaps one of the most important facilities that such an application should provide is the ability to read/write RFID tags. The ability to read/write RFID tags was achieved through a specialized software component that is performing the following main tasks (Cerlinca & et al., 2008):

WRITE operation:

- establishing a connection with the RFID reader;
- encoding the data to be written on the RFID tag;

- searching for an RFID tag in the proximity of the RFID reader;
- writing the encoded data to the RFID tag;
- closing the connection.

READ operation:

- establishing a connection with the RFID reader;
- searching for an RFID tag in the proximity of the RFID reader;
- reading all the data encoded in the RFID tag;
- decoding the data;
- closing the connection.

Next we present some considerations regarding the use of mobile devices within the RFIDHospitalTracker system. Mobile devices can be successfully used in places where there are no fixed RFID readers. Thus, the hospital staff members (doctors, nurses, caregivers and other employees) have a portable RFID reader (for example, a handheld device, like Pocket PC or mobile phone, equipped with an RFID reader) and possibly with a wireless connection to the web. The mobile application that runs on such devices performs the following main functions:

- read and write RFID tags (figure 5 and 6). The tag selection window presented in figure 5 is useful in cases when there are more than one RFID tag in the proximity of the RFID reader;
- work in stand-alone mode (independently of the main servers);
- store huge data;
- integrate and exchange information with other components of the RFIDHospitalTracker system. The tag template and database tags are transferred to Pocket PC devices or mobile phones through specialized web services that are located on the web server. All data transferred between mobile devices and PC server is first converted from the database format into the XML format;
- ensure maximum security;
- employ a multi-user and user-friendly interface.

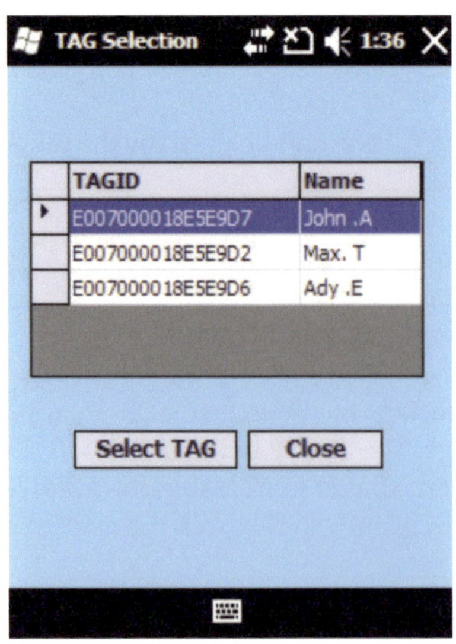

Fig. 5. Tags selection window

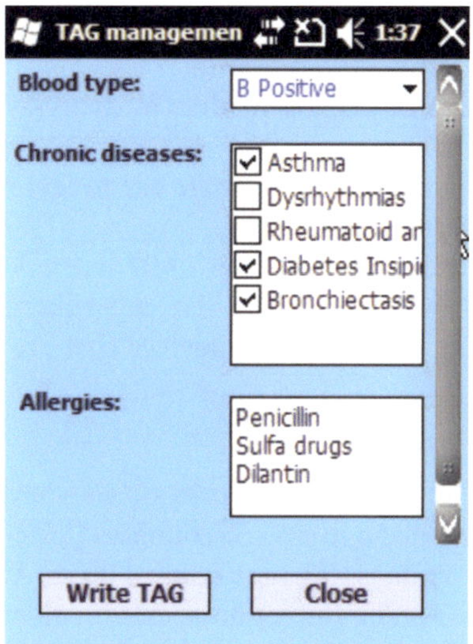

Fig. 6. RFID tags reading/writing

The mobile application cannot be used for tracking purposes or to add new patients.

Another important component of the RFIDHospitalTracker system is represented by the information terminals that will allow patients to check their own medical records by scanning their wristbands. A typical information terminal consists of the following components: an embedded PC, a display, a keyboard and an RFID reader. The application that runs on information terminals can be used for the following purposes:

- to read and display information stored on RFID tags;
- to display a patient's EMR;
- to locate hospital staff members.

RFIDHospitalTracker system is adaptable and scalable, enabling the hospital to deploy other tracking solutions for people (patients, doctors, nurses) and objects (medical devices and assets) without investing in new hardware or system architecture to support such solutions.

Any technology implementation in healthcare must deal with privacy and security issues. Our system provides several security enhancements and options to ensure the security of data and communication between applications:

- all RFID tag data is encrypted using AES algorithm (the private key is generated using tag unique ID and a hash function);
- password-based access to all web services used for communication and synchronization between mobile devices and the RFIDHospitalTracker system.

The main benefits of the RFIDHospitalTracker system are:

- error prevention of identifying wrong person by medical staff;
- elimination of some current medical errors, such as: not finding out whether a patient is allergic to a certain drug or giving a patient someone else's prescription;
- fast location of patients and doctors in the hospital.

Barriers to RFIDHospitalTracker adoption

We have to mention that the using of the RFIDHospitalTracker system raises some problems (issues), such as:

- privacy and legal issues;

- direct costs, for example, implementation costs, maintenance costs;
- difficulties in implementing RFIDHospitalTracker system within a hospital, for example, tagging persons and placing readers in doorways at hospitals will require an initial deployment;
- people acceptance, for example, hospital staff has to feel comfortable with the fact that they can be tracked and located every time.

Before the healthcare industry fully adopts the RFID technology there are still some open problems to be solved, such as determining the consequences and side-effects of radio waves on the exposed humans and electrical devices that are common in an hospital (like pacemakers, heart monitors).

HL7 (Health Level Seven) Compatibility

Health Level Seven, Inc. (HL7) is the global authority on standards for interoperability of health information technology with members in over 55 countries (*, 2007). HL7's vision is to create the best and most widely used standards in healthcare. HL7 is an international community of healthcare subject matter experts and information scientists collaborating to create standards for the exchange, management and integration of electronic healthcare information (*, 2010). HL7 promotes the use of such standards within and among healthcare organizations to increase the effectiveness and efficiency of healthcare delivery for the benefit of all.

In order to support other clinical software already developed by other companies or organizations we have designed a HL7 server. The main purposes of this server are:

- to acquire medical data about patients using the HL7 messaging protocol;
- to request clinical data about a patient from different servers and applications;
- to allow the clients of the HL7 server that are not HL7 compatible to understand the HL7 protocol and request/receive data using simple and intuitive ASCII commands.

The HL7 server will be primarily used to obtain the EMR of a patient based on his RFID tag. There are two different ways of getting clinical data:

- using the standard HL7 messaging protocol our HL7 Messaging Server connects to a list of medical applications;
- using simple and intuitive ASCII commands any non-HL7 application can connect to the Messaging Server and request data about a patient (Table 2).

Fig. 7 presents some examples of the use of these commands for:

- logging in using server address and port, user name and password;
- setting the current patient;
- getting patient name, SSN, mother's maiden name, birth place, etc.

Future development

We will try to develop our application to track the interaction between medical staff and patients. Also, we'll add a software module that will optimize the patient treatment process by prioritizing and tracking patients from admission to discharge, based on medical priority and treatment.

Our research will focus on the development of various software modules to use the medical information collected via RFID. For example, once information such as body temperature of a patient is received, the hospital control center can immediately analyze it. According to configuration settings, for example, in order to keep some diseases from spreading, the control center could send essential information to medical staff via mobile phone. Also, temperature-monitoring tags could be used for refrigerators containing lab specimens, blood bags and other sensitive items to ensure safe conditions.

Command	Description
login(<IP>,<Port>,<user>,<password>);	Login to a specified HL7 application or server using IP:Port and, for authentication, a user and a password.
usePatient(<SSN>,<Language>);	Setting the current patient using his Social Security Number and a language.
getName();	Getting the name of the patient.
getDateOfBirth();	Getting the date of birth for current patient.
...	
getLastObservations();	Getting all observations about current patient from the current server.
getSimplifiedLastObservations();	Getting all observations about current patient from the current server and return it in a simplified way.
getLastError();	Will return the last error in the language selected by the *usePatient* command.

Table 2. Some commands used to obtain data from the HL7 Message Server

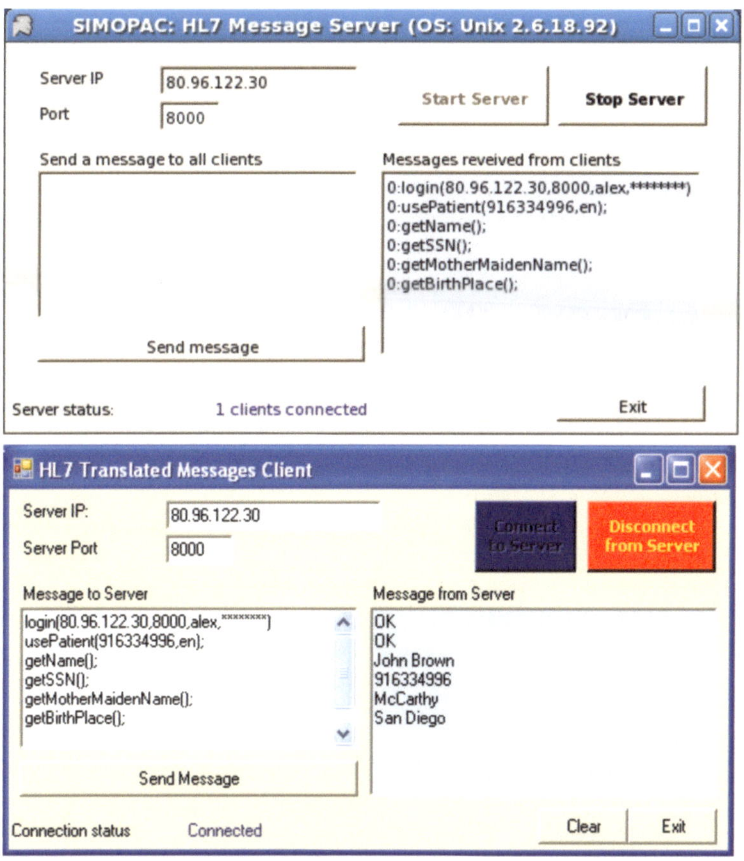

Fig. 7. The use of HL7 Message Server (Linux server and Windows client)

5. Conclusions

RFID applications can provide significant benefits to the healthcare industry to ensure patient safety and also to improve supply chain efficiency. In fact, healthcare is predicted to be one of the major growth areas for RFID. An analysis (Frost & Sullivan, 2005) estimates that the RFID in healthcare and pharmaceutical applications markets will reach $2318.7 million in 2011.
This chapter presents some applications that integrate RFID technology in healthcare domain. It also presents an RFID-based integrated system for patients and medical staff identification and tracking, called RFIDHospitalTracker. Our RFID-based system could be used in hospitals with large patients flow, allowing hospital staff to read patients' identification tags (RFID cards), which can help avoid some medical errors, like giving patients improper prescriptions or someone else's treatment.

6. References

BlueBean. (2007). The Benefits of RFID in the Healthcare Organization, *RFID Solutions for the Healthcare Industry*, available at http://www.rfidhealthcare.com/

Cerlinca, T.; Turcu, C. & Cerlinca, M. (2008). Integrating Mobile Applications into RFID Based B2B Systems, *22nd International Conference on Advanced Information Networking and Applications*, pp. 1341-1345, ISBN: 978-0-7695-3096-3, March 2008, Gino-wan, Okinawa, Japan.

Crounse, B. (2005). RFID: Increasing patient safety, reducing healthcare costs, 2005, available at http://www.microsoft.com/industry/healthcare/providers/businessvalue/housecalls/ rfid.mspx

Dyess, D. (2009). Medical Errors Cause Increasing Number of Preventable Deaths Each Year, *Health News*, August 2009, available at http://www.healthnews.com/family-health/medical-errors-cause-increasing-number-preventable-deaths-each-year-3559.html

Frost & Sullivan Research Service. (2005). World RFID in Healthcare and Pharmaceutical Markets, September 2005, available at http://www.frost.com/prod/servlet/report-brochure.pag?id=F548-01-00-00-00.

Harrop, P. & Crotch-Harvey, Trevor. (2008). RFID for Healthcare and Pharmaceuticals 2008-2018, *IDTechEx*, 2008, http://www.idtechex. com/research/reports/rfid_for_healthcare_and_ pharmaceuticals_2008_2018_000146.asp.

Lebowitz, J. & Mzhen, V. (2007). Four Patients Die After Improper Use and Prescription of Painkiller Drug FENTORA, September 2007, available at http://www.marylandaccidentlawblog.com/2007/09/four_patients_die_after_improp_1.html.

Sirivattha, P. (2003). Issues on patient identification related error, *Technical report*, University of Hawaii at Manoa, Information and Computer Sciences Department, 2003.

Vilamovska, A.M.; Hatziandreu, E.; Schindler, R.; Oranje, C.; Vries, H.; Krapels, J. (2008). Study on the requirements and options for RFID application in healthcare, July 2008, http://ec.europa.eu/information_society/activities/health/docs/studies/200807-rfid-ehealth.pdf

*. (2007). HL7, Introduction to HL7 Flash Tour, available at http://www.hl7.org/index.cfm

*. (2009). National Association of Boards of Pharmacy (NABP), About the issue, available at http://www.dangerouspill.com/about_the_issue.html

*. (2010). Wikipedia, Health Level 7, available at http://en.wikipedia.org/wiki/Health_Level_7

13

RFID-based Direction Finding Signage System (DFSS) for Healthcare Facilities

Nitin Sharma and Jong-Hoon Youn
Department of Computer Science
University of Nebraska at Omaha
USA

1. Introduction

As RFID technology matures, it has huge potential for numerous applications in disparate fields of study. Due to the potential benefit and transformational technology that RFID promises, the United States Department of Defense (DoD) is dedicated to adopting passive RFID technology (United States Department of Defense suppliers' passive RFID Information guide version 10.0, 2008). RFID has potential benefits for supply chain, inventory management, asset tracking and improved labor management and productivity within an organization. Therefore, we can see top retailers like WAL-MART adopting the technology to harvest its full potential. Likewise, Technology giants such as HP, IBM, Intel and Sun have considerable efforts and plans ahead for adoption.

There have been many considerable efforts by the research community in adopting RFID in different fields of study. Healthcare applications using RFID are of primary interest as well. Although our application can be thought of as generic, we point out that we can improve healthcare quality by reducing patient wait-time and providing better patient care. Our research deals with an application based on RFID technology. We focus on designing a system using cost-effective devices so that mature applications can be built using the existing technology.

We propose an application titled "Direction Finding Signage System (DFSS)". A DFSS is a system that directs a person to a desired destination in an unfamiliar environment. The need for such a system arises when people need directions to locate a particular destination within a building. We further strengthen our idea by taking an example of a hospital environment, where emergency services are provided 24 hours a day and the volume of emergency patients is huge in number. Such a situation would result in a long waiting time for patients to discover their destinations of interest. We propose an efficient RFID application with RFID tags, readers and display devices that can visually help patients take the shortest route to their destinations. This would help them save their precious time and help them get the desired medication and care at the right time without wait-time hassles. Furthermore, the application can help a hospital's security team be destination-aware of the patients or users in case of an accident or an emergency.

This application is written in a .NET environment in C# for Windows platform. The choice of the software is solely on the basis of simplicity of the Application Programming Interface (API) provided.

The rest of the chapter presents important aspects of the application. Related studies on current signage solutions are described in Section 2. Section 3 describes the proposed system architecture and technology. Section 4 describes the hardware and software aspects of the implementation of the system by suggesting cost-effective hardware solutions and briefing the algorithm used. Section 5 describes our deployment scenario. Section 6 discusses potential future improvements and insights, and we conclude the chapter in Section 7.

2. Related studies & background

This section discusses the current major technologies from the literature and how they are used for navigation purposes.

An RFID tag is a compact tag that can be fixed at one place or can be carried by a person. The tags store information regarding an object, place or a person. RFID readers then read these tags to get the information stored. The authors in (Miller et al., 2006) have described the indoor localization attempt by using the RFID tags on fixed locations and using a moving reader to provide the location for tracking the person or object carrying the reader.

Global Positioning Systems (GPS) is a rapidly advancing technology with applications in aircraft, automotive and in-building navigation. However, there are a few problems related to GPS, such as inaccuracy and inefficient operability in dense urban environments or indoor locations. A navigation system using embedded wearable GPS devices and a system providing verbal guidance and environmental information via speech and Braille displays has been proposed (Loomis et al., 2000). There is yet another approach using Differential Global Positioning Systems (DGPS) for developing a navigation system for the visually impaired (Hashimoto et al., 2001).

Infrared transmission-based devices can also be used by people with sight disabilities as well as the elderly. Such a mechanism involves infrared transmitters that transmit directional information to handheld receivers. The basic principle in developing this device is that infrared transmission is directional. The direction in which the person receives the clearest message is the intended direction. This technology is otherwise referred to as the Talking Signs.

A Verbal Landmark (VL) (Bentzen & Mtchell, 1995) based system is an inductive-loop based radio-frequency system which, when brought under the desired range of the signal, transmits the speech message to the intended receiver. The receiver is a handy device that can be comfortably carried by the user. The transmitters are also compact and can be installed at desired locations. Messages transmitted are in different forms suitable to people who are visually impaired or unable to identify the signs or are unclear about the visual traffic directions. This technology, though cost-effective, is not very efficient in terms of reliability.

A navigation system for the blind comprises a cane equipped with a color sensor, light emitting diode, a control unit and a vibrator. This has been developed and implemented in Japan (Magatani et al., 2001). The methodology involves optical beacon systems attached to the ceiling in the building that continuously send out infrared signals which are picked up by the respective receivers held by the user within the range of the beam. Some hospitals paint colored paths and the feedback from the color sensor vibrates the cane for navigation.

The System for Wearable Audio Navigation (SWAN) (Walker & Lindsay, 2006), designed at the Georgia Institute of Technology, includes a small, light-weight laptop or even a handheld device, a pedometer and tracking devices such as RFID tags, RF sensors, and inertial sensors. The sensors help evaluate the direction the user is currently facing and also the direction of the actual destination. After the direction is determined, an audio interface provides the user with a speech message indicating the directions along the path. This system also allows the user to learn about the surrounding environment, and thus transmits sounds indicating obstacles on the path.

Dead Reckoning (Roston & Krotkov, 1992) is another technique on which indoor navigation can be based. It refers to estimating the current position of a user based on past courses navigated and speeds from a known past position. The future position can be analyzed from the current position and past data. For this purpose, a DR graph is plotted and constantly updated with changes in the surrounding environment and user walking traits. This process only provides a rough estimation, and thus inaccurate.

Although there are many different technologies associated with various navigation systems, we believe that there is tremendous potential in RFID-based systems.

3. The proposed system

3.1 Radio frequency identification system

RFID works on radio waves, and operates in different frequency ranges. It has been widely used in inventory tracking, supply chain management, manufacturing, asset tracking and health care. A typical RFID system consists of a reader, an antenna and a tag. Generally, the reader is called the interrogator and the tag is called the transponder. The transponder is a microchip attached to the antenna. An interrogator (reader) typically consists of a transmitter and a receiver module and an antenna.

RFID operates on Low frequency (125/134 KHz), High frequency (13.56 MHz) and Ultra-high frequency (850 MHz to 950 MHz & 2.4 GHz to 2.5 GHz) modes according to the type of antenna used. The operating frequency is chosen according to the data rate and the range required for the application.

RFID tags can be further classified as active tags and passive tags. An active tag is a battery powered tag, which consumes energy from the battery to power up its circuitry and broadcasts the signal to the reader. Since it has the source of power within itself, it can transmit over a longer distance. A passive tag instead powers up its circuitry with the help of the radio signal emitted from the reader. At the same time, the data exchange takes place. Hence, it transmits over shorter distances than active tags. The semi-passive tag utilizes the functionalities of both active and passive tags. It has its own battery to power up its microchip as an active tag, but the RF signal is reflected back to the reader like a passive tag. Furthermore, the tags are also classified on the ability of the reader to read or write on the tags as "read-only" and "read-write" tags.

Although RFID has enormous capabilities, its use has been criticized because of security issues. When two or more readers overlap, the tag may not be able to respond to multiple queries at the same time. This is called reader collision. Also, when an area contains many tags, the reader may not be able to read a particular tag. This is called tag collision. Although there are plenty of techniques that address reader collision and tag collision, behavior plays a significant role in the adoption of RFID in any applications.

3.2 DFSS system architecture

The proposed system architecture is shown in Figure 1. The user scans the RFID tag at the information desk. This information is then sent over the network to the server and, in return,

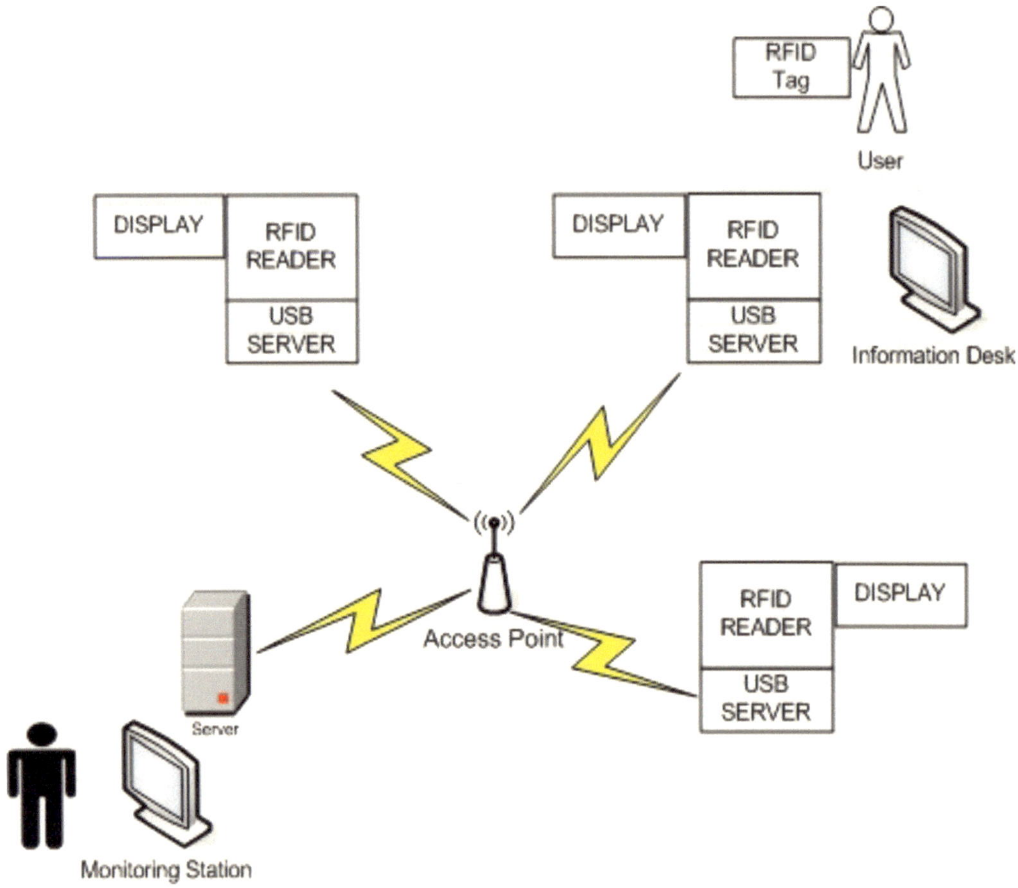

Fig. 1. DFSS System Architecture

the server sends back the direction information to the display device associated with the reader that scanned the tag. The system aims to guide the patient to a particular destination. The monitoring station also helps locate the troubled patient so that immediate assistance can be provided without wait-time hassles. The system proposes low cost hardware and has no cabling costs by using wireless communications for the infrastructure. The system is scalable and can be easily integrated into an existing wireless network setup.

4. Implementation

4.1 Hardware

We explored some interesting devices that could be used to design the system. As the tags would be scanned from just few inches away from the reader, we decided to use a low frequency, cost-effective RFID reader. We chose the Phidget RFID readers, which are designed to be used with any tag designed for EM4102 protocol. We chose passive read-only tags for our purpose. The Phidget comes with tags of different shapes and sizes. The orientation of the reader and tag in respect to each other affects the distance over which the tag is usable. For smaller tags, it must be closer to the reader to be read. The effective range for the credit-card size tag that we use is 5 inches.

Fig. 2. Phidget RFID reader, EM4102 tags, USB servers, 20X2 PhidgetTextLCD

The device used in this system has a few limitations. Multiple RFID readers within 1-2 meters distance could interfere with each other, which could be mitigated by enabling and disabling the antenna of the readers. Also, the Phidget RFID has no collision detection or avoidance capability. Hence, a reader would not be able to read two tags that are within close proximity. We believe these issues will not affect our system as we have low range for the read field, and there is a very low probability of two users trying to use the reader simultaneously. The Phidget RFID reader comes with an on-board LED; however, we decided to integrate yet another device for efficient display of the results. The device, PhidgetTextLCD, allows us to display messages on a 2 x 20 LCD screen. The PhidgetTextLCD has a 2 x 20 LCD (2 lines high, 20 characters per line) with each character having an arrangement of 40 pixels (8 pixels high by 5 pixels wide). The LCD display has support for both standard ASCII library and unicode characters. The onboard interface kit could also be used to integrate various sensors into the display device and open doors for pervasive computing.

Device Specifications

Antenna Output Power (max, far field)	< 10 µW
Antenna Resonant Frequency	125kHz - 140kHz
Communication Protocol	EM4102
Read Update Rate	30 updates / second
External +5V Supply Voltage	5VDC
External +5V Supply Current Limit	400mA
External LED Supply Voltage	5VDC
External LED Supply Current Limit	16mA
External LED Output Resistance	250 Ohms
Recommended Terminal Wire Size	16 - 26 AWG
Terminal Wire Strip Length	5 - 6mm (0.196" - 0.236")
USB-Power Current Specification	500mA max
Device Quiescent Current Consumption	16mA
Device Active Current Consumption	100mA max
Typical Read Distance - Credit Card Tag	11cm (5")
Typical Read Distance - Disk Tag	6cm (3")
Typical Read Distance - Key Fob Tag	7cm (3.5")

Table 1. Phidget RFID reader Device specification (PhidgetRFID Product Manual, 2007)

The Phidget Devices use a Universal Serial Bus (USB) to communicate with the computer, thus providing developers with an easy to use and robust API. The Phidget libraries are available for many languages such as: Java, C, C++, .NET(C#) and Delphi. We chose C# for our development.

These devices were sparsely deployed at many locations and were allowed to communicate to a central station using an existing Ethernet or Wi-Fi network. We focused on using wireless communications for the system to reduce cabling costs. We found two USB servers for this purpose. The KeySpan USB server has 4 USB ports, an RJ45 Ethernet connector and an embedded USB host controller with software drivers that would make the device appear as an additional USB host controller on the station. We used a wireless bridge to connect to an access point in our network. This requires two different power sources to each of the devices. Since we aimed at reducing power costs, we sought for yet another device with an integrated 802.11 support. The other USB server we used was the Silex hi-speed USB 802.11 wireless device server. This device server provided connectivity through an integrated wireless antenna, therefore for reducing the need for a bridge or any interim device. The Silex USB server required virtual link software to be installed on the station for connectivity with the USB devices.

4.2 Software

We chose to use .NET libraries provided by Phidgets to interface the readers and LCD kits. We maintain a software association between readers and the destinations of interest and also between readers and their respective display devices. This helped us to identify the display device to send the destination information to once we scanned a tag on the reader. We found the shortest path between locations using the Floyd-Warshall shortest path algorithm, as it computes the shortest path between all pairs of vertices in the system. The system requires weighted graphs of the map of setup building, and also the directionid (north, south, east, and west) between each pair of vertices.

The algorithm developed is as follows:

1. For the first time when a reader reads a unique tag, it provides the user with an interface (as shown in Figure 3) to input the destination of interest. Then it adds the (tagID, dest) tuple to the run-time hash scan table.

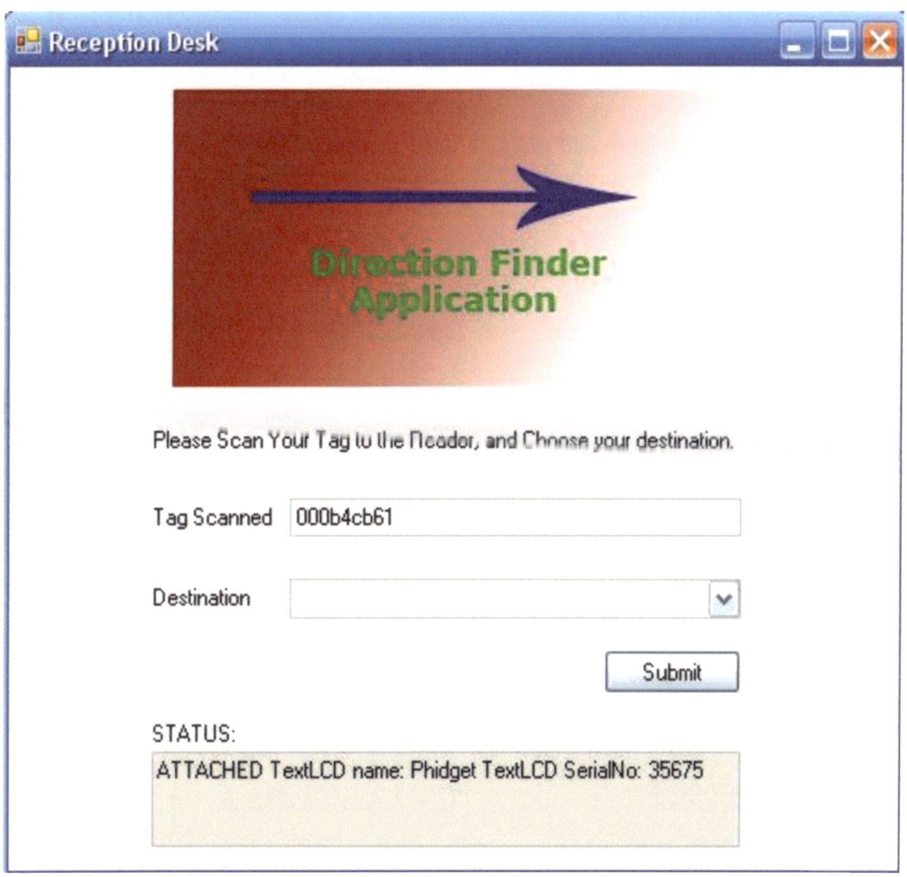

Fig. 3. GUI snapshot at information desk

2. Once the reader re-reads the tag which it finds on a run-time hash scan table, it reads the current reader's ID that scanned the tag and destination reader's ID from the scan table, and feeds the (currfid, destrfid) tuple to a function that returns the next-hop RFID reader. The function does so by sending the (currfid, destrfid) tuple to the Floyd-

Warshall algorithm, which returns the distance to be traversed to reach the destination and the next-hop reader ID along the shortest path.

3. With the next-hop reader ID, the system queries the hash for the (currfid, nxthoprfid) tuple to get the directionid, which is sent to the display device associated with the currfid for display purposes.

Thus the user can navigate to the direction of the next hop reader and repeat the scan process unless he reaches the destination of interest. The display device also shows how far the user is from the destination.

5. Deployment

For deployment, we chose the third floor of the PKI (Peter Kiewit Institute) building at UNO (University of Nebraska at Omaha). The RFID readers and LCD displays were located at different points and marked on the map. Then we generated the test scenarios to verify the shortest path to the destination. We believe our map of critical interest is similar to a maze-like hospital building.

We identified four locations on the floor as our primary points of interest. There is an information desk that would require the patients to scan their tags, and express their point of interest. The information desk can be attended by a staff member or remain unattended at the cost of the user submitting their destination of interest from a scroll on a PC. This information is required by the central server for computation of the shortest path and the next destination. On computation, the server sends the direction information to the display device and the user follows the direction.

Fig. 4. PKI third floor map for deployment.

As the users follow the direction specified and reach intermediate readers, the central station will be able to monitor and track the users as shown on the screen snapshot in Figure 4. This could help the staff monitor and track the patients. We believe this will be extremely useful in a medical environment as it helps nurses/staff to provide care and assistance to patients/users walking along the hallway, if needed.
In this chapter we have presented the accomplishment of our idea and are highly optimistic about the benefits provided by the prototype for tracking in medical environments where patient care is of utmost importance. We tested our setup with three tags, each with different destinations, and achieved our expected results.

6. Discussion and further work

The system has many benefits as compared to previous approaches in direction signage systems. We prove that our system is robust, cost-effective, scalable, and easy- to-integrate.
We show that we can support multiple tags with multiple destinations in the system with minimal concerns about tag and reader collisions. This proves the robustness of the system. The system is capable of supporting multiple patients/users from a single setup infrastructure. This proves the scalability of the design. Each patient would have a tag with their destination of interest on the system, and the direction information would only be sent on-demand, i.e., only when the patient scans the tag on a reader. Hence, the setup would not create much traffic in the air. Again, because the communication medium between the reader and the central station is Wi-Fi, it can easily integrate into any building with existing wireless network infrastructure. The choice of implementing low cost devices and relying on wireless media for communication to the central station in order to reduce cabling costs proves that our system is more cost-effective than other solutions previously proposed.
Furthermore, as every reader event can be observed from the central station, the system allows staff to track visitors and/or patients within the building and can provide alerts for staff to assist if needed. We can even expand the system to let the system administrator override the path computed and redirect a patient to a less congested route. We believe that our simple proof-of-concept can lead to many more exciting future developments and research into direction signage systems.

7. Conclusion

We have presented a cost-effective direction signage system that can be used in environments where waiting times in a maze-like building, such as a hospital, is of primary concern and of utmost interest. Our study focuses on using RFID technology to reduce such wait-times as minimally as possible. Therefore we proposed a DFSS that can help people in such environments find their destination with the best possible care and attention without having to change much of the existing infrastructure in setting up the system.

8. References

Loomis, J. M., Golledge, R. G., & Klatzky, R. L. (2000). GPS-based navigation systems for the visually impaired, *Symposium on Low-Vision at the Meeting of the American Academy of Optometry*, Orlando, FL.

L.E. Miller, P. F. Wilson, N. P. Bryner, Francis, J. R. Guerrieri, D. W. Stroup, and L. Klein Berndt, RFID-assisted indoor localization and communication for first responders, *Proceedings of the Int. Symposium on Advanced Radio Technologies*, 2006.

H. Hashimoto, K. Magatani, and K. Yanashima, The development of the navigation system for visually impaired persons, *in Proc. 23rd Annual EMBS Int. Conf.*, 2001, Istanbul, Turkey

Magatani, K.; Sawa, K.; Yanashima, K., Development of the navigation system for the visually impaired by using optical beacons, Engineering in Medicine and Biology Society, 2001. *Proceedings of the 23rd Annual International Conference of the IEEE*, vol.2, no., pp. 1488-1490 vol.2, 2001

Bentzen B. L. and Mtchell P. A., 1995, Audible Signage as a Wayfinding Aid: Verbal Landmark verses Talking Signs, *in Journal of Visual Impairment and Blindness, Ed. Shively S.*, Nov-Dec 1995 Vol 89 No 6, Page 494-505, AFB Press, USA

Walker, B. N., & Lindsay, J. (2006). Navigation performance with a virtual auditory display: Effects of beacon sound, capture radius, and practice. *Human Factors*, 48(2), 265-278

Gerald P. Roston and Eric P. Krotkov, "Dead Reckoning Navigation for Walking Robots". Raleigh, NC July 7-10, 1992

14

RFID Modeling in Healthcare

M. Laskowski, B.C.P. Demianyk, G. Naigeboren, B.W. Podaima,
M.R. Friesen and R.D. McLeod
University of Manitoba
Canada

1. Introduction

Increasingly, healthcare management systems include investment in and implementation of technology to track the status and movement of various entities within the healthcare environment, including patients, healthcare workers and physical assets. This is often a means of understanding patient flow, controlling inventory, tracking equipment usage, and thereby (ideally) assessing efficiencies in order to optimize resources and processes within that environment (Wang et al., 2006). The focus of this chapter is to highlight the contributions of RFID systems modeling, particularly in relation to an understanding of the nature and extent of system error that is often overlooked experientially. The healthcare environment was chosen as it is an increasingly complex and interesting application area for RFID, and in which a wide range of RFID-based applications and devices already exist and can be envisioned for the future. The insights gained through modeling provide a complementary set of data to those gained from the experiential knowledge of performance in existing installations. To that end, this chapter focuses on a case study of an agent-based model (ABM) of a hospital emergency department (ED), with extensions to modeling the provisioning of a real-time location system (RTLS) using RFID for patient tracking.

To contextualize this work, Section 2 reviews conventional and emerging RFID applications in healthcare. Section 3 introduces the agent based modeling technique, invoked to investigate system performance in an application for RFID-enabled patient tracking within an ED. The ABM was developed as a decision support tool oriented towards optimizing RFID placement (minimizing uncertainty) for an actual ED where healthcare managers are considering the deployment of such systems. Section 4 outlines the ABM simulation results, with a particular focus on the nature and extent of system error and uncertainty – both spatial and temporal – that modeling illuminates. Section 5 discusses implementation strategies for an RFID RTLS system reflecting a Service Oriented Architecture approach that leverages existing software systems and focuses on being IP-centric and application- and device-agnostic.

2. RFID healthcare applications and patient tracking systems

The current scope of RFID technology in healthcare includes conventional as well as emerging 'smart' RFID devices and applications. The overall objective of RFID applications in healthcare is to automate manual processes as well as to reduce the time required to track and locate assets, including vital supplies and people and thereby resulting in institutional and productivity gains (Kanyuk & Young, 2004). Ultimately, RFID applications in healthcare can support service providers in decision making functions and arduous tasks typical within a complex clinical environment. Over the range of applications, RFID in healthcare generally shows capacity to i) identify, track, and locate patients, healthcare personnel, medical files, medical equipment, medical supplies, medications, and laboratory and pathology samples; ii) detect incorrectly packaged medications or supplies and doses; iii) institute systems to recall counterfeit or contaminated medications and supplies; iv) correctly identify and locate potentially 'misplaced' surgical instrumentation - even those accidentally left within a patient during a surgical procedure; v) corroborate medical supply and surgical instrumentation sterilization; vi) ensure safe and proper handling and disposing of bio-hazardous materials, expended medications, medical supplies and devices, and waste; vii) and, even 'confirm' actual anatomical locations for surgical procedures. Thus, at the patient point of care (POC), RFID can serve to correctly identify patients to receive and care providers authorized to deliver a prescribed medical treatment or procedure. Overall, in the healthcare sector there are many emerging RFID inventions that can offer drastic improvements in processing times associated with medical records retrieval, efficiency of service and reliability of medical diagnosis, dispensing of medications, and personnel management (Solanas & Castella-Roca, 2008; Wicks et al., 2006). The overall goal of these technologies is to facilitate improvements to patient safety and quality of care, and to do so in a cost effective manner.

First-generation RFID applications have been in commercial use for well over 20 years. These conventional applications in healthcare are primarily based upon identification, for example technologies and methodologies for improving patient POC and reducing errors based on barcodes and RFID (Rao & Dighe, 2004; Bearing-Point Inc., 2006; Podaima & McLeod, 2006). In general, conventional applications enable systems to be built around inventory tracking and control. Extensions include pharmaceutical supply chain inventory and tracking for medical reconciliation. Tied into a hospital management system, they hold considerable potential to reduce adverse drug events at the patient POC. This is accomplished through corroboration of the patient ID with the drug prescribed by the physician. These technologies carry increasing potential as the supporting electronic technology improves and connectivity protocols become standardized (Dubois et al., 2001). One of the limitations with early adoption of both RFID and barcodes is that they are inherently submissive, allowing for identification with little or no support for interactivity and automation.

Second-generation or 'smart' RFID-enabled healthcare solutions offer an expanded functionality beyond conventional RFID tagging alone. While conventional RFID is typically reserved for supply chain management, incorporating identification, asset tracking and locating (Glabman, 2004), smart RFID-enabled devices incorporate the functionality of control - both sensor and actuation - facilitated by advancing microelectronics technologies such as system on chip (SoC). Within this expanded capability, applications incorporate an RFID enabled electromechanical lock or latch used in combination with a pervasive health informatics platform. In a medical compliance configuration, this serves to mitigate against adverse drug events and can be implemented in smart syringes, smart couplers, smart medical containers, smart pumps, smart clamps, smart valves, smart pipettes, and smart bandages, for applications requiring the proper identification and corroboration of IV

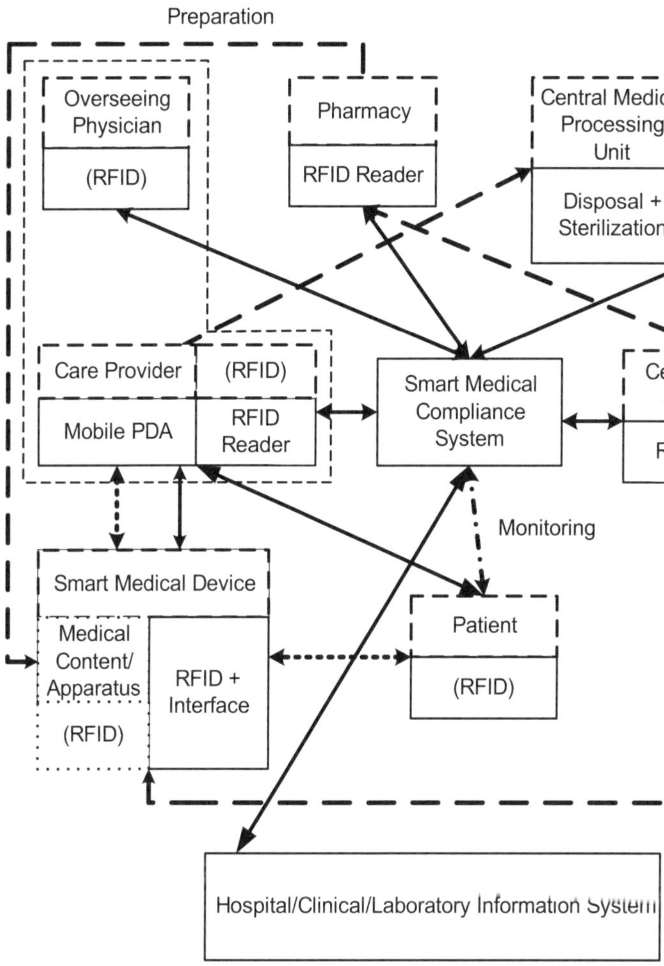

Fig. 1. Medical compliance platform - POC interaction components

infusion (smart pumps), and medication dispensing (Podaima, 2007). The RFID tags on these devices can be either active or passive, and the control and communication can be derived from the interaction of an RFID reader and tag in conjunction with the associated electronics and overseeing medical information management system. Smart RFID also introduces novel designs for integration with evolving and legacy POC systems. Figure 1 illustrates a conceptual overview of the POC interacting components within the medical reconciliation and compliance platform.

Smart RFID with its enhanced functionality lends itself to improved fail-safe or 'intelligent' medical devices, medical compliance and reconciliation systems encompassing medication prescription, transcription, administration, and pharmacological preparation, and physical and biological sensor-based functional-monitoring modalities. The benefits of smart RFID are compounded when used in conjunction with conventional RFID tagging (e.g. RFID badges, RFID wristbands) for identifying patients, healthcare providers, and medical supplies and equipment. For instance, a smart RFID-enabled hand washing hygiene surveillance and compliance system implemented across an entire healthcare facility can be a most important tool in preventing the spread of infection. This application is representative of the kind of capabilities of enhanced smart RFID-enabled devices, as it can reduce the risk of widespread (nosocomial) infections causing harm or even death due to staff and visitor non-compliance.

As another example, a Smart Coupler incorporating the functionality of smart RFID-enabled fail-safe locking technology is manufactured by Colder Products Inc. (Bacheldor, 2006). The two-piece smart coupler consists of a male and female sub-miniature plastic connector containing an integrated RFID tag and RFID reader, respectively, typically used for connecting medical and surgical equipment, including blood-pressure cuffs, blood analyzers and analytical instruments, and IV infusion bags. When two pieces of the connector are joined for coupling, the RFID reader scans the tag. If there is a mismatch or incompatibility, the error could trigger an alarm and/or the completely disable the coupling function. In addition to this smart technology ensuring that the equipment is hooked up properly, it also ensures that healthcare providers aren't reprocessing and reusing devices that are actually intended to be disposable. By programming the smart coupler system (RFID reader) via a computer interface, application specific data and rules can be customized for the particular application.

A further example of a smart RFID application in healthcare is the Smart Syringe currently under development by Virtuistix Inc. (Podaima, 2007). The Smart Syringe extends the basic injection principle of a standard syringe in that it offers much greater capability and purpose via integration with an RFID-enabled capability. Thus, a Smart Syringe has a bi-directional communication channel along with a microcomputer interface. The capability therein incorporates a controller that authorizes the operation of a manual or automatic mechanical injection plunger with a lockable-latch or pinch-off regulating mechanism, thereby modulating the flow of fluid through a syringe. The Smart Syringe incorporates an RFID tag and accompanying interface (*in situ* and/or external) in a method and system to control the mechanical operation of a plunger either i) manually enabling or precluding an instance of user operation of the syringe plunger regulating mechanism, or ii), automatically enabling or precluding an instance of power assisted (e.g., electromechanical) operation of the syringe plunger regulating mechanism. Hence, in addition to facilitating flow by activating a syringe plunger regulating mechanism (opening, closing, or modulating), this smart syringe assembly incorporates a lockable-latch which will prevent unauthorized, erroneous, or inadvertent operation. In the field, information will be available as to the status of operation (metrics, performance), maintenance, and serviceability (replacement).

The RFID-enabled Smart Syringe can be identified with an RFID reader juxtaposed with a mobile or handheld computer (computer on wheels or personal digital assistant), or a stationary communicating computing device. In this configuration, the RFID reader can thereby be used to interrogate the smart syringe status as well as identify patient (via RFID wristbands or badges), healthcare provider, and other medical supplies and equipment used in the treatment modality. The handheld computer can provide on or off-line audio and video status as to the authentication and compliance of the medical procedure or treatment in real-time. Thus, the communication and electromechanical control can be derived from the interaction of the RFID tag (and associated electronics) and the RFID reader and, in some instances, an overseeing healthcare information management system. Upon identification, corroboration, and authentication among patient, care provider, and medication preparation in the Smart Syringe, the interrogating RFID reader could be used to activate or preclude the Smart Syringe latch and/or plunger regulating mechanism according to the prevailing medication compliance protocol.

Future smart RFID-enabled medical devices will extend well beyond traditional uses, and as such will incorporate various sensors and actuators, likely to be widely available as implantable devices (Ross, 2004). The integration of RFID with bio- and chemical sensors is also of great interest, in that the combination can be used to create new devices for both diagnosis and even treatment (e.g., smart bandages) to function in areas previously impracticable. These sorts of hybrid RFID devices can be used practically and robustly - both *in vitro* and *in vivo* - and have far reaching clinical efficacy. Blood glucose monitoring is one example of an implantable RFID SoC device that can be used by diabetic patients without the requirement for taking blood samples subcutaneously several times per day. Intestinal and esophagus monitoring with pH and temperature sensors also fall into the same class of device; they offer a minimally invasive means of biopsy and diagnostics, facilitated as a direct result of venerable RFID-based SoC technology.

It should be recognized that both conventional- and smart-RFID are not panaceas to solve all healthcare management issues, as they have current technological limitations characteristic of many new technologies, and are fraught with privacy and security concerns. Additional considerations include the reuse (i.e., sterilization) of devices, which implies an additional constraint that may require the device to be subject to temperature, chemical, pressure, and/or electronic processes not otherwise needed in less sterile environments. As with other medical devices, Clinical Grade Smart Medical RFID-enabled devices will be required to meet the stringent standards and guidelines of various governing bodies and institutions of the healthcare industry. These include mandates regarding security, integrity (encryption), and privacy. Clinical grade RFID-enabled devices will also be required to meet rigorous EMI and EMC (electromagnetic interference and compatibility) guidelines (Ashar, 2007; Witters, 2009). Thus, smart RFID-enabled medical devices come with an associated overhead, but are not superfluous in deployment, and can be used within the framework of an engineered POC system (Ehrmeyer et al., 2005; Roberts, 2005). Currently, the capacity exists for seamless integration with purposeful function, and an evolutionary path to improved overall medical compliance.

Over the next several years, we expect to see a significant push for improved healthcare services and medical records adoption demanding quicker, better, and cheaper point of care and clinical biomedical testing and diagnostics. To fulfill this need, opportunities will continue to emerge for both conventional- and SoC- RFID-based smart healthcare technologies to play a viable role in delivering on this promise, toward the ongoing goal of improved patient safety and quality of care.

While this discussion demonstrates an increasing range and variety of RFID applications and devices prototyped and implemented in healthcare settings, much less work has been done on *modeling* RFID systems in healthcare, in order to gain insights into implementation parameters of RFID applications. In the following section, a case study is presented which investigates the modeling of a conventional RFID application - that of patient tracking in an ED - with a particular interest in the nature and extent of uncertainty and error in the system, which is often overlooked in real installations and real data.

3. Case study: agent based modeling of RFID placement for patient tracking in an emergency department

3.1 Introduction

Evolving RFID RTLSs are meant to augment and/or automate the existing data capture (electronic records systems) in place in many healthcare institutions. A typical patient trajectory capture system may include the collection of time of arrival, time of triage and registration, time and duration of treatment including specialty consultations, and time of discharge from the ED. In some cases, data are entered by healthcare workers for multiple patients in aggregate when there is a break in the workflow. As such, there may be considerable uncertainty associated with the data themselves, making it difficult for policy-makers to make statistically verifiable decisions as they attempt to optimize patient access and flow through the healthcare facility. In this context, RFID systems offer a complementary and more automated means of data collection. However, critical issues are again associated with uncertainty and consequent inferencing required to remove ambiguities in the data.

To preface a discussion on error and uncertainty in RFID systems, a brief discussion of the system hardware is introduced. For our purposes, RFID systems are considered to be tag-and-reader systems, comprised at minimum of a number of tags and at least one reader. These RFID systems operate over a variety of frequency ranges and with a variety of complexities associated with the tags as well as readers. In general, a tag is associated with an asset or patient and the reader is associated with a location in the physical environment. The tag can be either passive or active, the latter requiring a battery allowing it to power a transponder and allowing it to be interrogated by a reader. A passive tag obtains its power from the field of the reader, allowing it to effectively "transmit" its identification back to the reader (Finkenzeller, 2003). Passive tags are modeled in this study.

Several sources of uncertainty and error become immediately evident, yet are often overlooked by users. First, once a tag is read, it is primarily a proximity measure, meaning the tag is somewhere within the proximity of that reader. The low frequency passive tags (<20 MHz) use inductive coupling as opposed to propagating electromagnetic radiation, and as such are typically limited to fairly close proximities (~1-2 m). As a consequence, when attempting to create a patient trajectory through a healthcare facility, time-stamped historical data are required and rules of inference are used to estimate a patient's approximate location. Second, a particular reader and tag system will have time-varying non-isotropic capture areas, impeded by distance and other objects in the surrounding area. Third, in the case of the passive tag with virtually no processing power, a reader may read a tag at apparently random intervals or may not read the tag at all, depending on tag orientation and its environment. A fourth issue is interference, where closely-spaced readers interfere to the point where both readers fail to read a tag that passes through or stays within the interference zone. The problematic case is the lack of a read (missed read) by one or both readers. In a study estimating service queue lengths using RFID, Sanders et al. (2008) highlight the extent of uncertainty inherent even in a straight-forward RFID RTLS application.

While both passive and active RFID tracking technology will continue to evolve, and uncertainties in operation and quality will be mitigated, these uncertainties will likely never be completely eliminated. As such, modeling plays an important role in optimizing the implementation of any RFID RTLS patient tracking system. The next section overviews the simulation environment developed for planning and managing a RFID RTLS system.

3.2 Agent based modeling

Agent based modeling has emerged as a simulation technique which attempts to model a system in as much detail as available or as possible. ABM is systems modeling, approached from the ground up or from the perspective of its constituent parts, in order to build an aggregate picture of the whole. Systems are modeled as a collection of agents, their individual behaviours, and their interactions. Agents are autonomous decision-making entities (generally used to model human beings, but can also include inanimate objects) able to assess their situation, make decisions, and compete with one another on the basis of a set of rules. ABM's conceptual depth is derived from its ability to model emergent behaviour that may be counterintuitive or, at minimum, its ability to discern a complex behavioural whole that is greater than the sum of its parts. ABM provides a natural description of a system that can be calibrated and validated by representative experts, and is flexible enough to be tuned to high degrees of sensitivity in agent behaviours and interactions. ABMs are particularly well suited to system modeling in which agent behaviour is complex, non-linear, stochastic, and may exhibit memory or path-dependence (Bonabeau, 2002).

A deterrent to the acceptance of ABM is that, as a technique, it does not lend itself to sensitivity analysis or the modeling of steady state phenomena without a potentially unrealistic number of simulations and substantial statistical analysis being undertaken. Although we agree with that position in general, we also argue that the advantages of ABM complement other techniques. In particular, there is a significant body of literature on mathematical modeling techniques applied to healthcare management, which can be complementary sources of data and cross-validation to ABM (Karnon et al., 2009).

ABM also provides one of the most useful tools available in terms of knowledge transfer and requirements capture, independent of whatever other techniques may be also employed. An ABM has an as close as possible correspondence to the problem as understood by both the practitioner as well as the ABM implementer. An ABM models emergent complex behaviour by specifying rules for simple underlying interactions between agent entities. As such, the resulting model closely resembles the system description, which could come from business rules or some other description by stakeholders who need not be overly familiar with ABM itself.

Many ABMs are developed to gain a better understanding of operations through the use of what-if scenarios. An ABM initially developed for improving patient access to healthcare (Mukhi & Laskowski, 2009; Laskowski et al., 2009) is extended here to allow modeling a RFID RTLS augmented ED, with the objective to gain insights into the nature of error and uncertainty associated with patient tracking. Of primary interest to a RFID RTLS is the accuracy and precision of patient trajectory through the ED. As each agent behaves in an autonomous fashion under the control of the simulator, an exact trajectory is recorded for each patient, which is then available for comparison and validation to trajectory data captured via the RFID RTLS model.

The basic model allows for various configurations, including provisioning the number of healthcare workers (HCW), the number and characteristics of patients, as well as the topography of the ED. Operational parameters in the ED include registration, triage, waiting and treatment areas. The ED floorplan used here closely resembles that of a large, acute-care, metropolitan hospital (Health Sciences Centre in Winnipeg, Manitoba, Canada), although tailoring of floorplans is possible. A schematic of the ABM at an instance of time is illustrated in Fig. 2. In the figure, patients and HCWs are illustrated within the ED with different icons for healthcare workers, and patients at various levels of acuity. This representation also provides a visual animation as the simulation progresses. In effect, the animation is a tool used to transfer information between practitioners and modelers, whereas the extensive collection of data for statistical analysis proceeds without any type of visualization.

The RFID RTLS environment allows for the overlaying or placing of RFID readers which serve as inanimate agents within the ABM. Reader placement can be engineered, or simply distributed in a somewhat systematic manner. For the purposes of this type of simulation, the latter is used to allow for placement roughly based on the density of readers with a specified granularity. In practice, actual reader placement would be subject to access, mounting, and functional constraints, with read patterns also modified as a consequence of the local environment (walls, furniture, equipment, required clearances, as well as people, the latter being time-varying). Tag collision is not considered in this simulation; it is assumed that over the duration of tags being within a reader's range, any colliding or interfering tags would be resolved in a time period considerably shorter than the tags traversing the reader. The resolution of the collision would be a consequence of multiple reads of the close proximity tags. This would also not be an issue with more sophisticated tag technology with collision avoidance (e.g. more sophisticated active tags)

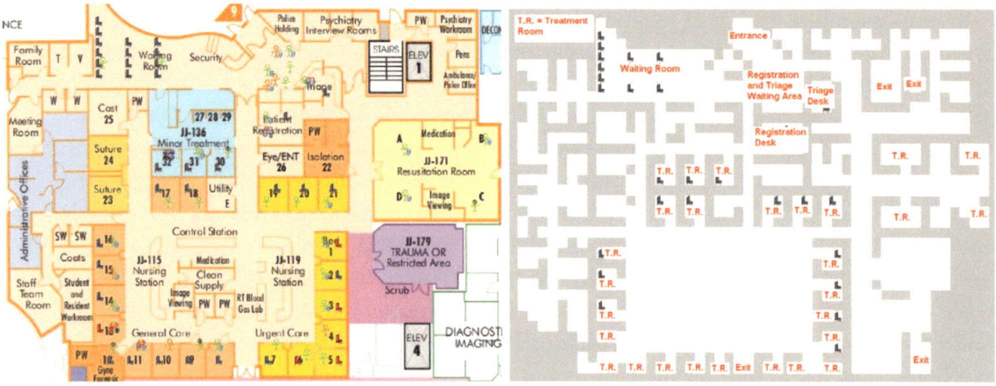

Fig. 2. Emergency department layout (actual and as seen by the ABM)

For the RFID RLTS ABM simulation, the exact trajectories of patients are compared with estimated trajectories inferred from RFID tag reads. Patient trajectories through the ED are largely governed by the triage score, by HCW resources, as well as by issues such as social distancing as aspects of agent behaviour. Several primary cases are considered: under-provisioning, intermediate-level provisioning, and over-provisioning of readers throughout the ED. The primary metric used for determining the quality of the RFID placement is the spatial error in the patients' actual position as well as the temporal error (duration of time when the patient is outside of a reader's range). Economic cost considerations have not been included, although one assumes the cost and maintenance to increase monotonically with an increased number of readers.

Fig. 3 illustrates the model running with a placement of readers illustrated, highlighting a read of a patient tag as the patients traverses the ED. In Fig. 3, patients are moving between registration, triage, waiting and treatment rooms in a hospital ED. The reader placement is seen as a coarse grid of concentric dots and circles, with the radius indicating the reader range. The darkest colored circles indicate that a reader has just read a tag, whereas the lighter colored circles indicate that a tag was read some time previously. Simulation parameters are set to resemble actual records of wait times for EDs of similar resource (number of HCWs, ED physicians, capacity). Although not exact, patient arrival rates and service times are adjusted to reflect times (several hours) spend in ED waiting areas and treatment rooms.

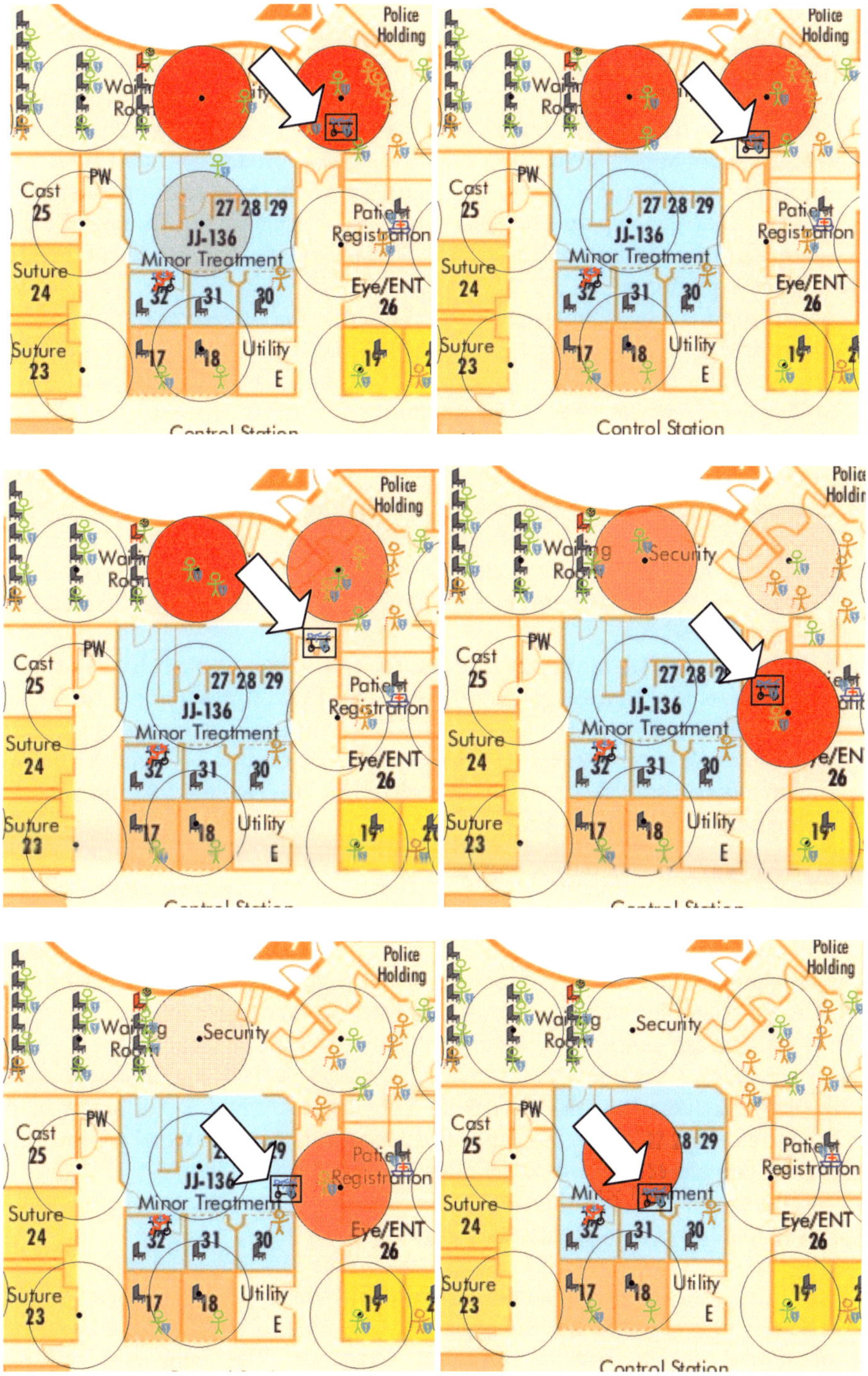

Fig. 3. Consecutive snapshots of patient trajectory (from entrance to treatment room)

3.3 Implementation details

The case study considered a modeling instantiation coded in an Object Oriented Language, C++ using Qt4 libraries running under the Linux operating system. For those interested in saving development costs, commercially available ABM frameworks/toolkits such as SWARM (SWARM, n.d.), Repast (Repast, n.d.), and Anylogic (XJ, n.d.) come with varying out-of-the-box functionalities. This object oriented ABM approach has natural extensions to the sort of spatial modeling called for by the spatial nature of the system under study. It also should be noted that developing an ABM within a object oriented framework from the ground up provides an additional degree of understanding the problem in contrast to using a more commercial platform.

The simulated world is a two-dimensional (X, Y) discrete Cartesian world of extremely high resolution (floating point). An implicit assumption is that RFID tags are always at the reader level, which allows for the elimination of the Z direction. This type of design is also closely related to how a floorplan would be drafted with RFID readers overlaid, making implementation of such aspects more natural. The model is not general purpose, but is a spatially directed ABM aimed at institution-level simulation of a healthcare environment. While a custom ABM generates results that are unlikely to be reproducible because of subtle design differences between the implementations of various researchers, any sufficiently complex system is unlikely to be reproducible in practice. An application-specific ABM is not filled with rarely used or obscure features and design choices to support those features of a more general framework.

Within the object oriented approach, all agents and other entities in the system are implemented as objects with a geographical hierarchical decomposition (world locations in world). For partitioning the world further, there is a world grid in locations; however, the underlying space is still "continuous" (floating point specification of X and Y coordinates). The simulation loop calls a tick() method, the semantics of which are that the object (agent) is getting their slice of simulation time where they can react to other agents around them. Agents can access a limited number of features of other agents and objects, which can be set by the programmer, depending on the model. They can also pass messages to other agents in order to achieve interaction. For example, they can sleep until bumped or acted upon. Another difference between this and other modeling approaches is the lack of a scheduler which drives the simulation. Agents spend a high proportion of their time reacting to one another, that the scheduler would effectively be running each agent at every time step. For example, a patient can assess the number of patients waiting before them, and decide whether or not to leave the ED without treatment, based on their condition and a model of how long they will have to wait. We acknowledge that a scheduler or script may be of benefit to the efficiency of our custom ABM, especially when extending the model to simulate more complex time-of-day variation of patient arrivals and staffing schedules. Partitioning the space of the world, has the tendency to reduce the number of agents that each agent has to interact with at each time step, speeding the simulation.

Fig. 4 illustrates an object inheritance model that closely resembles the model used in this study. The RFID RTLS model builds upon an ABM that was initially developed for studying patient access (wait times) and was subsequently extended for application in modeling the spread of nosocomial or hospital acquired infections. The bold text illustrate the modifications to extending the ABM to modeling the performance of RFID RTLS in an ED. In addition to the RFID reader class, extensions were also made to support the data collection in terms of the error model.

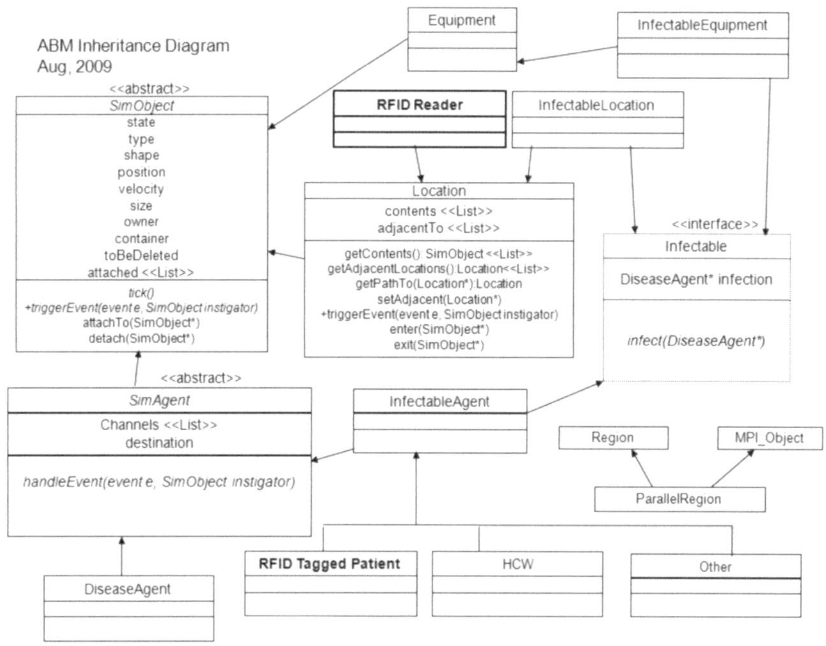

Fig. 4. ABM inheritance diagram with RFID extensions

Each agent acts in a set, predictable, but arbitrary order. That is, agents and all other objects are simulated in the order in which they were added to the simulation governed by the arrival rates of patients of various triage scores. Time is discrete with each time step having a resolution of 1 second (1 time step or tick = 1 second). Patients are modeled as occupying a circular space with a radius of 60 cm, representative of their physical person as well as a concept of personal space.

Using this technology, we model a hospital ED as a collection of interacting agents. Patients arrive and require treatment, and are tagged with an RFID tag upon entering the ED. Nurses and other agents that decide upon treatment paths for patients, for example an erController, direct patients through the system. Doctor agents treat patients. RFID readers are implemented in a similar way to agents, as a class extending the location class which read the tags of every patient that enters their reading radius.

Agents practice rudimentary path planning, typically choosing the shortest path to the destination location (e.g. travel from waiting room to treatment room). The agents also selectively practice microsocial distancing, where they will tend to distance themselves according to their local density. These are examples of localized agent decision making.

A simulated RFID reader notes when an agent (patient) enters its reading range, a roughly circular region with occlusions primarily due to walls. The RFID model requires a line of sight between the reader and tag; walls provide 100% attenuation in the simulation. The model can be modified to allow for propagation through walls, but would require an attenuation model to account for a reduced range in the previously occluded region. Furthermore we model regions of interference where multiple readers' ranges overlap. We employ a placement algorithm which attempts to place readers approximately equally spaced in a grid throughout the modeled ED floorplan. (The ABM also allows for readers to be placed manually as would be the case in an actual ED and more detailed planning). A heuristic places the reader on either side of a wall if the equally spaced grid placement algorithm should place the reader inside a wall. We call this a naïve grid placement strategy. Longer term goals of an ABM would be to further develop the feedback loop of an optimizer adjusting reader locations in a manner to minimize uncertainty. In doing so, one would also have to model the actual RF environment via a site survey, and model the performance of the RF system taking into account interactions in a system with a plurality of patients, visitors, HCWs, as well as portable equipment.

We do not receive a signal or notification when the tag leaves a reader's reading radius. Furthermore, we cannot expect better resolution from this system than the error resulting from the reader's own radius. That is, we know that a patient entered the radius of a circular reader area, and that the patient is known or thought to be in that particular reader's range until another reader event trigger. Therefore, we define error as 0 if read by a reader and the tag is within that reader's radius (line-of-sight does not need to be maintained). In general, with a tag within a reader's range, the tag may be read many times. These multiple reads in effect confirm the patient's proximity and are implicitly accounted for here. This is an optimistic error model as when an agent is within a read radius, the contribution to spatial as well as temporal error is zero.

Since ABM is very compute intensive, parallelism should be used wherever possible. Parallelism was implemented during the collection of data, as the ABM is tailor-made to not only exploit fine grained parallelism, but also exploitable at the coarse grain level (process). The latter parallelism was exploited here where individual ABMs would be run on a small compute cluster.

4. RFID patient tracking ABM simulation results

4.1 Variable reader configurations

The simulation parameters are summarized as follows:

- The emergency department layout is drawn on a 1464 by 2001 pixel grid, corresponding to an area of approximately 30m by 40m. While the scaling is not precise, it is chosen to be a reasonable representation.
- The RFID reader range was chosen as a circular region of 100 pixel radius (approximately 2m), roughly corresponding to the approximate reading radius of many newer passive readers.
- RFID reader spacing was varied from 100 pixels to 350 pixels between readers, in increments of 50 pixels. This created six reader configurations, hereafter noted as 100, 150, 200, 250, 300, and 350, respectively, and shown in Fig. 5. Readers are again indicated by black dots, with their ranges indicated by circles. Readers in the process of a read are highlighted. The placement scenarios are not optimal but are an attempt to capture patient flows through major traffic routes in the ED modeled.
- At each of the six reader configurations, the simulation was run 500 times, with roughly 200 patients visiting the ED in each simulation.
- The variable of interest in the simulations was uncertainty or error, defined as both spatial error and temporal error. The measure of spatial error used in this model is the difference in trajectory inferred from the readers, relative to a patient's actual position as known within the simulation. As such, when a person is recorded within the range of a reader, the spatial error would be zero. As the patient moves beyond the range of the reader, the spatial error is the Euclidean distance from the reader to the actual patient

position. The temporal area in this model is defined as the percentage of time that a patient is not within the range of a reader.

Other measures of error, such as Manhattan distances would be equally reasonable and well within the modeling alternatives. Once a patient moves within the range of another reader, both the instantaneous spatial and temporal error would return to zero. A complication occurs when modeling interfering readers (an over-provisioned scenario). Here a dead zone is modeled, reflecting the overlap of two readers' ranges. The difference to the error measure is that the read may be delayed as an agent enters a reader range which was reduced by a dead zone, contributing a longer period of proximity uncertainty or error.

Fig. 5. RFID reader configurations (top left to bottom right: reader spacing at 100, 150, 200, 250, 300, 350 pixels).

Fig. 6 illustrates the typical spatial error between reader-inferred location and actual location for a single patient, plotted over the duration of stay in the ED, for the six reader configurations. The trajectory error illustrated is for a the reader arrangements seen in Fig. 5 and for typical patient instances. The behaviour of the patient is highly stochastic, governed by a number of random variables. In Fig. 6, the spatial error is plotted as being the Euclidean difference between the last read reader and the patient location once outside the read range. When a patient is within a reader's range the error is set to zero.

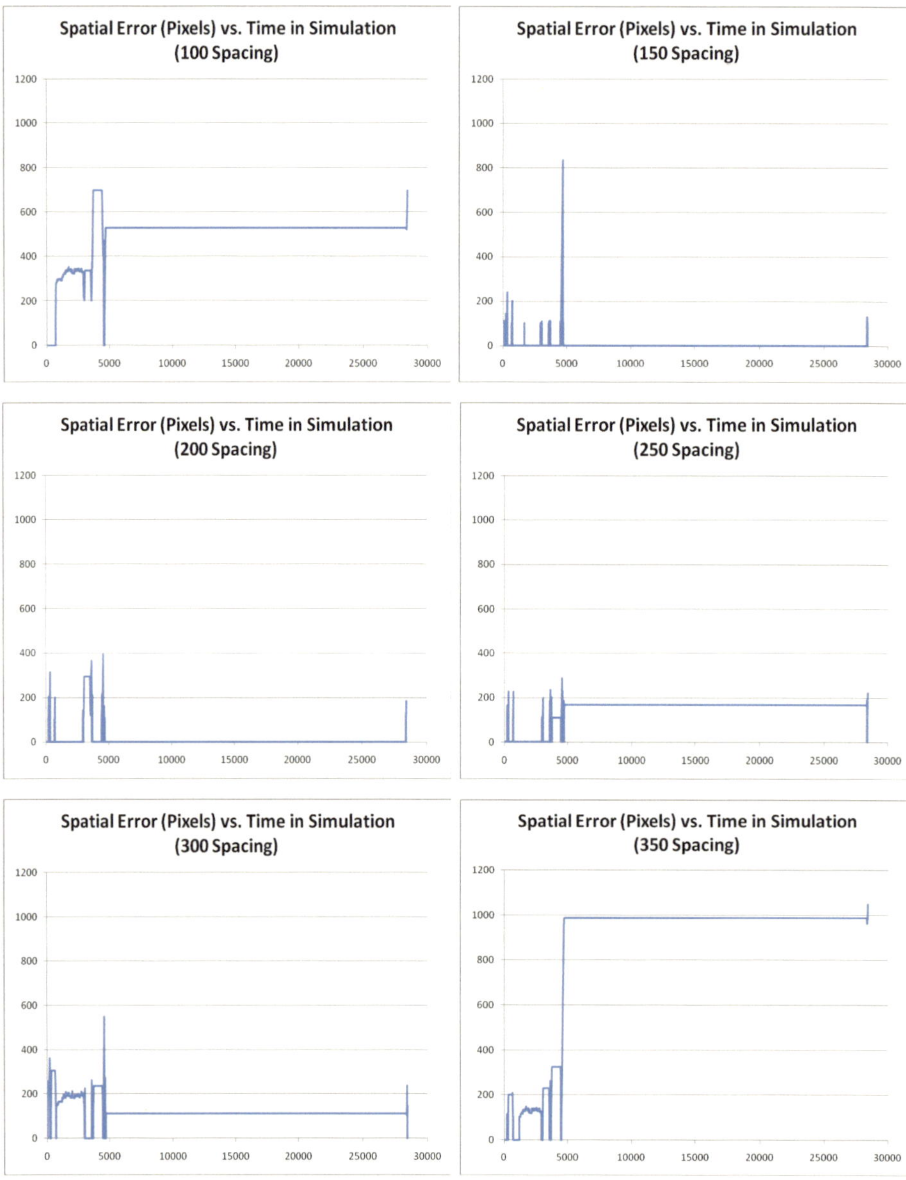

Fig. 6. Spatial error between RFID reading and actual patient location, for the same patient at each reader configuration

Fig. 7 and Table 1 summarize the normalized error for both spatial and temporal error. For example, in Table 1, (reader spacing of 350 pixels (sparse readers)), the patient would be in the read range of a reader 25% of the time, with a normalized error of 237 pixels or approximately 5 m. As the number of readers increase, the error or uncertainty is clearly reduced. The exception seen in Fig. 6 and Fig. 7 is the case of the reader configuration at 100 pixels, i.e. closely spaced readers with significant interference zones. Once a patient has been read by a reader, the patient is not considered to be in error if the patient moves out of the active read range and moves through or stays (waits) in an interference zone of that particular reader. However, if a patient moves into an interference zone of a reader and stays (waits) there, the patient is considered to be in error unless and until they move into the active read range of that reader and is read.

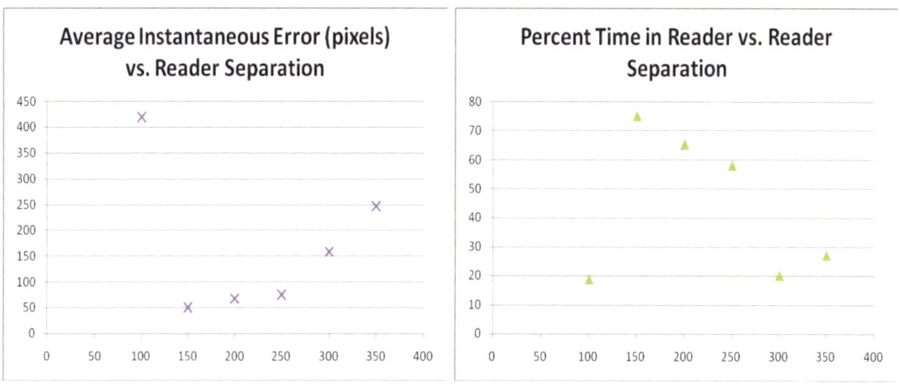

Fig. 7. RFID location errors (spatial and temporal)

Reader Separation	Probability of being in reader range	Variance of probability	Normalized Spatial Error (Pixels)	Variance of Spatial Error (St.Dev.)
100 (dense)	0.17 (17%)	0.08	407	38199 (195)
150	0.77	0.10	45	6212 (78)
200	0.64	0.11	71	6787 (82)
250	0.59	0.12	76	6225 (78)
300	0.23	0.08	156	6938 (83)
350 (sparse)	0.25	0.09	237	31099 (176)

Table 1. Summary of RFID location performance (spatial and temporal with variance)

One of the most interesting aspects of the simulation is the percentage of time that a patient would be in range of a reader. From these simulations, in an under-provisioned system, the patient's position may only be known approximately 25% of the time, whereas for the over provisioned case the patient's position is also known approximately 20% of the time. This highlights again the importance of strategic installation locations for the readers, to capture the major traffic flow patterns through the topography, taking into account that one can live with a high degree of uncertainty in certain areas with less in others. Depending on the RFID model, there likely is a near optimal configuration of readers that will minimize the uncertainty associated with an RFID RTLS.

It should be noted that although the uncertainty is generally decreased with an increasing number of readers, there is a point of diminishing return. Although not linear, the cost of provisioning and maintaining the ED with readers at 150-pixel separations is significantly greater than an ED with the readers at 250-pixel separations. In addition, we did not consider any inference of behaviour as a consequence of the location of the tag being read. For example, if the last read reader was located in the waiting room, one can likely infer that the patient is still in the waiting area until another reader reads the patient's tag.

4.2 Variable reader ranges

The case study further examined the performance of the readers in terms of their read radius. Readers and associated technologies tend to improve the range of RFID systems over time, whether passive or active. In this simulation, the readers' read areas are either decreased of increased by a factor of two from the baseline configurations outlined earlier.
The best case scenario from the previous placement of readers is denoted as 250, i.e. the readers being roughly spaced 250 pixels or approximately 5 meters apart. This best case scenario is argued as function of reader error and diminishing return if more readers are deployed (Fig. 7). Based on other metrics (Fig. 6), one could also argue that the reader configuration at 200-spacing is a best case scenario. As is typical, implementation-specific priorities and their associated metrics (cost, performance, etc.) define a best-case scenario for a given user. This section illustrates other variations of reader capability that are also easily modeled within the ABM framework. Fig. 8 illustrates the case where the reader range

Fig. 8. RFID coverage (red) and interference (blue) areas for small, medium, large coverage areas.

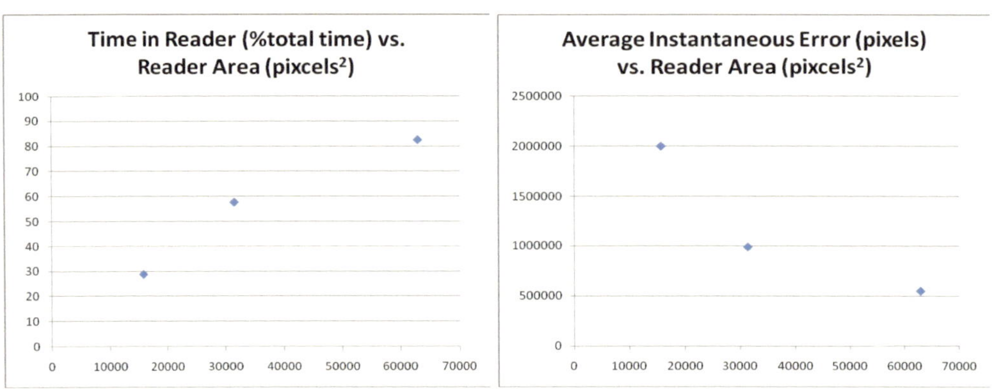

Fig. 9. RFID range performance (spatial and temporal)

(coverage area) is reduced or increased by a factor of ½ or two respectively. The middle illustration of Fig. 8 is the same as the previous case with readers spaced 250 pixels apart having a read radius of 100 pixels. An identical number of simulations were performed with the results presented in Fig. 9. Here one can see that the case of readers spaced at 250 pixels, with a 100 cm (1m) read range is likely reasonably close to the "near optimal" configuration. The trade-off again would be diminishing return if one were required to invest in the large coverage readers, as cost tends to increase with read range.

4.3 Error interpretation

It is difficult to illustrate explicitly the effect of the various errors seen with an arrangement of readers. It is however enlightening to attempt to graphically illustrate why some reader placements result is large error in estimating the patient's location. Fig. 10 illustrates the coverage areas, blind spots, and regions of modeled interference. Red illustrates coverage area, with blue representing regions of interference (the over provisioned case). Ideally the reader would have a circular red read area. Degraded read regions would be a result of walls and obstructions where the tags would not have a clear path to the reader and thus, in the passive case, be unable to acquire sufficient energy to function. Degraded performance also occurs in regions of reader interference where the readers are too closely placed. From this perspective, it is relatively easy to identify poorly performing patient trajectories. For example, from this illustration and observation of patient trajectories, one can envision a patient making their way to a treatment room not equipped with a reader and waiting there for treatment for a considerable period of time. In these cases (which may be pathological) the error can be considerable over hours of wait time. In Fig. 10 the arrows illustrate a path a patient may traverse to a treatment room (left). Overlaying the view seen by the ABM (right), one can see that the patient trajectory traverses a region of considerable interference (horizontal arrow), and ending at a treatment room without read coverage (occluded due to walls).

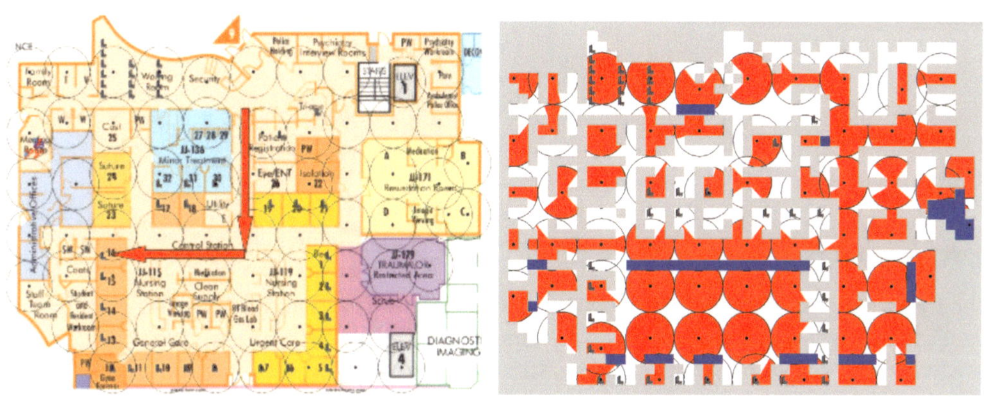

Fig 10. Trajectory of outlier patient with high temporal and spatial error (left), reader coverage (right), interference regions in blue, readable areas in red

Fig. 11 illustrates the difficulty with simply increasing reader density if interference is a problem. Red illustrates the percentage of floor space covered by readers while blue illustrates regions of interference. This corresponds to the high error seen in Figures 6 and 7 in the case of reader spacing at 100 pixels (over-provisioned), where interference is modelled as error.

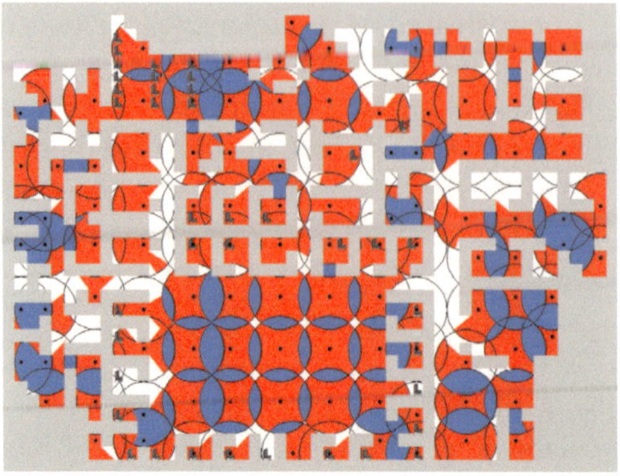

Fig. 11. Dense reader spacing leading to increased interference regions.

There are techniques (beyond the scope of this chapter) for reducing the error associated with estimation of patient trajectories. The simplest are those based on the Kalman algorithm as well as others based on conditional probabilities (Bhatia et al., 2007). Fig. 12 illustrates a histogram of a passive tag's maximum read distance from a GAO GP-90 passive reader using clamshell tags. The data was collected under ideal conditions and as such represents a bound to error anticipated in the read itself. It does however lend credibility to the ABM with fairly abrupt read ranges simulated. Any backend system deployed for collection of RFID RTLS data would presumably be also equipped with estimation software to further reduce error.

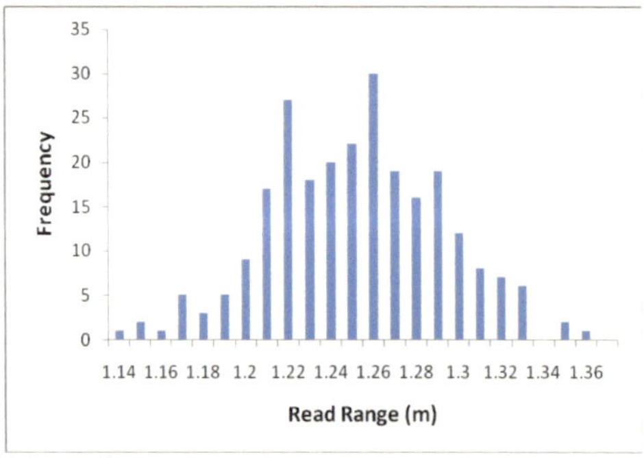

Fig. 12. Histogram of GP90 read range.

5. A simple prototyped RFID RTLS

The RFID RTLS application modeled using an ABM approach may be a conventional RFID application in healthcare, yet its implementation demands strategic consideration of hardware and software details and integration. This chapter closes with a brief discussion of a desirable architecture, emerging best practices, and available technologies to create an RFID RTLS using Supervisory Control and Data Acquisition (SCADA) loosely based on Service Oriented Architecture (SOA). The prototyped framework represents an integration of technology, architecture, software and RFID devices that leverages existing software systems, with an emphasis on being IP-centric and application- and device-agnostic.

The basic notions are derived from the requirements of a framework incorporating the best practices from existing data acquisition systems as well as underlying communication systems. The current generation of RFID RTLS are purpose-built hierarchical network architectures and tend not to leverage more modern advances in SOA. SOA is meant to overcome the difficulties of creating distributed applications by making the data and programmatic communications between entities over a distributed network transparent to the developer. The overall architecture of an RFID RTLS is illustrated in Fig. 13.

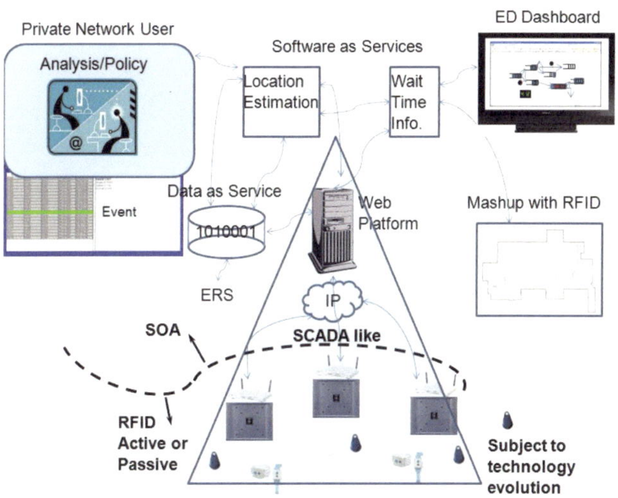

Fig. 13. Overall SOA SCADA RFID RLTS architecture

5.1 RFID RTLS system components

The prototyped RFID RTLS installation uses the principles of a SCADA system. The primary objective of SCADA is to remotely monitor and/or control remote devices over a communication channel and present the data back to the user in a friendly manner. Another possible feature of SCADA is the capability to not only acquire data but also to record it for future analysis. Furthermore, SCADA systems can log information about the status of the remote equipment. As such, SCADA is well suited as a part of a data collection platform for an RFID RTLS. A minimal SCADA system is composed by at least one Master Station (MS) computer, at least one communication channel to the Remote Transmit Units (RTU), and the Remote Transmit Units connected to a sensor or device that collects data.

Typical industrial applications of SCADA include system control of electrical networks, gas pipelines or weather stations across a region or country. The prototyped application incorporated SCADA technology with an improved architecture, thereby leveraging much of the legacy of the data acquisition and industrial proven stability of SCADA systems. While SCADA systems are well suited to data collection, their evolution to networked applications has been slow. SCADA applications lag well behind many internet applications which typically do not require any type of interface to data collection systems but more typically are terminated at a terminal or computer with a human user interface. It is the integration of SCADA and SOA that is compelling as an architecture for an RFID RTLS.

The prototyped application collects the location of the traceable individuals (patients) and equipment in a personal care home setting, with each element equipped with an RFID tag. The system stores the information relative to the time and location of the last identification of a particular element. The definition or granularity of the system depends on the number

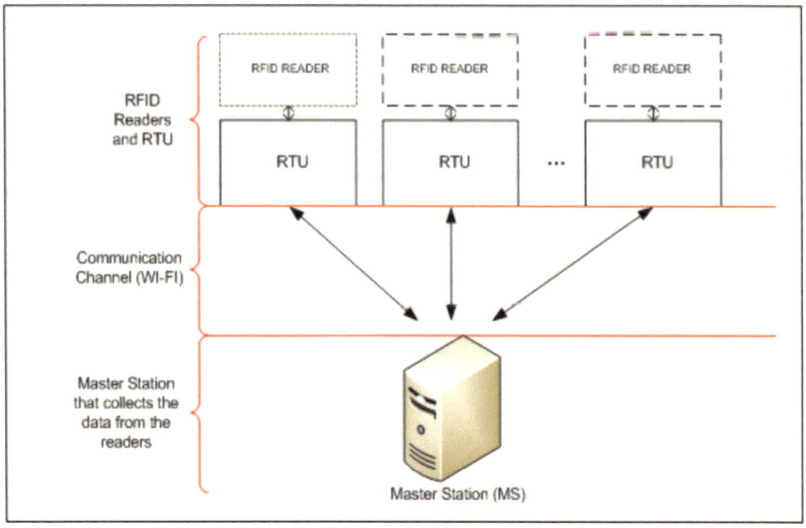

Fig. 14. A Minimal SCADA system (IEEE, 1994).

of RFID readers installed, each equipped to send the data to the MS. Each reader is equipped with an RTU that interfaces the serial communication port from the RFID reader to the existing wi-fi in the building. Finally, data is collected by a computer and presented to the user using a simple web interface or GUI application.

Ideally RFID readers should be installed in different locations within a building in a somewhat optimal configuration, guided by planning tools such as the ABM discussed in previous sections. The system should have as many sensors as the level of accuracy and precision desired in the positioning and tracking within a building. When an element with a tag is in the read range of a reader, the reader will detect the tag ID and it will transmit it to the MS. The readers used here were the GAO GP90 readers for passive tags. The tags used were the long range (1-2 m) passive clamshell type.

The communication module is responsible for sending and receiving the data to/from the readers and the MS. The options to establish a communication channel between the readers and the MS are multiple. The one selected for this implementation is wi-fi (802.11). The reasons for the decision are: the wi-fi infrastructure exists already in the building where the RTLS will be installed; cost savings in the installation because there is no need to install single cables per reader; considerable bandwidth to transmit tag IDs; and, open standard with low or no interference (well developed medium access control protocols).

Each reader has a RTU built with two ports: a serial port to communicate with the RFID reader and a built-in wi-fi connection to communicate to the remote system. The module used is the RCM5400W from Rabbit Semiconductors. This module has a Rabbit 5000 series microprocessor. The software can be easily written in Dynamic C, based on ANSI C. The small size of the board make is simple to mount in a small enclosure or even within the RFID reader.

In the development of the prototype, the MS was assigned a pre-defined and fixed IP address. Each RTU was assigned a unique non routable IP address from the MS which served as a DHCP server. Ideally as part of a networked environment, the MS would also be assigned a dynamic IP address. The RTU serves as an interface between the RFID reader and the wi-fi communication channel. The RTU has two functions: 1) send the data to the

MS addressing the packages to the MS IP, and 2) the MS can PING each RTU to know if it is live or not. The other elements of the wi-fi network are the access points necessary to carry the signal to each RTU and the wi-fi board in the MS. The overriding principle is that even as the physical layer RFID technology changes, much of the backend can remain the same.

The SCADA software is an innovative approach based on years of in-field use and research. The SCADA software typically has three modules. The modules can run on the same or different systems based on the load and resources available. The first module is the communication module. This module handles the communications between the MS and the RTUs. Each time a RTU has data to send, the communication module is responsible for receiving and storing it in a database. Also, this module checks that the wi-fi network is working and that each RTU is live every preset number of seconds. The second module is the database module. This module receives the data from the communication module and stores it in a database. It also stores configuration information for the system to run. The third module is the presentation module. This is the module allows the user to see the location of each tag associated with an individual or equipment, in a user-friendly manner. There are two possible alternatives: 1) a web based application; or 2) an application that can be installed. Each module has a specific task, and the interaction between them facilitates the system functionality.

To interact with one another, each module is implemented as a SOA. There are multiple definitions for SOA, but the OASIS group (2006) defines SOA as a paradigm for organizing and utilizing distributed capabilities that may be under the control of different ownership domains. It provides a uniform means to offer, discover, interact with and use capabilities to produce desired effects consistent with measurable preconditions and expectations.

Each module has functions that can be called or invoked by remote modules. For example, the presentation layer can ask the communication module for a location for a particular individual or equipment. Because the communication module remembers the last reading, it will provide the information without need to query a database. In the same way, each time a reading arrives at the communication modules, it stores the information into a database, calling a database storage procedure located in the database SOA module. Each module is independent of one another and each can be local or remote. However, reliable communication links are necessary for proper functionality.

In general, the prototyped application is built on a modularized approach, with each module governed by modern SOA principles. To a large degree, the modules are independent of one another, allowing for demonstrable rapid prototyping, an emphasis on reuse, and the utilization of existing infrastructure. Although this application does not explicitly utilize a formal SOA, it illustrates many of the benefits. Specifically, it is - to a large degree - access technology agnostic: wi-fi or cellular can be replaced without interfering with the functionality of other modules. It is also largely RFID technology agnostic: the technology or vendor can be interchanged with little code modification. For example, the transition from a passive to an active tag can be relatively easily accommodated. The SOA layer at the application level gives each module independence and flexibility. Moreover, it allows the presentation layer to access information about the readers through data services. As algorithms are developed for improved estimation, for example, the location estimation module would be replaced or modified without further disruption. Additional data must also be supported in a flexible manner. For example, if parameters associated with signal reception strength were to be used as input to an estimation module this should only be a change in the XML that is used to support the data encapsulation. An additional benefit of an SOA for an RFID RTLS within a healthcare facility is that these systems are required to integrate or dovetail with other healthcare systems at some level, be it other RFID applications or the existing hospital electronics records system. Although not described in detail here, it should be noted that as a consequence of being heavily IP-centric, all of the existing security advancements of the IP community can be readily deployed here.

6. Conclusion

RFID based RTLS systems hold great potential to play a significant role in augmenting electronic records systems in healthcare. However, they are not without their limitations. One of the more challenging obstacles to overcome is dealing with the inherent uncertainty of relatively low cost RFID systems, as illuminated by the modeling and simulations in this case study. As such, the agent-based modeling techniques are particularly well suited to modeling RFID-enhanced healthcare systems. In mitigating uncertainty, some backend processing is required. As RFID systems continue evolve and more sophisticated tags are deployed, it is reasonable to expect additional levels of physical layer triangulation to be employed, as well as more intelligent protocols that may also query tags in an attempt to yet further reduce location estimation error. Many of these issues will be resolved in evolving medium access control protocols as well as at the application layer for the RFID RTLS system. Nonetheless, this chapter illustrates the role that ABM has in provisioning a RFID based tracking system, and further modeling extensions to refine the characteristics of the topography and agents will enhance the results.

7. Acknowledgements

The authors thank the Thorlakson Foundation for financial support, as well as Dr. A.S. Alfa, NSERC Industrial Research Chair in Telecommunications at the University of Manitoba. We also thank Dr. S. Mukhi at the Canadian Network for Public Health Intelligence for initiating the agent based model for patient access.

8. References

Ashar, B.S. (2007). Radiofrequency identification technology in health care: Benefits and potential risks. *JAMA*, Vol. 298, No. 19, pp. 2305-2307.

Bacheldor, B. (2006). Medical-tube couplings get RFID technology. *RFID Journal*, (June 13, 2006).

Bearing-Point Inc. (2006). *RFID in Healthcare: Poised for growth, Point of view: Technology infrastructure & integration, emerging technology*, pp. 8.

Bhatia A.; Mehta B. & Gupta R. (2007). *Different Localization Techniques for Real Time Location Sensing using passive RFID*, http://kict.iiu.edu.my/integ/docs/LocalizationTechniques.pdf

Bonabeau, E. (2002). Agent-based modeling: Methods and techniques for simulating human systems, *Proceedings of the National Academy of Science* [Online]. 99 (Suppl 3), pp. 7280-7287. http://www.pnas.org/content/99/suppl.3/7280.full#xref-ref-3-1

DuBois, J.A.; Dunka, L; Allred, T. *et al.* (2001). Point-of-care connectivity; approved standard. *POCT1-A, NCCL Exec. Off.* (CIC): Wayne, PA, USA, Vol. 21, No. 24.

Ehrmeyer, S.S.; Hausman, P. & Lebo, R. (2005). Using technology to improve patient safety at point of care. *Point of Care Testing Journal*, Vol. 4, No. 4, (December 2005), pp. 146-149.

Finkenzeller K. (2003). *RFID Handbook: Fundamentals and Applications in Contactless Smart Cards and Identification*, 2nd edition. John Wiley & Sons, Inc.

Glabman, M. (2004). Room for tracking: RFID technology finds the way. *Materials Management in Health Care*, Vol. 13, No. 5, (May 2004) pp.26–38.

Kanyuk, P. & Young, J. (2004). *RFID in Healthcare: Novelty or Mass Market?*, Report Ref. Code: BFTC1095, Datamonitor USA.

Karnon, J.; Mackay M. & Mills T.M. (2009). Mathematical modeling in health care, *18th World IMACS / MODSIM Congress*, http://mssanz.org.au/modsim09

Laskowski, M.; Demianyk, B.; Friesen, M.R. & McLeod, R.D. (to be published). Modeling an RFID tracking system in an emergency department. *IEEE Workshop on Healthcare Management*, Venice, Italy, February 2010.

Mukhi S. & Laskowski M. (2009). Agent-based simulation of emergency departments with patient diversion, in *Electronic Healthcare*, D. Weerasinghe, Ed. Berlin: Springer, 2009, pp. 25-37.

OASIS Consortium. (2006). Reference model for service oriented architecture *1.0 OASIS Standard*, 12 October 2006 [Online]. Available: http://docs.oasis-open.org/soa-rm/v1.0/

Podaima, B.W. (2007). Smart Medical Compliance Method and System, *Patent: WO/07/041843*.

Podaima, B.W. & McLeod, R.D. (2006). Point of care engineering and technology - An overview. *29th Conference of the Canadian Medical and Biological Engineering Society (CMBE29)*, Vancouver, BC, Canada, No. CE-2.5, June 2006.

Rao, A.C. & Dighe, A.S. (2004). Radiofrequency identification and point-of-care testing. *Point of Care Testing Journal*, Vol. 3, No. 3, (September 2004), pp. 130-134.

Repast; Recursive Porous Agent Simulation Toolkit; http://repast.sourceforge.net/index.html

Roberts, A. (2005). Point-of-care IT: improving patient care. *Enhancing Clinical Care*, http://www.HCTProject.com, pp. 352-354.

Ross, P.E. (2004). Managing care through the air, Aging + Technology. *IEEE Spectrum*, (December 2004), pp. 26-31.

Sanders D.; S. Mukhi M.; Laskowski M.; Khan M.; Podaima B.W. & McLeod R.D. (2008). A network-enabled platform for reducing hospital emergency room waiting times using an RFID proximity location system, *19th International Conference on Systems Engineering*.

Solanas, A. & Castellà-Roca, J. (2008). RFID technology for the health care sector. *Recent Patents on Electrical Engineering*, Vol. 1, No. 1, (January 2008), pp. 22-31.

SWARM; http://www.swarm.org/index.php/Main_Page

The Institute of Electrical and Electronics Engineers, Inc. (1994). IEEE standard definition, specification, and analysis of systems used for supervisory control, data acquisition, and automatic control, *IEEE Standard C37.1-1994*, pp 10-44..

Wang S.W.; Chen W.H.; Ong C.S.; Liu L. & Chuang Y.W. (2006). RFID application in hospitals: A case study on a demonstration RFID project in a Taiwan hospital, *Proceedings of the 39th Annual Hawaii International Conference on System Sciences.*

Wicks, A.M.; Visich, J.K. & Li, S. (2006). Radio frequency identification applications in healthcare. *Int. Journal of Healthcare Technology and Management*, Vol. 7, No. 6, pp. 522-540.

Witters, D. (2009). *Medical Devices and EMI: the FDA Perspective.* http://www.fda.gov/cdrh/emc/persp.html

XJ Technologies; AnyLogic, http://www.xjtek.com/

15

RFID-based Disaster-relief System

Osamu Takizawa
National Institute of Information and Communications Technology
Japan

1. Introduction

In the 1995 Great Hanshin-Awaji Earthquake, bills were widely posted as a means of distributing information in the disaster-afflicted area. Resident safety and evacuation information and emergent risk assessment results were directly posted on damaged buildings, and in this manner information was communicated within the disaster area. This example highlights the importance of establishing a system that can be used in rescue operations for the rapid collection of information scattered throughout an affected area; such a system could rely on manpower, rescue robots, or other elements that are independent of existing means of communication. This feature would prove particularly important in the event of a large-scale disaster that would likely cripple the communication infrastructure.

An RFID tag (for Radio Frequency Identification, a type of electronic tag) is a small device that can store, input, and output data through non-contact means. In addition to its wide use as a non-contact IC card, the RFID tag is on its way to becoming commercially feasible for attachment to merchandise and cargo in the logistics industry. In addition to logistics applications, a range of uses in other fields, including firefighting and disaster prevention, were recently highlighted in a report by a study group (MIC, 2004Mar) organized by the Ministry of Internal Affairs and Communications.

Given the anticipated arrival of a "ubiquitous information society" in which RFIDs are embedded in large quantities in house walls, and in traditional utility poles along the road for normal use, the authors are moving ahead with development of an RFID writer/reader designed to write or read rescue-related information to or from an RFID. This device could serve as an information resource for rescue work in the disaster area and allow disaster victims or rescue workers outside of the disaster area to collect needed information instantaneously, in large quantities, and in a non-contact fashion, informing those outside the afflicted area of the conditions within the region. This chapter describes such RFID writer/reader, followed by a discussion of an information sharing system using the device and thoughts on the further potential of an overall damage information collection system.

2. Development of an RFID writer/reader to collect damage iInformation

This section outlines the RFID and provides a description of the development to date of the RFID writer/reader device based on anticipated application to the collection of damage information.

2.1 Outline of the RFID (Nebiya & Uetake, 2003)

The RFID consists of a small IC (Integrated Circuit) chip capable of storing information and responding to commands from an interrogator (the writer/reader), and a metal antenna. The interrogator can read information stored in the RFID in a non-contact fashion using electromagnetic waves or through electromagnetic induction. Fig. 1 presents an example RFID.

Fig. 1. Example RFID

The basic IC is 0.1 to several square millimeters, with its own storage capacity ranging from around ten bytes to several tens of Kbytes. The IC also has memory and logic circuitry as well, allowing it to perform processes such as computation, authentication, and encryption. In Japan, two frequency bands—13.56 MHz and 2.45 GHz—are the main radio frequency bands assigned to RFID. The maximum communication distance between the RFID and interrogator is roughly 70 cm for the 13.56-MHz band and roughly 1.5 m for the 2.45-GHz band, in accordance with regulations under Japan's Radio Wave Law. In Japan, frequencies from 950 MHz to 956 MHz have been available for use in 2005, which has enabled the design of RFIDs capable of communication over longer distances.

RFIDs are classified into two types: an active type incorporating a battery, and a passive type that does not require a battery. A passive RFID modulates the carrier wave sent from the interrogator with the information written in the RFID's storage area and returns the signal to the interrogator, transferring the information. With a rectifying circuit in the antenna, this type of RFID receives the power to reflect the signal by rectifying the electromagnetic wave received from the interrogator. While in areas such as logistics, the passive type is the mainstay device due to its lower cost and maintenance-free design, battery-driven RFIDs can extend the distance of communication with the interrogator and can actively transmit information—functioning as a beacon, for example.

An additional type of RFID has been developed that allows not only reading of data previously written to the RFID (as in read-only devices) using the interrogator, but also permits write operations to the RFID using the same interrogator.

RFIDs are characterized by better reading efficiency per unit time (relative, for example, to barcodes) thanks to the ability to read multiple RFIDs at once (so-called "multi read"). Moreover, the manufacturing cost of a single RFID has dropped to approximately several tens of yen, paving the way for further cost reduction through increased production volumes.

2.2 Outline of the RFID writer/reader under development

The authors have performed on the development of a 2.45-GHz passive RFID writer/reader. We have already developed three types of writer/readers: a cart-mounted device, a backpack-mounted device, and a handheld device.

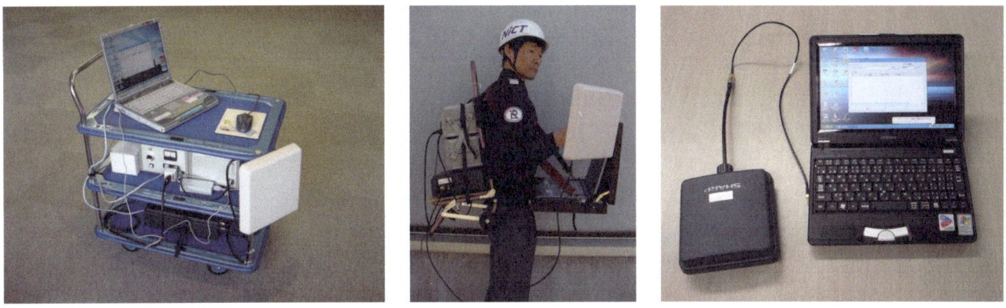

Fig. 2. RFID Writer/Reader for Collection of Damage Information
(Left: cart-mounted type, Center: backpack-mounted type; Right: handheld type)

Designed to be battery-driven and portable under the assumption that it will be carried into disaster-stricken areas, the writer/reader under development is comprised of an antenna section, handling write and read operations to and from an RFID; a main body; a notebook PC controlling the main body; and batteries supplying power to these devices. While some RFID writer/readers are already commercially feasible for use as hand-held inventory terminals, most such terminals can only read an RFID upto several centimeters away, as with barcodes. However, assuming the necessity of reading a difficult-to-reach RFID in the event of a disaster (such as one buried under rubble), extending the readable distance is an obvious necessity. The possible communication distance of the authors' system is roughly 2 m at present. At the beginning of the development process, a high-output stationary writer/reader requiring a private radio station license was modified and rendered portable by adding batteries, in an attempt to secure the longest readable distance with a passive RFID available in Japan today with a portable device. However, since reduced size is critical for mobile activities in disaster areas, a low-output device was adopted, at the expense of readable distance. We should note here that although they allow extension of readable distance, active RFIDs require periodic tag-battery replacement. This poses the difficult challenge of replacing large quantities of tag batteries to prepare for a disaster that could occur at any time. Further, many RFIDs are read-only (that is, they send only a fixed tag ID), rendering them unsuitable for collection of damage information.

RFIDs are generally used in a client/server configuration in which the RFID is commonly employed as an identifier (ID), and the server retrieves information from a database under its control via a network using the read ID as a key. The authors' system, in contrast, is designed to use the RFID for data storage, with all necessary information written to the RFID based on the assumption that the client/server system will not function at the time of a large-scale disaster.

The following sections provide an outline of developments.

2.3 Cart-mounted type

The cart-mounted type has the following basic functions in writing and reading information to and from the RFID:

- Writing Japanese character strings to a single RFID (simplified write function)

- Reading Japanese character strings from an RFID and saving these to a control PC (read function)
- Voice synthesis of Japanese character strings read from an RFID in real time (read-out function)
- Automatic location of an empty tag among multiple RFIDs and writing of information to the tag (write function)
- Clearing a tag to an empty state by deleting read data from the RFID (retrieval function)

Early in this development, we restricted the data to be exchanged to Japanese character strings (text data), assuming that damage information would be written and read in natural language. Fig. 3 illustrates an example of a screen for the simplified write function, and Fig. 4 shows an example of a screen for the read function.

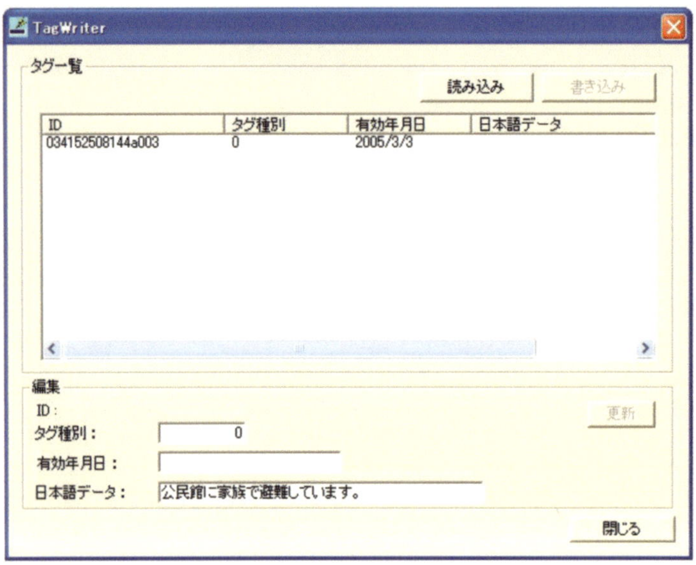

Fig. 3. Screen Example for Simplified Write Function

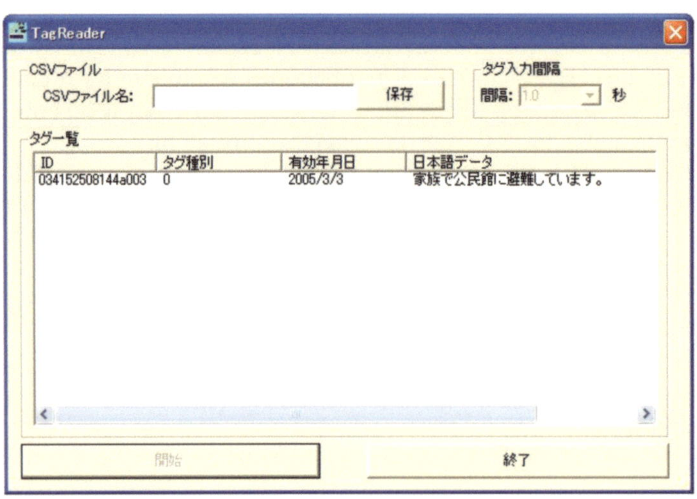

Fig. 4. Screen Example for Read Function

For the successful deployment of this system in society, RFIDs must be ubiquitous, or present everywhere in high concentrations. To this end, it is important to be able to use this system for commercial purposes (e.g., proving store information) in ordinary periods and then to switch to damage information collection in the event of a disaster. Voice synthesis of information from RFID would represent an expansion beyond the range of ordinary commercial applications; accordingly, a function in which Japanese character strings are read in real time via voice synthesis was incorporated.
The RFID ("Intellitag" from Intermec) memory consists of 128 bytes, and can be broken down as follows:

System ID		8 bytes (not rewritable)	
Manufacture ID (manufacturer type information)		2 bytes (not rewritable)	
Hardware tag type (tag type information)		2 bytes (not rewritable)	
Software tag type (tag identifier 02 h, 53 h, 48 h)		3 bytes (not rewritable)	
Software tag type (NICT global code 02 h, 00 h, *)		3 bytes (rewritable)	
User area		110 bytes (rewritable)	
	Tag type		2 bytes
	Date of expiry		8 bytes
	Japanese data		100 bytes

Thus, 100 bytes of Japanese character strings are writable to each tag (50 characters in two-byte character format).

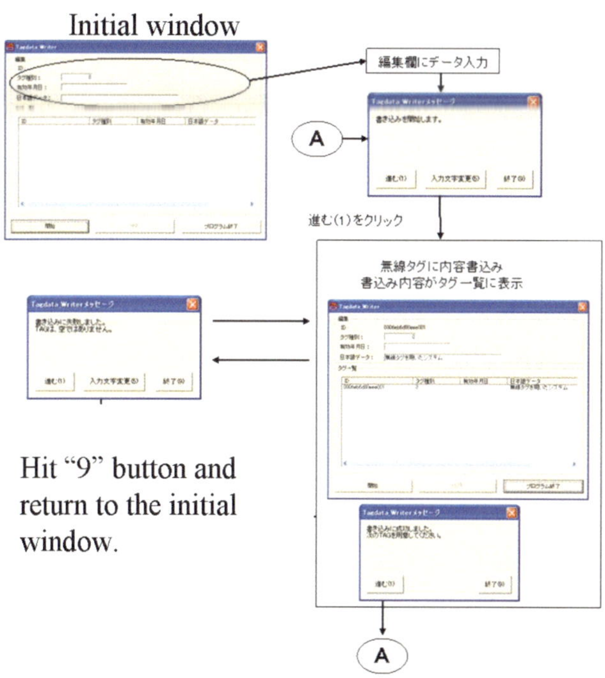

Fig. 5. Screen State Transition Diagram of the Write Function

Write function consists of automatic selection, from within the antenna's field of view, of an empty tag (i.e., with no written information) followed by writing of information to the tag. Fig. 5 presents a screen state transition diagram of the write function.
A specific example of processing is given below.
When the user enters "Test Writing," a character string to be written in the Japanese Data field, from the keyboard and presses the Start button, the system enters the waiting state for numeric key input. Fig. 6 illustrates the screen in this state. Here, the user uses a numeric key to select Proceed (to proceed with writing), Change Input Character String, or End.

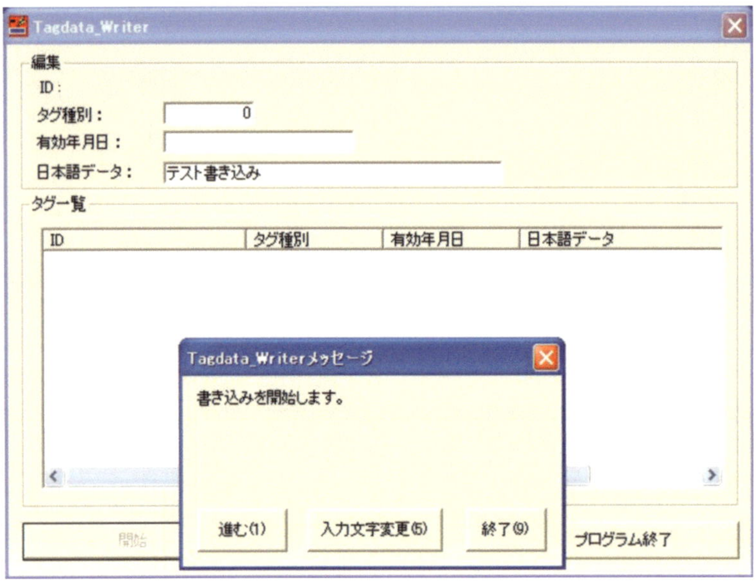

Fig. 6. Write Function Screen (start of write operation)

When the user selects Proceed (to proceed with the write operation), the system reads tags within the antenna's field of view (Fig. 7 shows that there were four tags in the field of view) and automatically selects one empty tag (i.e., a tag with a blank Japanese Data field) (Fig. 7 shows that the tag with the ID "03c312508144a000" has been selected), writing the character string "Test Writing" to that tag. Fig. 7 illustrates the screen after the character string is written.
Proceeding from the screen in Fig. 7, the user returns to the screen in Fig. 6. Here, if the user wishes to enter another character string, this is executed from the keyboard, and the write process begins again. The system automatically selects one empty tag from among those in the antenna's field of view (Fig. 8 shows that the tag with ID "0385b1508144a001" has been selected), and writes the character string "Test Writing 2" to that tag. Fig. 8 shows the screen displayed after the character string is written.
Retrieval function consists of clearing a tag to an empty state by deleting previously read data from the tag; new data can then be written to the RFID. Fig. 9 shows a screen state transition diagram of the retrieval function.

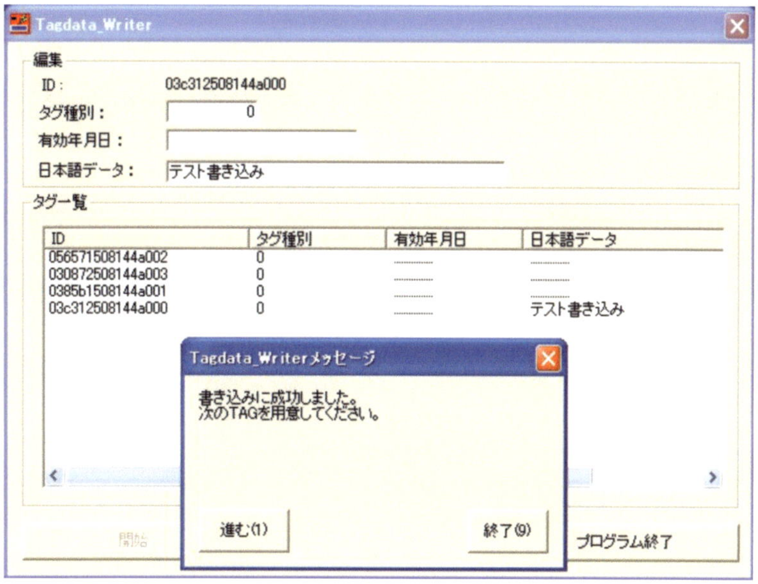

Fig. 7. Write Function Screen (After Writing to Tag)

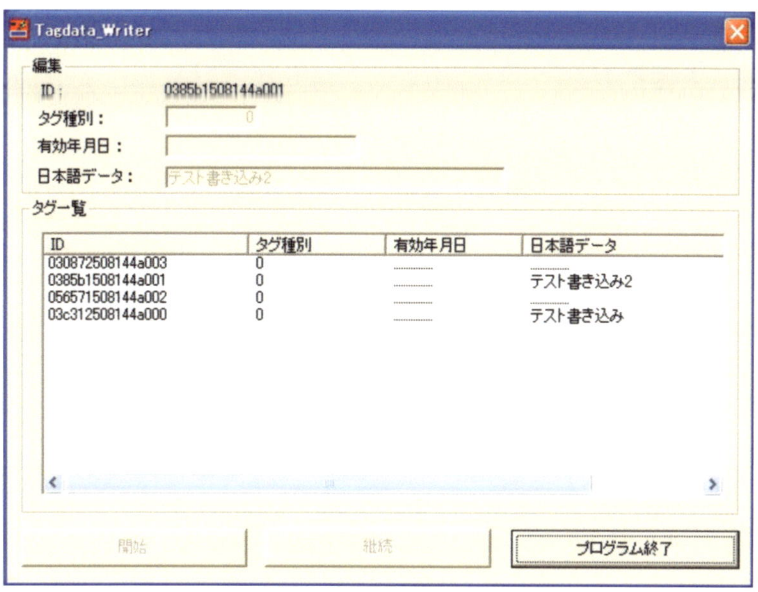

Fig. 8. Write Function Screen (after writing different character string to second tag)

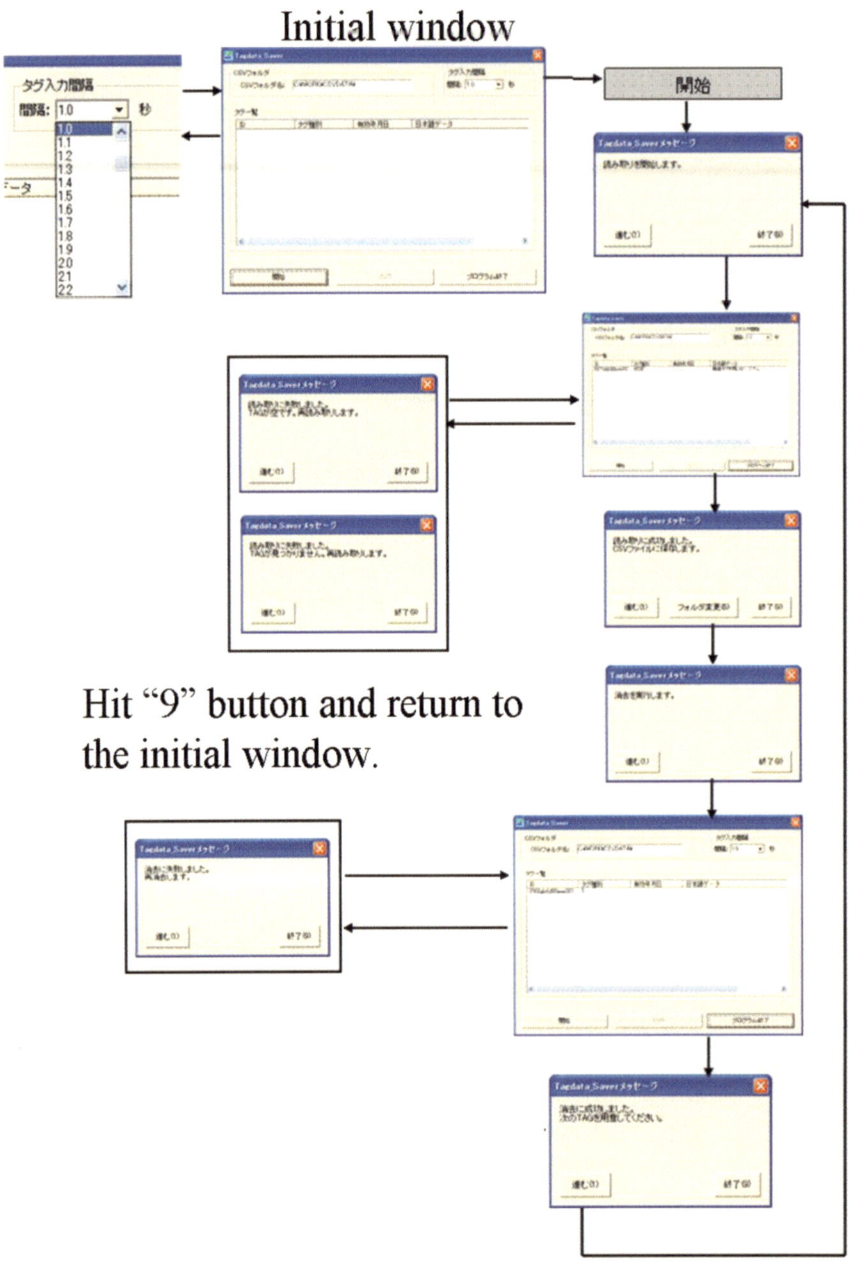

Fig. 9. Screen State Transition Diagram of the Retrieval Function

A specific processing example is given below.
When the user presses the on-screen Start button from the keyboard, the system enters the waiting state for numeric key input. Fig. 10 shows the screen in this state. Here, the user selects Proceed (to retrieve data) or End, via numeric key.

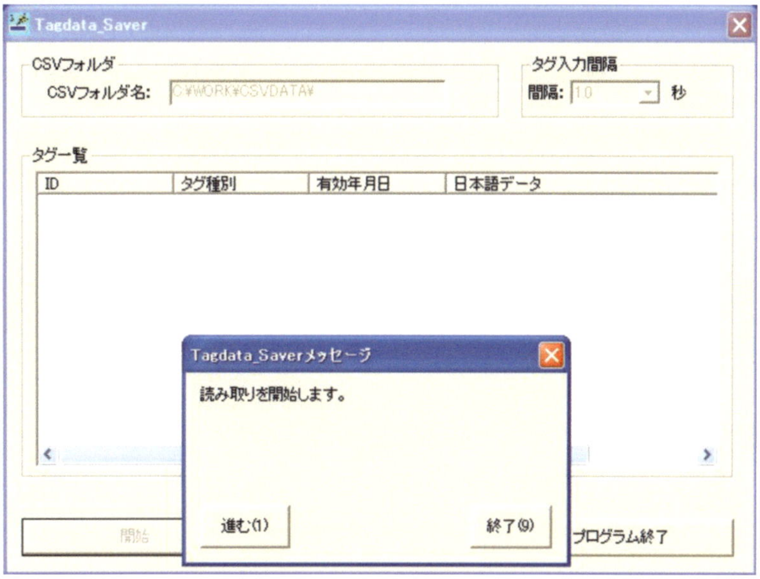

Fig. 10. Retrieval Function Screen (at the start of retrieval)

Fig. 11 shows the system as it waits for the read data to be saved to the control PC, after having read four tags within the antenna's field of view. The storage format is CSV (comma-separated values).

The user selects Proceed (to save), Change Folder to Save, or End.

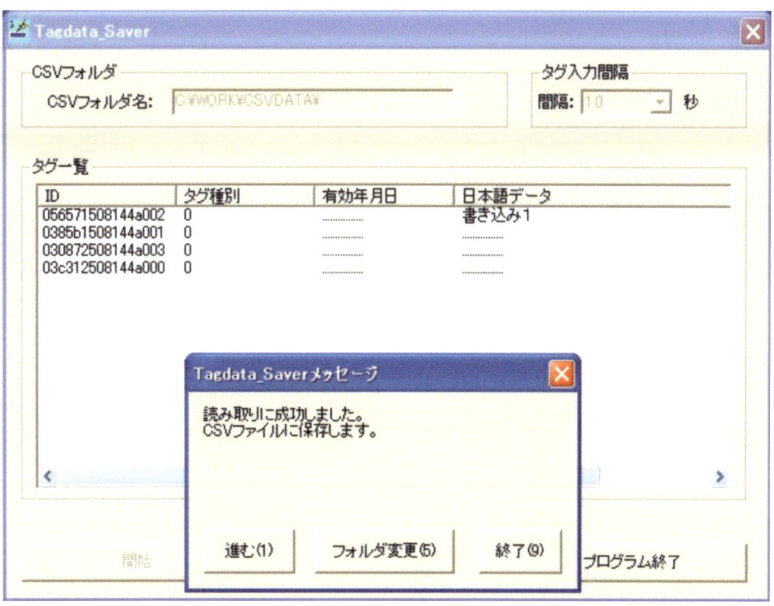

Fig. 11. Retrieval Function Screen (after reading tag)

When the read data is saved, the system enters the waiting state for a request to delete the data from the read tag. Fig. 12 shows the screen in this state. The user selects Delete or Stop (end without deleting).

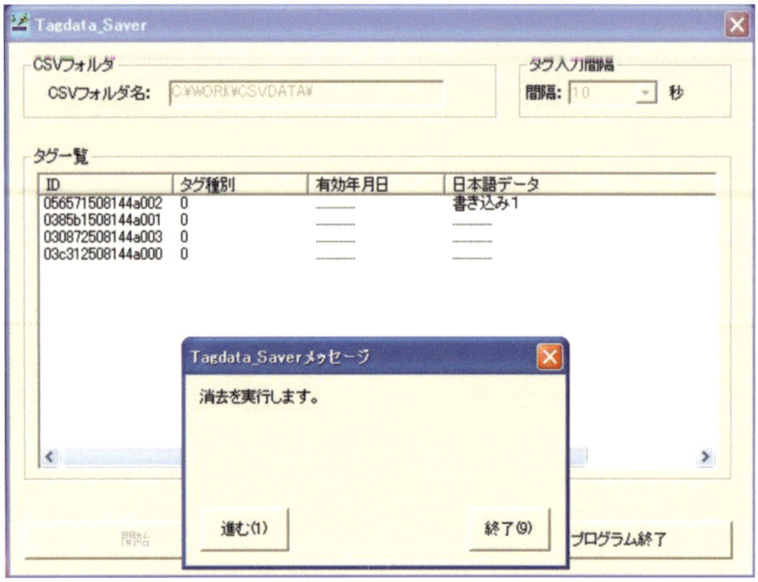

Fig. 12. Retrieval Function Screen (waiting for deletion processing)

When the user selects Delete, the data that has been saved will be deleted from the tag with the ID "056571508144a002", emptying the tag. Fig. 13 shows the tag in the empty state.

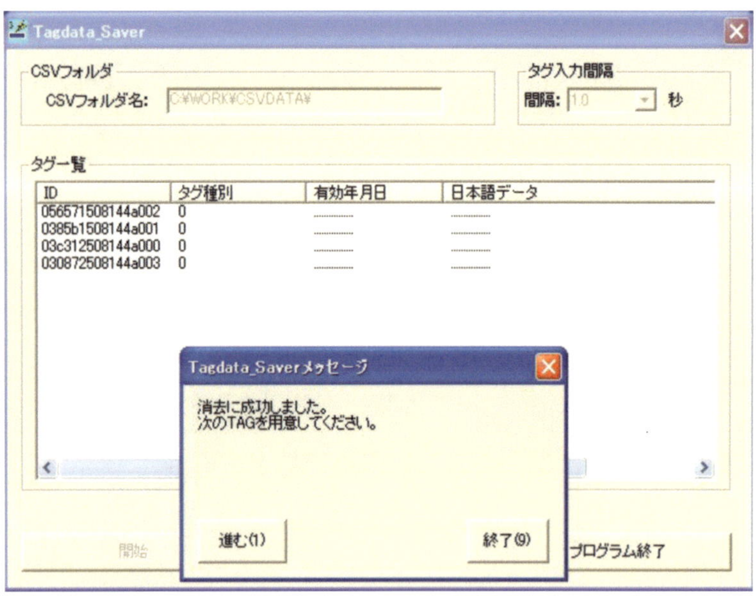

Fig. 13. Retrieval Function Screen (tag in empty state)

2.5 Backpack-mounted type

The conventional cart-mounted system has been redesigned to be wearable by someone on foot. The wearable system is shown at the center of Fig. 2. With this system, the RFID writer/reader and the batteries are piggybacked, with the control PC suspended from the shoulders like a drawing board.

The following four achievements were additionally realized in the backpack-mounted type:

- Announcement of operating conditions via voice synthesis
- Binary data division write and retrieval functions
- Control function via TCP/IP
- Security function

Explanations are as follows:

1. Announcement of operating conditions via voice synthesis

The developed system was geared toward eventual miniaturization to handheld size. The aim was therefore to enable the basic write and retrieval functions to be employed without viewing the screen and through operation of the buttons at hand. A function to announce screen status via voice synthesis was therefore added to those developed. Since the function of reading out data via voice synthesis had already been developed, the announcement function was realized through the incorporation of this read-out function.

2. Binary data division write and retrieval functions

Functions have been developed to write binary data, available in advance in the control PC, to an empty RFID, together with a file name; binary data read from an RFID in the control PC can then be saved as a file. Here we should note that the data capacity of 100 bytes may be insufficient for binary data when this data becomes available. For this reason, a function has been developed to divide and write binary data automatically to multiple empty RFIDs within the system's field of view if target binary data size exceeds the capacity of a single RFID. Conversely, another developed function reads and merges partial data items stored in

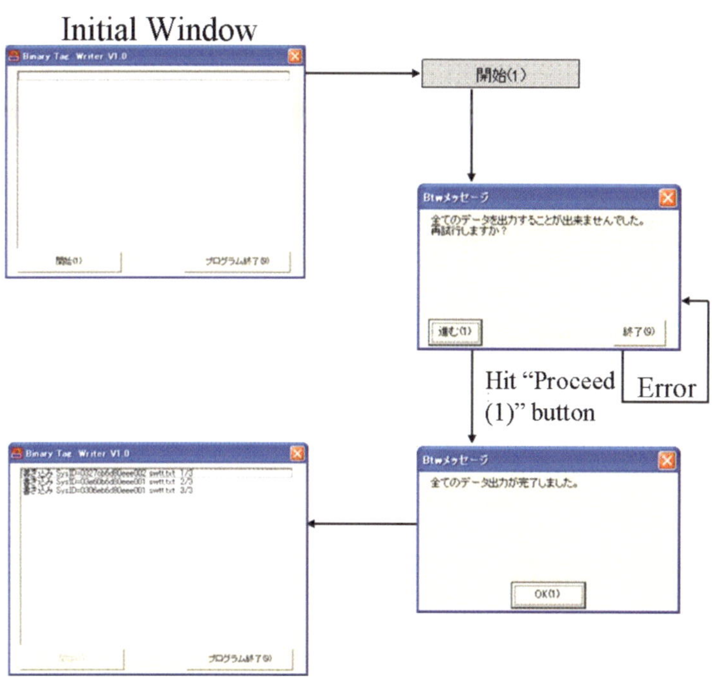

Fig. 14. Screen State Transition Diagram of the Binary Data Division Write Functions

a divided manner in multiple RFIDs, thus restoring the data to the original file. These developments lessen the limitations of RFID capacity within the system.
Fig. 14 shows the screen state transition diagram of the binary data division write functions. Fig. 14 shows an example of dividing and writing binary data under the file name "swtt.txt" to three RFIDs. The user specifies the file name of the data to be written in advance in the initial setting file.
Next, a screen state transition diagram of the binary data retrieval and merge functions is illustrated in Fig. 15.

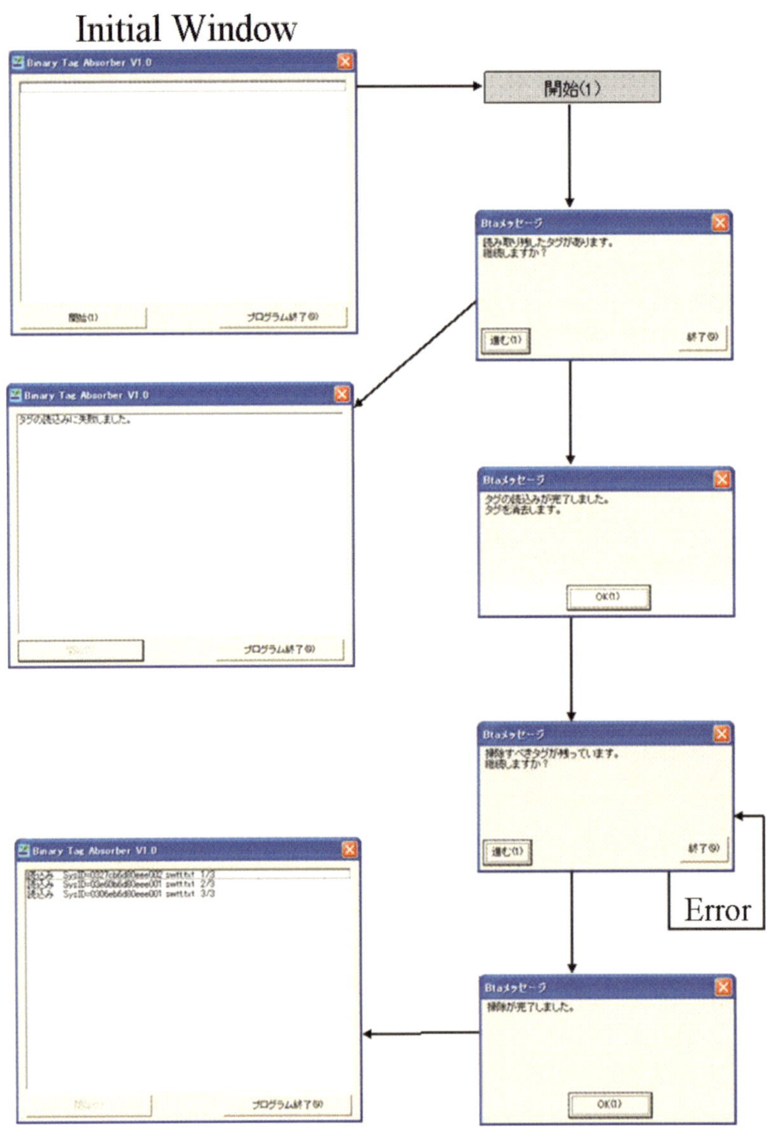

Fig. 15. Screen State Transition Diagram of the Binary Data Retrieval and Merge Functions

Fig. 15 shows an example of reading binary data that has been divided and written to three RFIDs. This data is read one element at a time, with each data item deleted in sequence (in a process referred to as "clearing") when all three pieces of data have been read. The three binary data items are merged to generate a binary file named "swtt.txt"; the data is then saved to the control PC. The user specifies the folder in which to save the generated binary file in advance in the initial setting file.
The tag memory for the binary data division write and retrieve functions described in the present section can be broken down as follows:

System ID	8 bytes (not rewritable)
Manufacture ID (manufacturer type information)	2 bytes (not rewritable)
Hardware tag type (tag type information)	2 bytes (not rewritable)
Software tag type (tag identifier 02 h, 53 h, 48 h)	3 bytes (not rewritable)
Software tag type (NICT global code 02 h, 00 h, *)	3 bytes (rewritable)
User area	110 bytes (rewritable)
File name	8 bytes
File name extension	3 bytes
Number of divided files	1 byte
Division sequence number	1 byte
Write size	1 byte
Binary data	96 bytes

The present functions require that the file name of the original data before division, the division sequence, and additional data be written to each tag. The size of the binary data writable to a single tag is therefore 96 bytes, smaller than that of the Japanese character strings described in Section 2.3.
The functions to handle binary data described in this section will be employed to develop applications designed to take advantage of RFIDs in various applications. Demonstrating the usefulness of RFID in a variety of applications in normal periods and at the time of a small-scale disaster (such as a normal fire), as well as in the event of a large-scale disaster, is expected to spur widespread use. For example, if binary data encrypted in advance—such as building position (precise GIS [Geographical Information System] coordinates), construction type, room arrangement, resident information (e.g., number of residents, contact information) and presence/absence of hazardous substances—is compressed and registered in an RFID attached to a building, then this information will be accessible from the RFID in the rescue team's command vehicle in the event of a normal fire, enabling rapid fire response and rescue. These application examples will be discussed further in Section 3.

3. Control function via TCP/IP

Interfaces controllable via TCP/IP have been provided in the writer/reader, allowing the device to be activated and operated from other applications within the control PC; this will also enable operation of the device from a remote PC via network.

4. Security function

As the use of RFID tags becomes more and more widespread, countermeasures against tampering (security issues) and unauthorized reading (privacy issues) will increase in importance. Accordingly, we also studied the possibility of access control through information encryption and using the unique IDs of RFID tags. (Takizawa et al., 2004Jan)

2.6 Handheld-type

Although wearable, the backpack-mounted type was too large for mobile activities at a disaster site. We therefore developed a smaller writer/reader allowing use of the same RFID technology. The smaller device is illustrated at right in Fig. 2. The main body of the writer/reader is of a size enabling insertion into the card slot of a notebook PC, with the antenna width reduced to roughly 1/4 of the original.

In the handheld-type, a function has been developed to write and read assessment results to or from an RFID tag in coordination with an application that displays emergent risk assessments and accepts assessment results in binary data form. This function is intended to check previous building data against the current status for a rapid assessment of damage. Assessment results are electronically deposited on-site for use among other rescuers at the time of a large-scale disaster, thus avoiding duplicative investigation, streamlining relief subsidy and other formalities, and contributing to the timely establishment of a detailed disaster database. Fig. 16 shows the entry application screen for the emergent risk-assessment results.

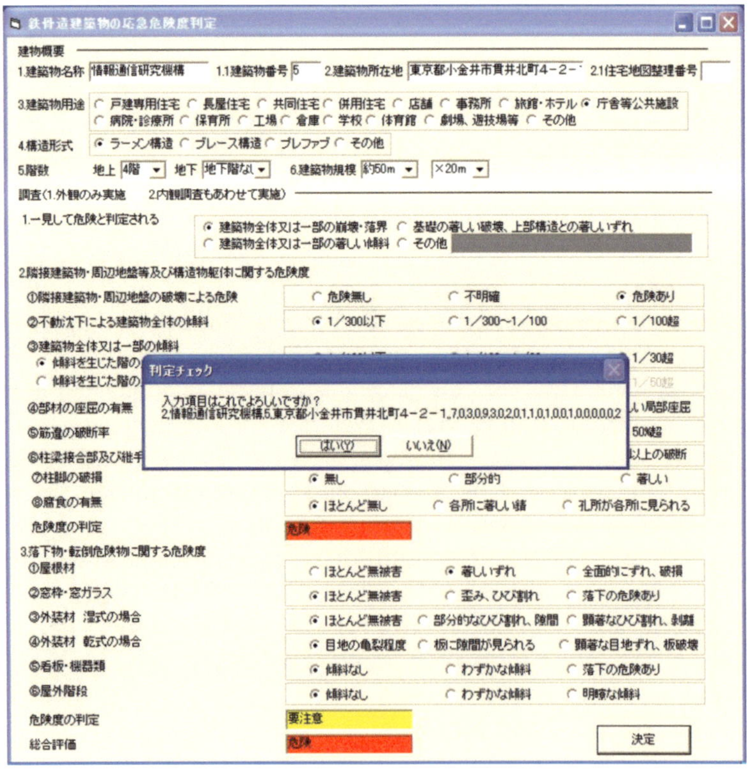

Fig. 16. Entry Application Screen for Emergent Risk-Assessment Results

Here we must note that the communication distance possible with the RFID shrank due to the weaker radio wave output of the writer/reader (relative to the backpack-mounted type). An RFID system in the UHF band is expected to provide communication over longer distances. In the next section, we describe the real performance of RFID system in the UHF band

2.7 RFID system in the UHF band to support firefighting

In FY2000, the Fire and Disaster Management Agency (FDMA) Japan developed an information system to support firefighting (FDMA, 2004). This system is mainly designed to locate firefighters deep underground or in other areas in which firefighting is difficult, where GPS and wireless communications are unavailable. Inertial navigation system devices worn by firefighters obtain location information, which is conveyed via ad-hoc communication or other means to the field command center. Here, firefighter locations are displayed on a 3D numerical map. Absolute location information transmitted from RFIDs attached to the evacuation lights indicating the escape route is used for error correction of this location information, and RFID reader/writers worn by the firefighters are used to receive and correct the absolute location information. In this system, the RFIDs attached to the evacuation lights are battery-powered active tags that emit trace radio waves on the 300-MHz band. This type of RFID is expensive, and because it emits radio waves like a beacon, the battery must be replaced regularly. These RFIDs cannot be used for other purposes, as rewriting the requisite data afterward is difficult. Thus, several factors must be overcome before the system is likely to become popular. In addition, signals from active tags can be received from approximately 10 m away, a distance at which location information is much more accurate than data from the inertial navigation system devices; this has led to problems with location correction. Replacing these RFIDs with passive tags (which do not require batteries) would make the system more affordable and easy to maintain. It would also enable the addition of information and support logistical management of the evacuation lights. Firefighting surveillance information could be recorded after installation. Replacing the tags in this manner would thus prove an effective way to popularize the system. However, one concern with passive RFIDs is that their effective range is shorter than that of active RFIDs. A shorter range may result in unsuccessful location correction when firefighters are working near the evacuation lights, which defeats the system's main purpose. For passive RFIDs, the frequency band yielding the longest range is said to be the UHF band (950–956 MHz). Thus, the author, who has participated in the development of this system, verifies whether replacing the RFIDs with passive RFIDs transmitting in the UHF band would support location correction.

It is assumed that alternative RFIDs will be used just as in the current system (that is, attached to the evacuation lights) for interaction with readers/writers worn by firefighters. Accordingly, to measure the range, we attached RFIDs to evacuation lights in a fixed position and determined the areas in which reading and writing was successful or unsuccessful when the reader/writer was moved. The assumed mode of use is shown in Fig. 17.

As of March 2005, domestic restrictions on RFID systems operating on the UHF band are still in effect, so measurement of the characteristics was contracted to a professional with an anechoic room who was licensed to operate experimental radio stations. The measurement setup in the anechoic room is shown in Fig. 18. The mode of use was as shown in Fig. 17.

High-output reader/writers (maximum: 1 W) were used, with evacuation lights and RFIDs at a height of 2.4 m and the antenna at a height of 1.5 m. Under these conditions, we varied the distance between the evacuation lights and antenna and the angle of measurement to determine the area in which RFID reading and writing was possible.

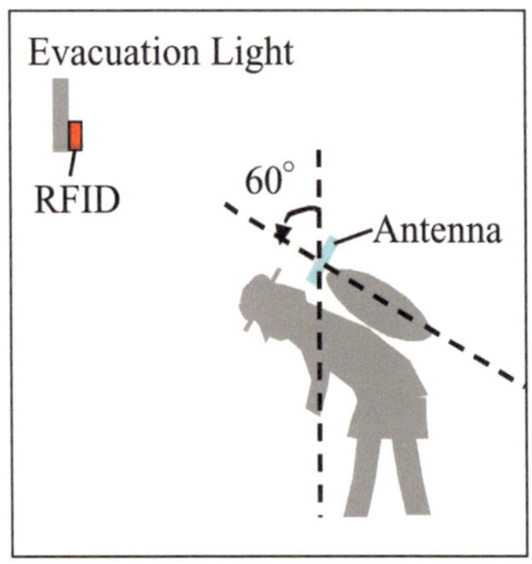

Fig. 17. Assumed mode of use

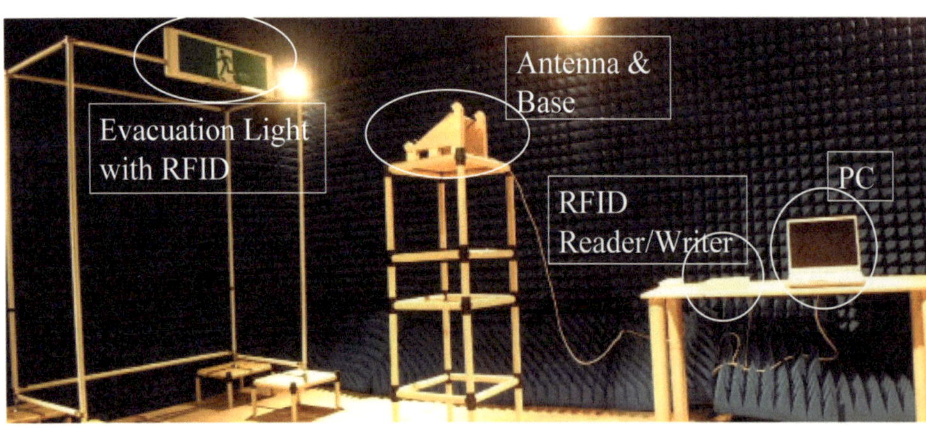

Fig. 18. Measuring the characteristics of passive RFIDs operating on the UHF band

Figure 19 shows an example of the relationship of antenna position to successful read rate, with the evacuation lights and RFIDs at the starting point. The unit of the scale in Figure 14 is meter.We determined that the RFIDs could be read at a distance of approximately 4 m from the evacuation lights, if read in the area in front of the lights. Location correction requires reading at a distance of less than 4 m from the evacuation lights. Thus, because the experimental mode used did not differ in this regard from the current system using active RFIDs, we can conclude that replacing active RFIDs with passive ones will pose no difficulty in terms of location correction.

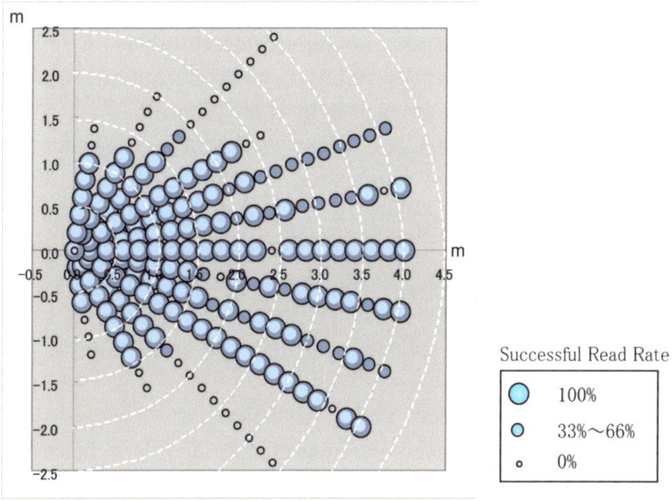

Fig. 19. One result of measuring characteristics

Measurement of the characteristics involved high-output (maximum: 1 W) reader/writers, as reported in part by the Information and Communications Council (MIC, 2004Dec). However, taking size, power consumption, and radio station license requirements into consideration, it is more appropriate to use low-output types (maximum: 10 mW) that do not require a license, such as the reader/writers worn by firefighters. Technical requirements governing reading and writing performance by these low-output units will be established in time, and verification will be required after experimental devices are developed. Details of the results of these measurements have been published in an FDMA document; specifically, a FY 2004 report by a development committee working on information systems to support firefighting in challenging underground areas (FDMA, 2005).

3. Information sharing system and damage information collection system using RFID (Takizawa et al., 2004Jun)

3.1 Introduction

It is widely recognized that rescue efforts of fire departments, the police, the Self-Defense Forces, and others were not smoothly executed in the Great Hanshin-Awaji Earthquake. Among the causes for this unsatisfactory response were delays in the collection of damage reports and resultant improper assignment of personnel, an overall shortage of personnel, and delays in support activities of personnel outside the afflicted area (due to a lack of understanding of the geography of the disaster region). In contrast, examples were seen in which many people were rescued by the local community—neighbors, volunteers from disaster-prevention organizations, and volunteer firefighters—without the assistance of the fire department, the police, or the Self-Defense Forces, as witnessed in Hokutancho, Tsuna-gun, Hyogo. The efficiency of the search for survivors was enabled by information: these neighbors and volunteers knew how many people lived where, where the elderly were located, and so on. In urban areas, however, where community ties are weaker, few know their neighbors. To gain an accurate grasp of those in need of rescue under these circumstances, there must be some way to determine numbers of family members, the whereabouts of the elderly, and more. Even in local communities, there are limits to volunteer relief and rescue efforts in times of large-scale disaster.

Similarly, as we learned from the Great Hanshin-Awaji Earthquake, gathering on-site damage information is essential for public organizations such as fire departments, the police, and the Self-Defense Forces if they are to initiate strategic relief and rescue efforts and promote recovery through the appropriate deployment of personnel. At present, however, it is difficult to ensure sufficient numbers of personnel to cover an entire disaster area, leaving no alternative but to rely on estimates of earthquake ground motion and damage, as well as on information collected through the circulation of a small number of personnel.

As a possible solution, the present section discusses an information sharing system (involving individual building structures) theoretically in place before an incident and subsequently applied in large-scale disasters to recovery operations. We then discuss a damage information collection system based on the existing information sharing system. As described in Section 2.2, the authors' system is designed based on the concept that all necessary information must be written to and stored in an RFID. Therefore, we had to develop many unique functions unavailable in conventional RFID systems designed for logistics and similar applications. Once developed, these functions paved the way for deployment in the information sharing system and the damage information collection system described in the following.

3.2 Information sharing system using RFID

3.2.1 Outline of the information sharing system

The information sharing system using RFID is designed to store information—on buildings, family identification, numbers of family members, and hazardous substance in the building—in normal periods. Each building would be equipped with such a system, enabling the sharing of information (both in normal periods and in times of disaster and recovery), ensuring efficient collection of information, and preventing on-site errors. The system will serve at normal times as a source of information for economic activities, such as those related to delivery, and as an information source for relief and rescue workers in the event of a disaster.

With client/server-based information sharing using conventional information communication networks, it is impossible to call up and acquire needed information if a large-scale disaster leads to breaks or congestion in the communication network. On the other hand, equipping the buildings themselves with RFID technology, a known reliable source of information, allows for the acquisition of information in the field and enables the establishment of a highly disaster-resistant system. Such a system, based on information gathered and managed in the field, will prove significantly useful in firefighting, disaster prevention, large-scale earthquakes, and other disasters that require an urgent response.

Moreover, in contrast to the client/server-based information sharing method, which raises the concern that an individual may not be able to control all of his or her information, including privacy information (in other words, the method is based on trust in the database administrator), an RFID system maintains information under the physical control of the individual, in effect assigning access control rights to the user. This is a considerable advantage in terms of privacy management. Enabling a choice between keeping information private (in normal periods) and disclosing it voluntarily (as in times of emergency or disaster) will provide ample reassurance to most users concerned with privacy.

3.2.2 Use of the information sharing system

In the following we describe a series of ways in which the RFID information sharing system can be used, from building construction to times of disaster and disaster recovery.

1. Installation of "electronic nameplates"

An RFID is attached as an "electronic nameplate" to a new or existing building. This refers to an existing nameplate embedded with an RFID that stores a range of information, from building information to details on resident families. Examples of data that may be stored in an electronic nameplate are shown in Table 1. This data can be put to common use, both in normal periods and in times of disaster, as described later. Residents of the building can make any necessary changes to the stored information as the need arises. On the other hand, if the residents do not wish to disclose this information, they can forego installation of the electronic nameplate or restrict the information stored. On the other hand, measures will be taken to limit the number of the people having access to the information.

Item	Information Stored
Building position information	Address, latitude and longitude, etc.
Building information	Building floor plan, construction type, etc.
Hazardous substance information	Hazardous substances subject to mandatory registration under the Fire Services Law
Life-saving information	Family members, blood types, presence/absence of the physically disabled, information about elderly residents under care, etc.

Table 1. Examples of Information Stored in Electronic Nameplates

2. Normal periods

In normal periods, the positional information stored in the electronic nameplate is used for mail deliveries and courier services. On the other hand, RFID can also be used as a guidepost for the visually disabled, thanks to its ability to convey information in a noncontact manner. Moreover, reading the owner's name, address, device numbers, and other data stored in the electronic nameplate will ensure efficiency in utility meter readings, such as those for gas and electricity. Further, if electronic nameplates and equipment such as gas and electricity meters can function in a coordinated fashion, it will be possible to perform meter-reading process simply by accessing the RFID.

3. In the event of fire

In the event of a fire in the building in normal periods, the information read from the electronic nameplate by a firefighter—such as the building's floor plan, resident family details, and information of hazardous substances—can improve life-saving in firefighting efforts and strategy. Meanwhile, if a resident is alone and injured, stored information on blood type, allergies, and other medical details will prove extremely useful during transportation of this person to hospital.

4. In the event of an earthquake

In the event of an earthquake, a number of people may be trapped under collapsed buildings and in need of rescue. Yet most urban dwellers don't know their neighbors, and the resultant lack of information can unnecessarily impede rescue efforts. Searches can be accelerated, however, using the information from electronic nameplates. This feature will prove essential in the effective absence of a local community, listing residents who may be trapped within collapsed buildings. Further, if successful evacuees are registered as such on the electronic nameplate, those in need of rescue will be more readily identified, further reducing search times.

An alternative application would involve coordinated operation with a health monitoring system adapted to perform continuous assessment of the building at the time of an earthquake, determining damage to the building simply by reading the electronic nameplate and increasing efficiency in the survey of overall damage.

5. Building damage assessment

Following an earthquake, various parties perform damage assessments to aid in recovery and reconstruction, including surveys to determine the distribution of government aid. Using electronic nameplates for such surveys would streamline the flow of needed information. Included among these earthquake-related surveys are emergent risk assessments for damaged buildings, particularly important in preventing life-threatening secondary disasters (JCQIEDB, 1998); damage assessments for scientific purposes (i.e., assessment of whether a building is partially or fully collapsed); damage classification for the assessment of required repairs and reinforcements and the continued use of the building (MLIT, 2001); and victim's certificate (classification of victims) for purposes of awarding assistance. While these surveys and assessments share many common items, at present this data is not structured to flow smoothly among the various parties. Survey details stored in electronic nameplates, however, can be used by the next survey team. This ensures reduced assessment times, speeding recovery and reconstruction. Further, with surveyors sharing a variety of survey results, differences in focus and oversights will be eliminated, both offering surveyors more material for judgment and providing for greater accuracy in assessments.

6. Recovery and reconstruction

During recovery and reconstruction, information relating to repair work can be added to a building's electronic nameplate, which would prove useful in future remodeling or a future disaster.

Given the foregoing considerations, it is clear that information sharing using electronic nameplates can form the basis for a system that will maximize efficiency in the field.

3.3 Damage information collection system using RFID

3.3.1 Outline of the damage information collection system

The damage information collection system is designed to collect damage information efficiently in the disaster-stricken area during and after an earthquake based on the information sharing system described above, in combination with IT-based digital information terminals and RFID writers/readers. Further, the system can be used to convey damage information to disaster response headquarters and elsewhere over existing communication networks and can help in the assessment of building damage for the phases of recovery and reconstruction. Moreover, the system can be used not only to help in the collection of damage information but can also be applied to information collection in normal periods.

Additional functions are possible: display of geographical information on a 3D screen to assist in surveys, fire simulations for firefighter reference, display of local geographical information and water resources (particularly the availability of water for firefighting), and simulation/analysis of ground motion and possible damage, for use in prediction or in the event of an actual earthquake or other disaster.

It should be noted that although the present damage information collection system may be built independently of RFID writers/readers, the system will provide a much wider range of functions when operated in coordination with these devices.

3.3.2 System configuration

The damage information collection system is comprised of a GPS (Global Positioning System), a notebook PC to operate the system, an RFID writer/reader, and a field damage information collection system (Shibayama & Hisada, 2003); (Shibayama & Hisada, 2004) incorporating a simplified GIS developed by the authors. In addition to the above configuration, the system allows hands-free surveying when combined with a head-mounted display (HMD).

3.3.3 Methods of use of RFID systems

1. Identifying building position and collecting damage information using "electronic nameplates"

A number of systems have been researched and developed to support the collection of damage information in the field (Zama et al., 2001); (Fukuwa et al., 2001), with nearly all systems using a GIS, as this system enables entry of damage information while viewing a map. However, even with a map entry of building position information entry can be slow if the user is not aware of the local geography. On the other hand, even GPS positional information is accurate only to several tens of meters, making it difficult to identify a target building among a series of adjacent buildings. To save time and effort in entering positional information, therefore, information is drawn from the electronic nameplates within the information sharing system. Matching surveyed building information on latitude and longitude, owner's name, address, and so on to a target building provides failsafe identification, improving the efficiency of collection of damage information. In addition to positional information, other building details stored in the electronic nameplate will likely prove useful in the gathering of information at the time of a disaster.

2. Using the RFID as a temporary storage device in the event of a disaster

RFID devices can be used to collect damage information even outside of the information sharing system. Specifically, the system can be applied to a successive flow of damage information tasks. The assessment of building damage is a process that is already in place, and is used after earthquakes. The flow proceeds from initial survey to emergent risk assessment to damage classification to victim's certificate (victim classification), with many items of shared information among the steps. Under these circumstances, providing an

RFID device for storage of damage assessment results in the field will allow common information to be written or read, drastically reducing mistakes, eliminating duplication, and ensuring greater accuracy and efficiency in surveys.

3. Information collection using "electronic nameplates" in normal periods

Above we discussed the uses of electronic nameplates to determine positional information and to collect disaster data at the time of an earthquake. In addition to these applications, electronic nameplates can find numerous applications in normal periods. For example, municipalities are required to conduct urban planning or urban infrastructure development surveys, to update maps, or widen roads, for example, or to perform field surveys for building or rebuilding permits. Furthermore, fire departments must carry out surveys in response to changes in available water resources, in addition to field surveys for disaster prevention. These surveys can be streamlined using the damage information collection system and electronic nameplate information.

3.4 Strategies for widespread use

The information sharing and damage information collection systems discussed above will become functional as the use of electronic nameplates becomes more pervasive. As things stand, however, instant widespread use would be difficult to implement. In this section we will examine methods of RFID use prior to widespread adoption of electronic nameplates, in addition to strategies to encourage such widespread use.

3.4.1 RFID utilization method prior to widespread adoption of electronic nameplate

RFIDs do not necessarily need to be available everywhere in a city in normal periods; empty RFIDs may at these times be kept in locations such as public meeting halls or elsewhere around the city for use as temporary storage devices in the event of a large-scale disaster. These temporary RFIDs could then be activated to facilitate information sharing following a disaster.

3.4.2 Strategy for widespread use

In order for electronic nameplates to become pervasive, they must be highly beneficial both for the ordinary households in which they will be installed and for the entities and public organizations that will use them. Several advantages were cited above; however, despite these advantages, widespread use will not be easy to implement. The road to widespread implementation will thus lie in the specific advantages of a system in which they are employed.

Electronic nameplates can store information such as building position and resident family details, and this information can be entered by residents on their own. In contrast, the ability to input building information (such as the floor plan and construction details) is generally limited to those with architectural expertise. This limitation can be overcome by allowing non-specialists to take advantage of building-related systems to acquire building information, thus eliminating one hurdle in the wide adoption of electronic nameplates. In the case of a newly built house, for example, building information can be obtained through the Government Housing Loan Corporation, (GHLC, 1950) which provides financing for home construction, or from a seismic performance evaluation system such as the House

Performance Indication System. (MLIT, 1999) In the case of an existing house, on the other hand, building information can be acquired through details of building permits for additions or remodeling. If these existing systems were to reward the installation of electronic nameplates through tax reductions or otherwise, then installation would become beneficial for both residents and public organizations, thus leading to more widespread implementation.

3.5 Security Issues in the information sharing system

While the information sharing system offers the convenience of acquiring and rewriting RFID information in conjunction with a portable terminal, the risk of abuse or information tampering cannot be ignored. Accordingly in this section we will address various RFID security measures.

The first possible measure would involve incorporating a locking feature on the RFID side. The RFID circuitry can be configured to remain silent; i.e., it can be programmed to ignore calls from the interrogator. Therefore, one possible scheme would consist of two coexisting storage areas in the RFID; one that responds to arbitrary calls and the other that does not (unless the lock feature is canceled), thus allowing only publicly disclosed information such as positional information and owner name to be read in normal times. Conversely, all information can be read once the lock feature is canceled in the event of a fire or earthquake. The challenge in this case lies in automating the trigger that will cancel the lock.

In terms of software, encryption of information within the RFID system is essential. Requiring user authentication before reading the RFID will ensure that only those authorized may operate the writer/reader. Alternatively, if the writer/reader is managed with great care, then authentication may be possible on a device-by-device basis, rather than on a user-by-user basis. In this case, the writers/readers would be ranked according to the security level of readable data. For example, writers/readers capable of handling the most sensitive information would be limited to public safety organizations such as fire departments.

One possible countermeasure against unauthorized writing to the RFIDs would consist of locking the active field at the moment the data written to that field is processed and written to the RFID.

It should be noted that in light of concerns over destructive attacks on RFID hardware, possible countermeasures include some very basic ones, such as installation of the RFID in a location that is not easily accessible. Such countermeasures are particularly practical in light of the RFID's noncontact design.

RFID security countermeasures are considered to represent the most important of all challenges facing widespread implementation of electronic nameplates; accordingly, the authors are currently studying the necessary security requirements and countermeasures. (Takizawa et al., 2004Jan)

3.6 Other applications

In addition to the applications discussed in the present section, RFID may be utilized in a variety of additional ways in the event of a large-scale disaster, as follows:

1. Individual identification of victims

When used as an ID card for individual identification, RFID can be used, for example, in entry/exit control in shelters, delivery of rescue goods and determination of required amounts, and resident access control in off-limit areas (as part of burglary-prevention efforts). These applications are immediately feasible with the use of a commercial IC card reader, without requiring the installation of the RFID writer/reader developed by the authors.

2. Information supply from utility poles

RFIDs may be attached to utility poles rather than to buildings, holding information such as shelter location information (location and map data), utility pole control, and positional information. In cases in which utility lines are underground, the RFID may be attached to a tree instead.

At normal times, the installer of utility poles (e.g., the power company) can use the RFID for utility pole regulation. Alternatively, when calling an ambulance or fire engine, location information can be read from the RFID on the nearest utility pole for transmission via email or telephone. Further, RFID technology can be used as a guidepost for the visually disabled. Lighter uses can be imagined: an RFID attached to a tree, for example, could hold information on the tree itself, for use in school field trips or the like.

In times of disaster, evacuees can search for shelters by acquiring information from utility poles. On the other hand, if utility poles fail or collapse, information related thereto can itself be useful for the power company.

In these uses, the information source must be installed outdoors, highlighting the advantage of RFIDs over two-dimensional barcodes, which are liable to become dirty.

4. Conclusion

This chapter discussed progress in the development of RFID writers/readers for use in the collection of damage information, as well as the deployment of this technology in information sharing and damage information collection systems. Going forward, we will need to conduct verification testing at a test site, pursue greater convenience in design, and study ways to achieve widespread use of electronic nameplates while pushing ahead with system development. We anticipate the incorporation of the RFID writer/reader into mobile phone terminals in the future. Commercialization of mobile phone terminals with an RFID read capability has already begun, and realization of write capabilities and improvement of the communication distance of the passive devices are among the technical challenges we must face if we are to ensure broad application in the collection of damage information.

5. References

Fire and Disaster Management Agency (2004). *FY2003 report of a development committee on information systems to support firefighting in challenging underground areas,* Fire and Disaster Management Agency , Tokyo. (In Japanese)

Fire and Disaster Management Agency (2005). *FY2004 report of a development committee on information systems to support firefighting in challenging underground areas,* Fire and Disaster Management Agency, Tokyo. (In Japanese)

Fukuwa, N.; Taikai, H. & Tobita, J. (2001). Intercommunications System "AnSHIn-system" ANS Mobile Disaster Information Unit "AnSHIn-KUn", *Architectural Institute of Japan Journal of Technology and Design,* No.12, Jan. 2001, pp.227-232. ISSN 1341-9463 (In Japanese)

The Government Housing Loan Corporation (1950). *The Law of the Government Housing Loan Corporation,* The Government Housing Loan Corporation, Tokyo. (In Japanese)

Japan Council for Quick Inspection of Earthquake Damaged Buildings (1998). *The Manual of Post Earthquake Quick Inspection of Damaged Buildings,"* The Japan Building Disaster Prevention Association, Tokyo. (In Japanese)

Ministry of Internal Affairs and Communications (2004). *Technical conditions for high-power passive tag systems using the 950MHz band (partial report from Information and Communications Council),* Ministry of Internal Affairs and Communications, Tokyo.

Ministry of Internal Affairs and Communications (2004). *Final Report of Investigative Study Group on High-Level Use of Electronic Tags in the Age of Ubiquitous Networks,* Ministry of Internal Affairs and Communications, Tokyo. (In Japanese)

Building Guidance Division, Housing Bureau, Ministry of Land Infrastructure and Transportation (Supervised) (2001). *Damage Assessment of Damaged Buildings and Restoration Technique Guideline,* The Japan Building Disaster Prevention Association, Tokyo. (In Japanese)

Ministry of Land Infrastructure and Transportation (1999). *Enforcement Ordinance of the Law of Promotion of Quality Reservation of a Residence etc.,* Ministry of Land Infrastructure and Transportation, Tokyo. (In Japanese)

Nebiya, H. & Uetake, K (2003). *Ubiquitous Radio Engineering and Micro RFID,* Tokyo Denki University Press, ISBN4-501-32280-2, Tokyo. (In Japanese)

Shibayama, A. & Hisada, Y. (2003). An Efficient System for Acquiring Earthquake Damage Information in Damaged Area, *the Journal of Social Safety Science,* No.5, 2003, pp.95-103, ISSN 1345-2088. (In Japanese)

Shibayama, A. & Hisada, Y. (2004). An Efficient System For Acquiring Earthquake Damage Information In Damaged Area, *The 13th World Conference on Earthquake Engineering,* No.1121, Vancouver Canada, Aug. 2004.

Takizawa, O.; Tanaka, H. & Yamamura, A. (2004). A Consideration on Security and Privacy Protection of RFID-based Disaster Relief System, *Proc. of the 2004 Symposium on Cryptography and Information Security (SCIS2004),* 2C5-4, Sendai Japan, Jan.2004. (In Japanese)

Takizawa, O.; Shibayama, A.; Hosokawa, M. & Hisada, Y. (2004). Research of Disaster Information Collection Support System using RFID, *Proc. of the Symposium on Application of Real-time Information in Disaster Management*, pp.191-198, Tokyo Japan, Jun.2004, Japan Society of Civil Engineers. (In Japanese)

Zama, S.; Endo, M; Hosokawa, M.; Hatayama, K.; Shibata, Y. & Harada, T. (2001). Development of Disaster Information Collection Terminal -as a n Element of Information System for Support of Fire-fighting Activities-, *Proc. of The Annual Conference of The Institute of Social Safety Science*, No.11, pp.113-116, Shizuoka Japan, Nov, 2001. (In Japanese)

16

Low Cost Identification Applications in Traffic Vehicular Environments

Jaume Segura, Juan G. Jordán, Miguel A. Jaen,
Francisco R. Soriano and Antonio Soriano
Institut de Robòtica - Universitat de València, Valencia
Spain

1. Introduction

The world is becoming wireless. In contrast with wired technologies, wireless technologies are widespread in several sectors and they are more and more present in many aspects of life. Wireless includes any technologies which uses no wire. They can be applied for specific applications and have been standardized. (Bluetooth or IEEE 802.15.1, WiFi or IEEE 802.11, ZigBee or IEEE 802.15.4, RFID standardized as ISO 18000, etc).

One of the most valuable reasons for the growth of wireless technologies for communications is the requirements for the mobility of modern applications. These requirements, and also security, make wireless technologies one of the best candidates for these applications and for establishing secure communications in traffic vehicular environments, so for vehicle to vehicle (V2V) as well as for vehicle to infrastructure (V2I).

Electronic Registration and Identification (ERI) of vehicles is a way to identify a vehicle univocally by means of some kind of wireless technology for communication. This protocol allows a wide range of interesting ITS applications, which involves secure identification of vehicles using symmetrical and asymmetrical techniques. These applications could be applied in private and public services (tolling systems, access to parking lots, information services, etc).

Figure 1 shows a schema which explains the whole architecture in order to understand how ERI standards are settled within an AVI/AEI architecture proposed in ISO 14814.

A specific application programming interface of wireless communications developed for Electronic Registration of Vehicles is explained in this chapter as an example of application of this technology. The philosophy of this concept is gathered in the ISO/TS 24534 and ISO/TS 24535 standards, also named ERI standards. This family of standards establishes the architecture of reference to Vehicle Identification independently of the physical technology used, so they do not specify a particular wireless technology to develop any kind of system application (the standard just suggests DSRC technology for deployment of the protocol). In other terms, these standards are linked to the Automatic Vehicle Identification (AVI/AEI) family of standards, which are ISO 14814 to 14816. This family of standards is the framework in which ERI is included.

Taking into account what is explained in the standards, ERI protocol allows establishing secure communications between the road infrastructure and the moving vehicles. This is an issue in which several Traffic administrations have been interested in. In this sense,

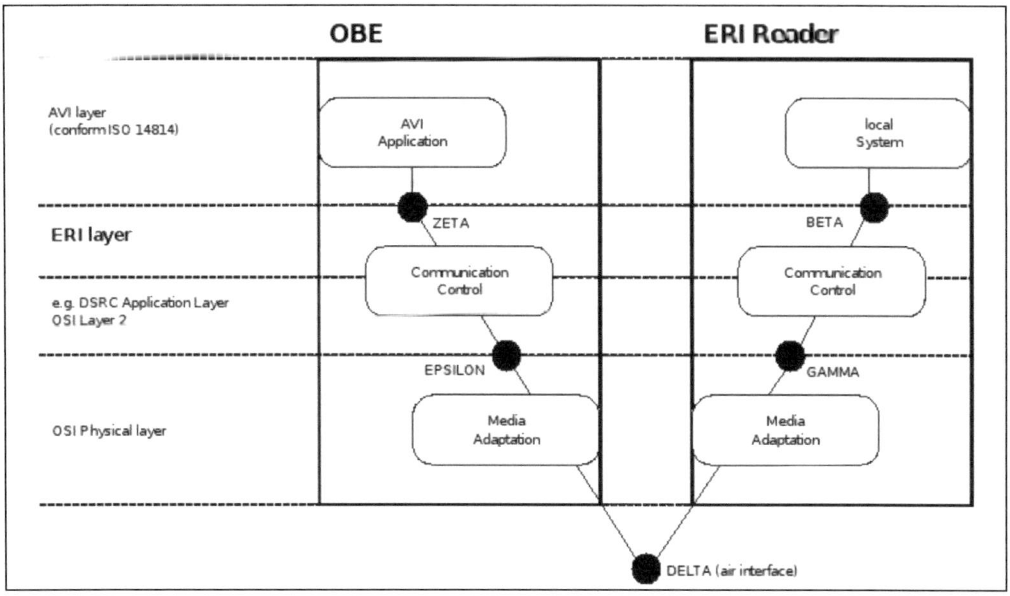

Fig. 1. ERI within the AVI/AEI reference architecture framework.

many countries have developed projects related to electronic identification of vehicles (some of them have taken into account feasibility studies for national or international application, e.g. EVI by ERTICO). Some of these projects are related to electronic plate (e-Plate), like the ones developed in Canada and United Kingdom. In the state of the art of this study, have been analysed some of these projects which show the administrations interests to study the requirements, costs and benefits of this kind of systems in order to implement identification systems for vehicles that serve as a proof of their identity, mainly for legal purposes.
The content of this chapter will be developed as follows. Firstly, a State of the Art of the applications developed in a worldwide framework will be outlined. After this, a summary of the ERI standards and the architecture of our application for several technologies (RFID, Bluetooth and WiFi) will be explained. Finally, the tests made on the developed system will be explained and the results for each one of these tests will be analysed.

2. State of the art of electronic vehicle identification systems

The study of the projects and initiatives in a worldwide context helps to determine the technological needs of EVI systems and serves to find out about the way in which problems have been solved in each application of the identification technology.
The main reference project studied within this state of the art was the EVI project [ERTICO_D2], [ERTICO_D3], [ERTICO_D4], coordinated by ERTICO and funded by DGTREN. The main aim of this project was the assessment of the decision makers in the European Comission and Member States to contribute to the establishment of a European Electronic Vehicle Identification system.
Other projects which have been studied were:

- EVI project was funded by DGTREN and coordinated by ERTICO. This project analysed the requirements for a Pan-European system for electronic vehicle identification. [Inno_Report, 2007]

- RFID within VIKING Euror-regional Project was based in EVI project by ERTICO, and studied the suitability of passive UHF RFID tags for monitoring applications. This project finished in 2004.
- e-Plate project in the UK and Canada which made an implementation of RFID system for vehicle identification. The project began in 2004 and finished in 2006.
- In Japan, there are some related projects for smart plates using DSRC and RFID [SMART_PLATE].

All these projects plot a worldwide sample of the development of electronic identification for vehicles.

2.1 EVI by ERTICO

The EVI project was an initiative of the European Commission, coordinated by ERTICO and funded by DGTREN to make a viability study for a Pan-European system for Electronic Vehicle Identification. This study aimed to identify and evaluate the main technical and non-technical issues for deployment, considering legal issues, institutional, operational and socio-political aspects.

This project has established framework architecture for an electronic vehicle identification system. Figure 2 shows a graphic of components of the EVI system.

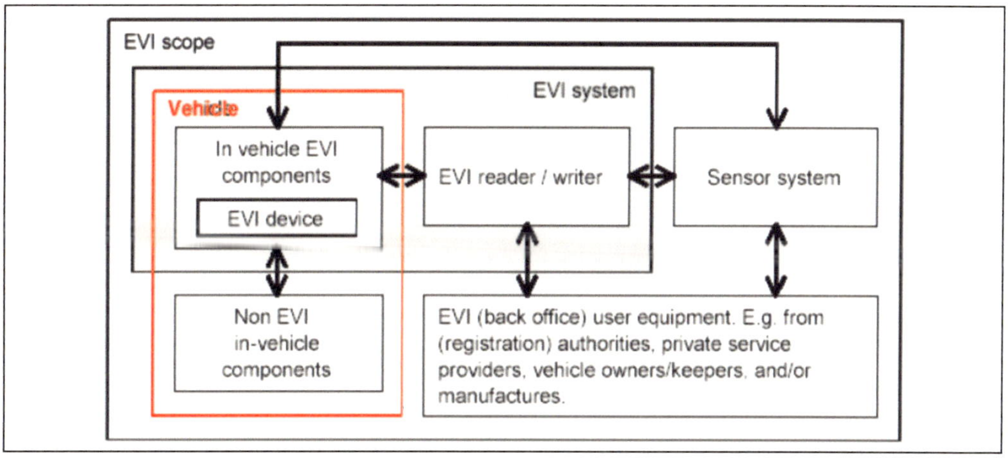

Fig. 2. Schema of components in the EVI scope

EVI system is composed by on-board vehicle components (sensor system, EVI and non-EVI in-vehicle components) and reader and/or external writers. Reader and/or writers are used to interchange EVI information with the different components on-board, they are used by registration authorities, private service providers, vehicle owners and manufacturers.

2.2 RFID from VIKING

In 2004, the Euro-regional project VIKING (Finland) developed an activity related to traffic monitoring using RFID technology called "Suitability of passive RFID tags for monitoring". This project was aimed to study the use of passive RFID technology for road monitoring. The study concluded that passive RFID tags in the UHF band (ISO 18000-6) were useful for traffic monitoring applications.

2.3 e-Plate

e Plate is a plate with an on-board chip which transmits a unique identifier for each vehicle. A small detector 'reads' the encrypted information and the output of the detector can be used locally or to transmit the encrypted message to a remote host. It was developed by Hills Numberplates Ltd in Birmingham - UK.

This project was initiated in 2004 by the British Government. It used active RFID technology embedded in the plate for the automatic identification of vehicles in the country. This project was developed and validated with trials during three years. Each e-Plate has an embedded tag with an encrypted (128 bits) and unique identifier number, which is transmitted by the tag and is detected by the RFID reader. The system allows simultaneous readings of multiple tags.

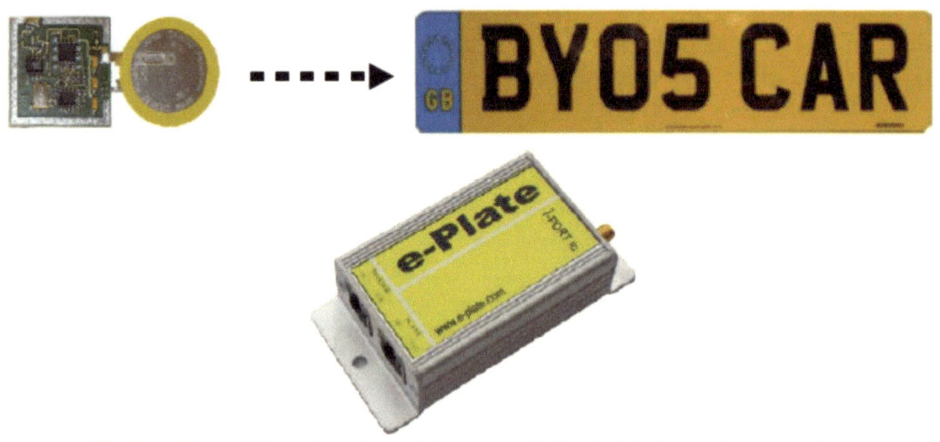

Fig. 3. RFID tag by Hill Numberplates Ltd and e-Plate reader by i-Port

2.4 Smart License Plate in Japan

A Smart License Plate is an automobile license plate equipped with an IC chip that stores information such as a license number and car specifications. A Smart Plate communicates with roadside units (antennas) to inform what type of vehicle is passing. [SMART_PLATE]

In 2000, the Ministry of Land, Infrastructures and Transport (MLIT) began the activity of a committee to study a system based on electronic plates (smart plates). These experiments were done thinking of a future commercialization of these smarts plates to gather information of private vehicles. In the textile district of Chojamachi in Nagoya, a trial to test the system was developed. In this district, a system for parking booking by using smart plates was adopted and it gave priority to freight goods vehicles in loading areas out of the road and on the road as parking spaces. It was concluded that the system improved the flux of freight goods. In the same way, vehicles with smart plates were firstly tested in the Japanese streets for a verification experiment.

In January 2006, the MLIT considered Smart License Plates development and commercialization using RFID technology. The embedded chip in the plates has information about the number plate, type of vehicle, etc. These systems were expected to solve traffic jams and to increase security allowing routing or re-routing to specific locations by using antennas installed on the road, etc. In April 2006, a pilot of the electronic plate system was deployed in several taxis of Chiba city.

3. ERI standards

ISO 24534 and 24535 [ERI_1], [ERI_2], [ERI_3], [ERI_4], [ERI_5], [ERI_BAS] describe all the required information for the establishment of identification and registration of vehicles by means of secure communications between vehicle and infrastructure. These standards describe the protocol for Electronic Registration and Identification. The first one is organised in 5 parts which establish the architecture of the system by means of the specification of its parts: ERT (tag) and ERR (reader). It also specifies the requirements of the system, establish operational parameters, defines functionalities, describes a data model and defines two security modes: by using symmetrical techniques and by using asymmetrical techniques.
Figure 1 shows a graph which represents the architecture of AVI/AEI (Automatic Vehicle Identification), where the ERI layer which allows communication and control capacities can be distinguished.
Figure 4 shows the different streams which both standards can define for different applications. The parts of the standard implemented are shown in orange colour. In the implementation of this standard some developments have been made in the AVI and ERI layers and the control and communication sublayer.

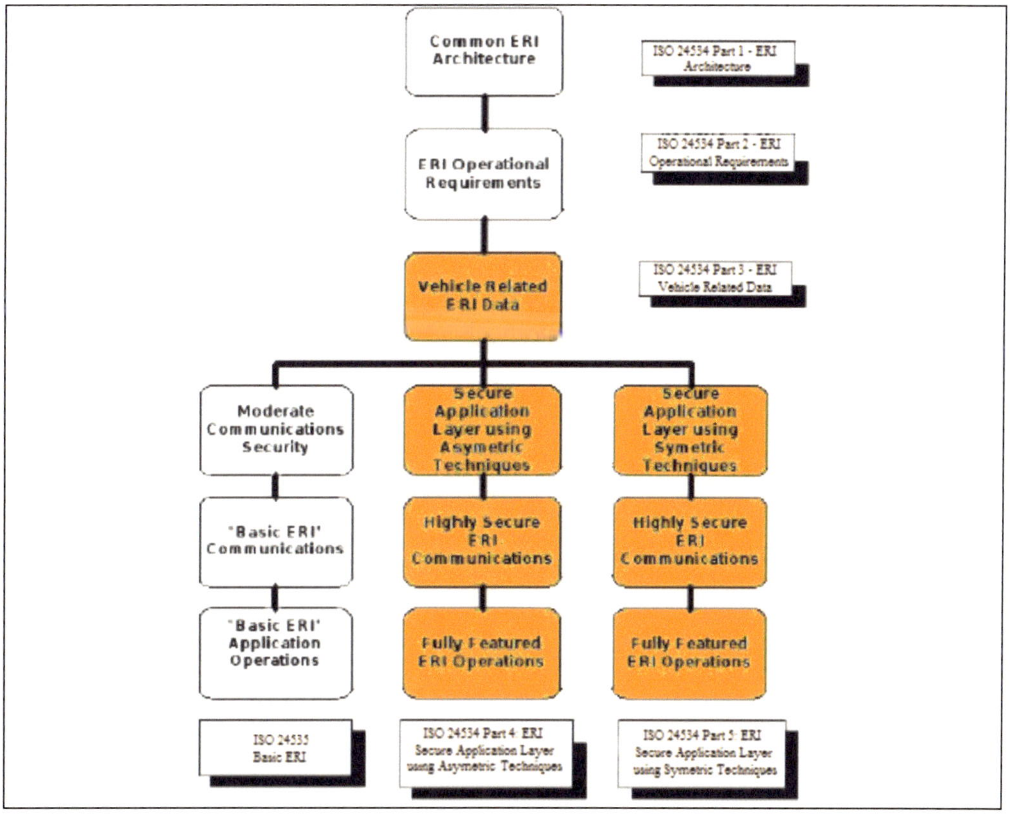

Fig. 4. Functional pile of the ERI system (includes different standards)

The protocol was implemented for each application scenario, originally planned as a set of messages to be used in each application and sequence. In order to apply the protocol, the ERI standards suggest DSRC technology, but several wireless technologies more and their

associated standards have also been studied to be used in this implementation, such as WiFi (802.11 b/g), Bluetooth (802.15.1), RFID (ISO 18000-4), ZigBee (802.15.4) and others.
Figure 5 shows a layered representation of this implementation. This implementation was made in Java, using J2ME for mobile devices on-board the vehicle (PDA) and J2SE for devices in the infrastructure (fixed and outside the vehicle).

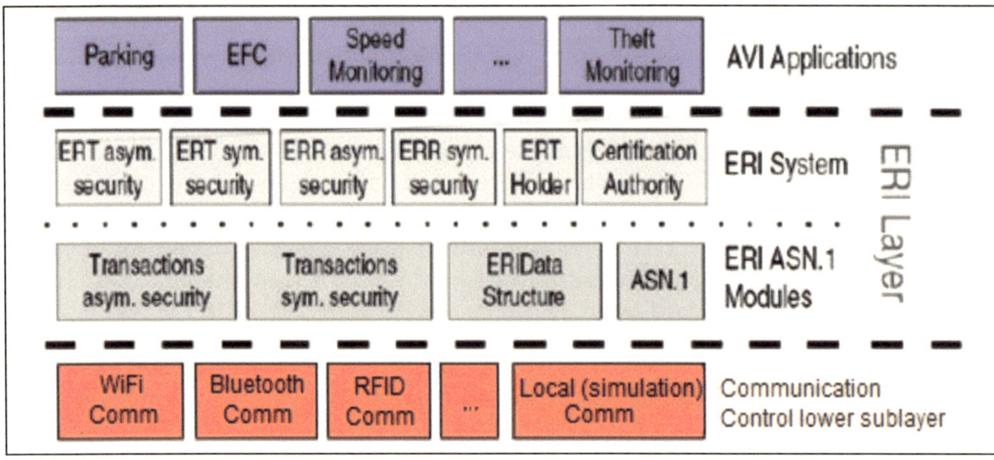

Fig. 5. View of the layered structure of the implemented software

3.1 AVI application layer

In the AVI application layer, several simple applications were implemented, which made use of the ERI functions. This layer was oriented to allow the implementation of the tests for the ERI protocol in different conditions.

3.2 ERI layer

The ERI layer as a two-layered structure has been implemented. The lower sub-layer is the layer for the ASN.1 modules as described in the ISO 24534-3 and 4 [ERI_3], [ERI_4] which is based on the ASN.1 vocabulary and transactions. The upper sub-layer implements the functionalities described in the standard [ERI_5], defining the actors in an electronic registration scenario: ERT (tag or device which stores identification data for the vehicle), ERT Holder (device used to access the owner of the tag locally), ERR (reader or device which allows remote access to the ERT for the service providers or authorities) and CA (certification authorities which generate and manage ERI certificates for remote secure connections).

3.3 Communication and control layer

This layer allows the wireless communications (by Bluetooth, WiFi, RFID) and local communications for simulation purposes.
Figure 6 shows the logical relationships which each module establish with each actor within the whole ERI system. The different colours indicate the different virtual machine which controls the module. The yellow modules are designated to J2ME devices and the orange ones are designed to run in backend service, therefore they work on J2SE, although they are J2ME compatibles.

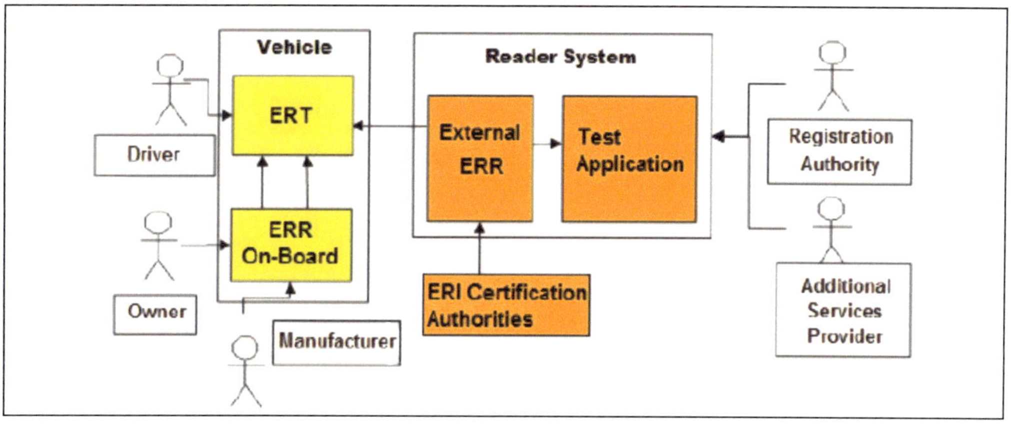

Fig. 6. Schema of the identification system.

Each one of the five parts of the logical system has its functionalities. The description of each logical part is as follows:

- ERT Module (Electronic Registration Tag): it is the Tag installed in the PDA. It has some classes to provide several functionalities and an interface to show data.
- ERT Holder Module: it is in charge of the local tag reading inside the car.
- ERR Module (Electronic Registration Reader): it is in charge of external readings, or any other operation programmed over the tags in the zone.
- Test Application Module: it receives data coming from the reader and makes a data process. It is for demonstrations.
- Certification Authorities: it is the system to generate certificates for security matters.

4. Test of the system

In order to test the API developed for Electronic Registration and Identification, several applications were developed. The tests for these applications have taken into account the range of velocity, distance and the inquiry time.

The communication staffs used in the tests were:

- WiFi router (802.11 b/g) and a pigtail
- Bluetooth dongle (802.15.1) and a pigtail
- RFID reader and 3 RFID tags (ISO 18000-4) at 2.45 GHz
- Patch panel antenna at 2.45 GHz for using with all technologies considered
- PDA containing ERI applications, which is used as ERT
- PC used as ERR with back-office applications (for monitoring the system)

The characterization of the test antenna was done at the anechoic chamber at the University of Valencia. Figure 7 shows a photo of the measurement process.

The measurement of this antenna was made by using the method of two antennas for the measurement of the antenna gain. Figure 6 shows the results for horizontal and vertical polarization.

The communication zone for the wireless technologies between tag and reader was also determined in an outdoor urban site for each one of the technologies used. All of them use the same frequency band at 2.45 GHz, so the covering area of the test antenna has been measured. Figure 7 shows a plot of the power density distributed in the test area.

Fig. 7. Antenna for the reader in the Electromagnetic Anechoic Chamber at the University of Valencia.

Taking into account reading distance and velocity, the EVI applications implemented present different profiles according to ISO 24534-2 [ERI_2]. The applications used:

- Theft vehicles control: static readings and at short distances. The profiles used are ERT-D6 for the velocity and ERT-C7 to C9 reading distance.
- Parking access control: static readings or with low velocity (less than 20 km/h) and at medium distances. The profiles used are ERT-D5 the velocity (the velocity for accessing a parking lot should not be greater than 25 km/h) and ERT-C6 to C9 (distance under 10 metres).
- Average velocity control: dynamic readings (different ranges of measurements) and variable distances.

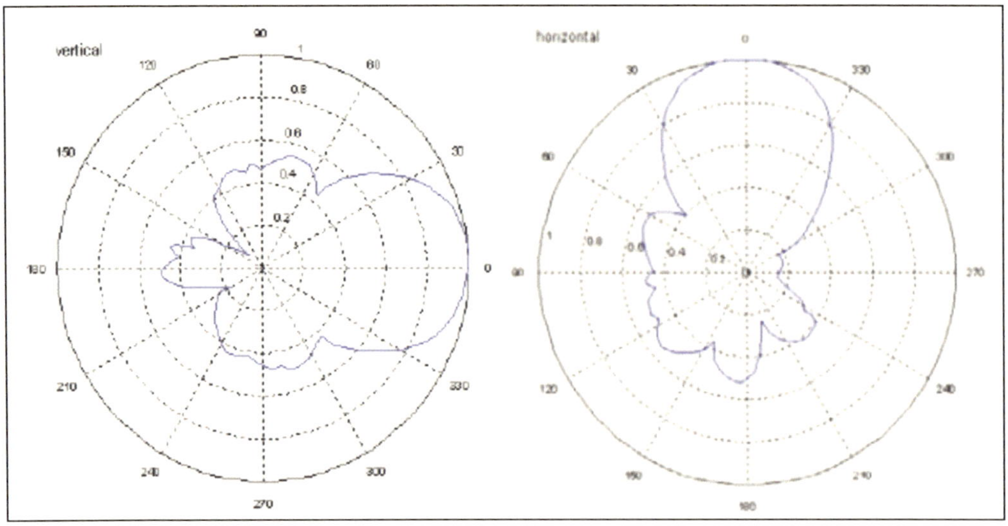

Fig. 8. Radiation pattern for the test antenna in horizontal and vertical polarization.

Tests with the vehicle in movement allowed testing within the communication zone. Communication between tag (inside the vehicle) and the reader was established. (located at the Institute of Robotics).
The tests were made at three communication distances: at 10 metres from the reader antenna (black point), at 22 metres (green point) and at 45 metres (white point). Also several velocities were used: 20 km/h, 40 km/h and 60 km/h (due to velocity limitations in the urban circuit). Figure 9 shows the communication area for the tests.

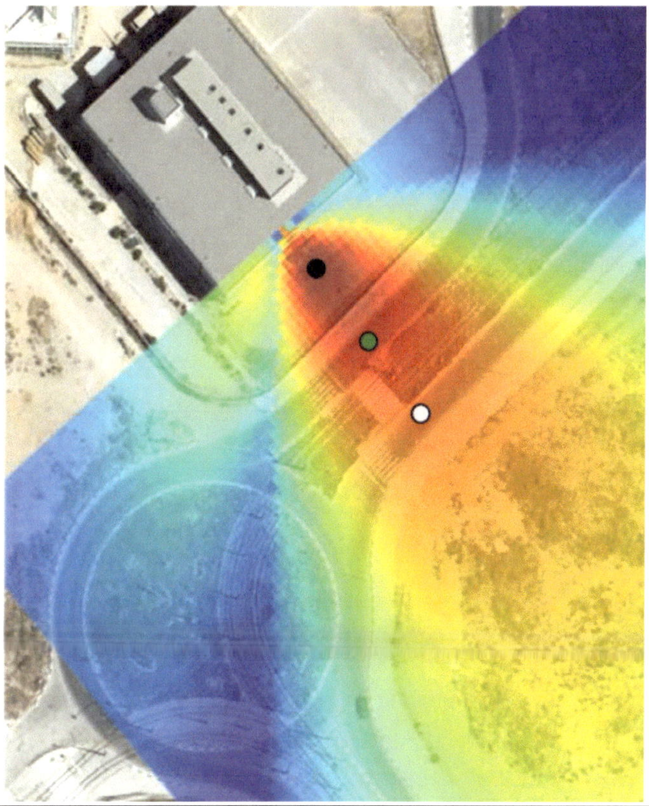

Fig. 9. Power distribution within the communication zone in the test area.

Figure 10 shows a capture of the data transmitted, processed and collected by the reader. The interchanged information is extracted from the ISO 24534-3. The information sent to the ERR was PDU, vehicle ID, chassis number, brand, model and colour of the vehicle.

4.1 Test of applications with WiFi

The applications tested for WiFi technology [802_11] has showed good results in the readings between tag and reader. Each application has been launched and transaction times have been measured with the reader. In accord with other authors, measurements have shown that the discovery time for WiFi is higher [Pering, 2005] than in Bluetooth technology. Figure 11 shows a histogram with these results. Also table 2 shows a summary of the test results for the ERI applications using WiFi technology at different distances and velocities.

```
Initializing
ERR: EriData created
Waiting for connection
Connection stablished from: /147.156.222.234:5555
ERR : initiated GetEriData
ERR : Sent PDU = 64 1 1 -128 1 64 108 101 99 116 111 114 65 49 4 78 106 -48
123 -128
ERR : Received PDU
**************************--ERI DATA--*******************************
** VIN = VS6DW990LK000007
** CHASSIS NUMBER = X500-100-24-7566-A-6454
** BRAND  = RENAULT
** MODEL = 19 XN
** COLOUR = RED
*********************************************************************
```

Fig. 10. Data transmitted and collected by the reader.

	20 km/h	40 km/h	60 km/h
22 m	Works	Works	Some errors in transactions
45 m	Works	Works	Some errors in transactions

Table 2. ERI tests with WiFi

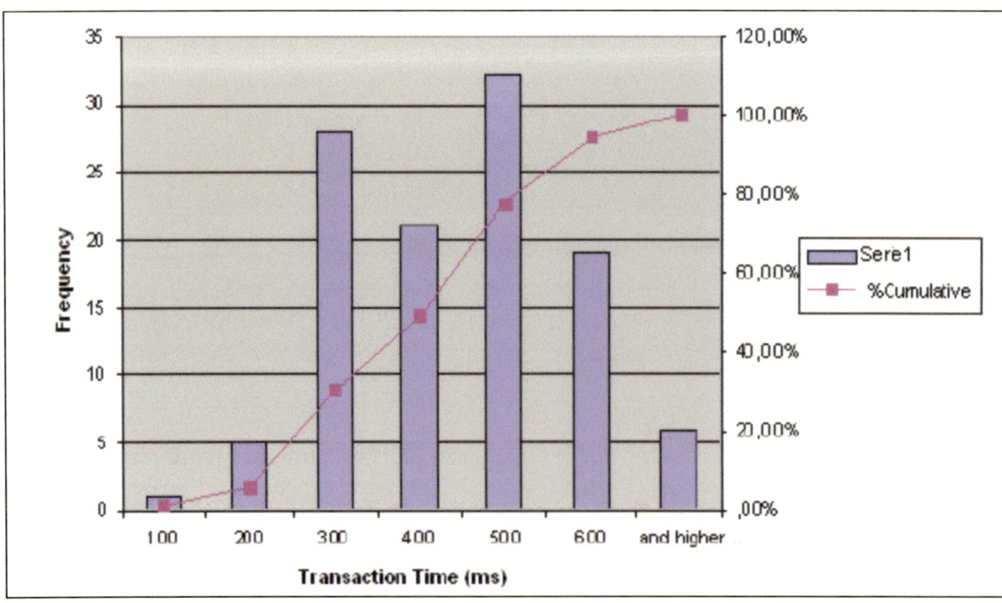

Fig. 11. Histogram with transaction times for WiFi technology.

4.2 Test of applications with Bluetooth

The applications using Bluetooth technology [802_15_1] have also given good results in the readings. Although Bluetooth devices have higher discovery times (between 10 and 40 seconds) than WiFi technology, the transaction times are lower [Pering, 2005] [Jiang, 2004].
Table 3 shows a summary of the tests results for the ERI applications at different distances and velocities.

	20 km/h	40 km/h	60 km/h
22 m	Works	Works	Some errors in transactions
45 m	Works	Works	Some errors in transactions

Table 3. ERI tests with Bluetooth

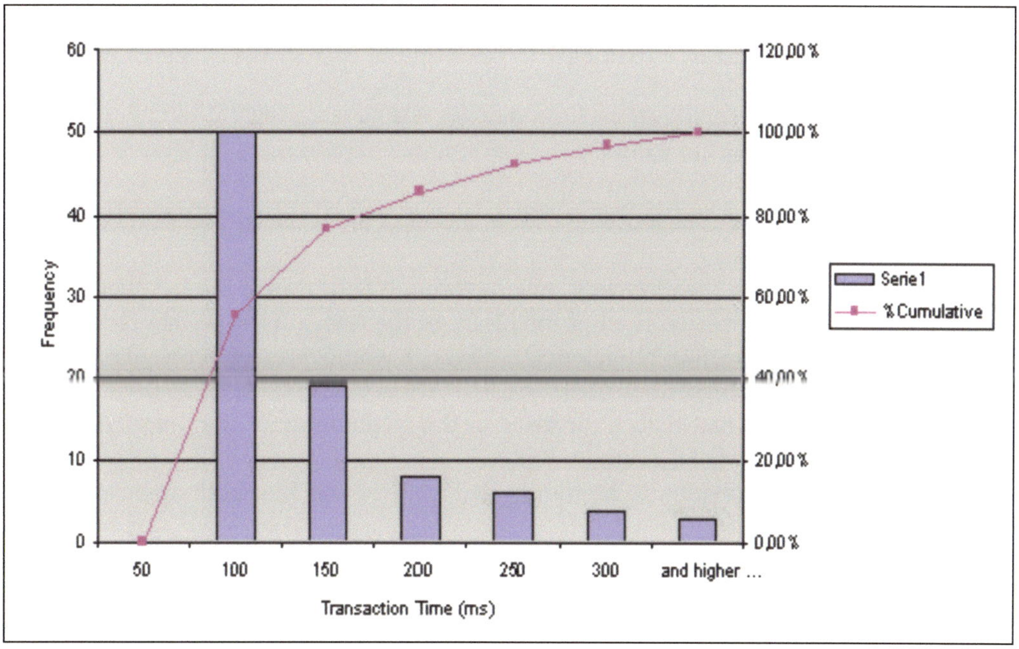

Fig. 12. Histogram with transaction times for Bluetooth technology.

4.3 Test of application with RFID

The RFID technology has limitations regarding the quantity of information to be sent in the communication between tag and reader. It also has limitations regarding security. For these reasons the implementation of the standard, can only be done with the ISO 24535. [RFID_4]
Table 4 shows a summary of the tests results for the RFID technology with the ERI applications at different distances and speeds.

	20 km/h	40 km/h	60 km/h
22 m	Works	Works	Works
45 m	Works	Works	Works

Table 4. ERI tests with RFID

5. Conclusions

This work has produced an API for the development of applications for Electronic Registration and Identification of vehicles has been implemented. This API has been published under GNU/GPL license. (API and documentation can be downloaded from http://smagris3.uv.es/preri).

Several applications for test have also been implemented. Three wireless technologies (WiFi, Bluetooth and RFID) have been proposed. These applications have been installed on the PDA and have been tested in a road next to the Institute of Robotics at the University of Valencia.

The tests of the developed applications have showed that [Segura, 2008]:

- The used technologies are valid for the applications implemented for slow uses (up to 60 kph).
- The discovery time in Bluetooth devices is too high (between 10 and 40 seconds) for dynamic applications.
- WiFi has problems in the discovery process through SSID due to the fact that it is not implemented in the version used of the J2ME. In the tests a fixed private range of IPs was used to simulate this behaviour; however it would present problems in a real implementation.
- WiFi and Bluetooth had high error taxes in the communication for velocities higher than 60 km/h. This is not the case for RFID.
- The amount of information to be transmitted in WiFi and Bluetooth applications are higher than in RFID.
- In case of a high humidity (or in a rainy day), the frequency of these three technologies used (2,45 GHz) has a high power attenuation.

6. Future research

This research has allowed establishing a framework to allow programmers to develop applications for electronic registration of vehicles. From the results obtained, this API can be applied to secure communications between vehicle and infrastructure by using asymmetric and symmetric security.

For future research, we are going to finish implementing Basic ERI (as shown in figure 4). We are also going to implement discovery process in WiFi technology by using a new version of the J2ME and assigning a wide range of IPs in order to avoid problems in a real deployment. Other technologies, as ZigBee (802.15.4) or WAVE (802.11p), are going to be implemented and tested.

7. References

[Inno_Report, 2007] Innovations Report (30.03.2007) "New long-range RFID technology from VTT for Chinese road tolls". <http://www.innovationsreport.de/html/berichte/verkehr_logistik/bericht-81876.html>. Last visit: 30.07.2009.

[ERTICO_D2, 2004] EVI ERTICO: D2 - Final Requirements

[ERTICO_D3, 2004] EVI ERTICO: D3 - High-level architectures, technology & realization options

[ERTICO_D4, 2004] EVI ERTICO: D4 - Final Assessment

[ERTICO_D5, 2004] EVI ERTICO: D5 - Conclusions & Recommendations

[SMART_PLATE] "Smart License Plate", http://www.apec-tptwg.org.cn/new/Archives/tpt-wg28/Intermodal/2006_TPT-WG-28_IEG_018.ppt. Last visit: 03.08.2009

[ERI_1] CEN ISO/TS 24534-1 Road transport and traffic telematics - Automatic vehicle and equipment identification - Electronic Registration Identification (ERI) for vehicles - Part 1: Architecture

[ERI_2] CEN ISO/TS 24534-2 Road transport and traffic telematics - Automatic vehicle and equipment identification - Electronic Registration Identification (ERI) for vehicles - Part 2: Operational requirements

[ERI_3] CEN ISO/TS 24534-3 Road transport and traffic telematics - Automatic vehicle and equipment identification - Electronic Registration Identification (ERI) for vehicles - Part 3: Vehicle data

[ERI_4] CEN ISO/TS 24534-4 Road transport and traffic telematics - Automatic vehicle and equipment identification - Electronic Registration Identification (ERI) for vehicles - Part 4: Secure communications using asymmetric techniques

[ERI_5] CEN ISO/TS 24534-5 Road transport and traffic telematics - Automatic vehicle and equipment identification - Electronic Registration Identification (ERI) for vehicles - Part 5: Secure communications using symmetric techniques

[ERI_BAS] CEN ISO/TS 24535 Road transport and traffic telematics - Automatic vehicle and equipment identification - Basic electronic registration identification

[Pering, 2005] T. Pering, V.Raghunathan, R. Want; "Exploiting radio hierarchies or power-efficient wireless device discovery and connection setup". Proc. 18th International Conference on VLSI. Kolkata, India. Jan. 3-7, 2005.

[Jiang, 2004] JR Jiang, BR Lin, YCh Tseng; "Analysis of Bluetooth device discovery and some speed-up mecanisms". Bluetooth White Papers. National Chiao-Tung University. Jul, 2004.
http://whitepapers.silicon.com/0,39024759,60143957p,00.htm

[802_11] IEEE 802.11 IEEE Standard for Information technology-Telecommunications and information exchange between systems-Local and metropolitan area networks-Specific requirements - Part 11: Wireless LAN Medium Access Control (MAC) and Physical Layer (PHY) Specifications.

[802_15_1] Bluetooth Core Specification Versions * Version 2.0 + Enhanced Data Rate (EDR), adopted November, 2004

[802_15_4] IEEE 802.15 Wireless Personal Area Networks

[RFID_4] ISO/IEC 18000-4 Parameters for Air Interface Communications at 2.45 GHz

[Segura, 2008] J.Segura, J.G.Jordán, M.A.Jaén, F.R.Soriano; "Implementation of the ERI standard and Evaluation of Applications with several low-cost technologies". The 11th International IEEE Conference on Intelligent Transportation Systems. October, 12-15, 2008. Beijing, China. pp. 490-494

17

Integration of Data Across Disparate Sensing Systems Over Both Time and Space to Design Smart Environments

Peter Bajcsy and Rob Kooper
National Center for Supercomputing Applications (NCSA), University of Illinois at Urbana-Champaign (UIUC), 1205 W. Clark, Urbana, IL 61801, USA

1. Introduction

Wireless sensing devices are frequently used in smart spaces, ubiquitous and proactive computing, and situation awareness applications (Satyanarayanan 2001), (Vildjiounaite, Malm et al.), (Weiser), (Ilyas & Mahgoub). One could list a plethora of applications suitable for the use of wireless sensor networks and other sensing instruments, for instance, health care (wellness system for aging), environmental monitoring (pollution of air, water, and soil), atmospheric science (severe weather prediction), structural health monitoring (equipment or material fatigue detection), military surveillance (vehicle movement detection), facility monitoring (security and life-cycle of a facility), wild life monitoring (animal migration), or intelligent vehicle design (obstacle detection) (Dishman), (Gupta & Kumar), (Mainwaring, Polastre et al.), (Wang, Estrin et al.), (Roush, Goho et al.), (East), (Rom'an, Hess et al.), (Abowd), (Dey), (Kidd, Orr et al.). The list of on-going projects that include wireless sensor networks and other sensing instrumentation is also growing every day (see NSF, NIST and DARPA projects such as NSF NEON, LOOKING, SCCOOS, ROADNet, USArray, TeraBridge, ORION, CLEANER, NIST SHIELD or DARPA Active Networks, Connectionless networks, DTT). All projects have in common the fact that they represent multi-instrument and multi-sensor systems that can be characterized as smart outdoor, indoor or embedded spaces. The challenge is to build smart spaces that can intelligently sense environments, gather information, integrate information across disparate sensing systems over time, space and measurement, and finally detect and recognize events of interest to trigger event-driven actions. We have been interested in the hazard awareness application scenarios (Bajcsy, Johnson et al. 2008) (Bajcsy, Kooper et al. 2006) that concern humans due to (a) natural disastrous events, (b) failures of human hazard attention or (c) intentional harmful behaviors of humans. Our focus is on the problems related to building hazard aware spaces (HAS) to alert innocent people, similar to the problem related to swimming pool surveillance systems to prevent human drowning (e.g. Poseidon developed by Vision IQ).

While building a real-time HAS system, one has to address the issues of (1) setting up the system to achieve desired accuracy and (2) operating it to achieve reliable performance with or without human intervention. In order to setup a HAS system, one ought to find ways how to deploy sensors, synchronize them, localize sensors and other instruments in the environment, and calibrate measurements coming from wireless sensors and instruments to obtain values represented in engineering units (for example, a raw value of temperature has to be converted to degrees of Celsius). These steps for HAS preparation allow us to answer questions about when, where and what hazards occur during the operation of a HAS system. In addition, one has to understand the limitations of smart wireless sensor networks (WSN), such as low-power, broadcast range, available on-board memory and CPU, to optimize the layout of sensor networks in terms of minimal wireless loss, minimal energy consumption and maximal information content received from the network. From the perspective of operating a HAS system, the objective is to perform reliable proactive data acquisition, hazard detection, human alert, hazard confirmation and possible understanding of specific hazard characteristics, and finally hazard containment. These building steps have been reflected in our research and development, and are illustrated in the overall HAS schema in Figure 1.

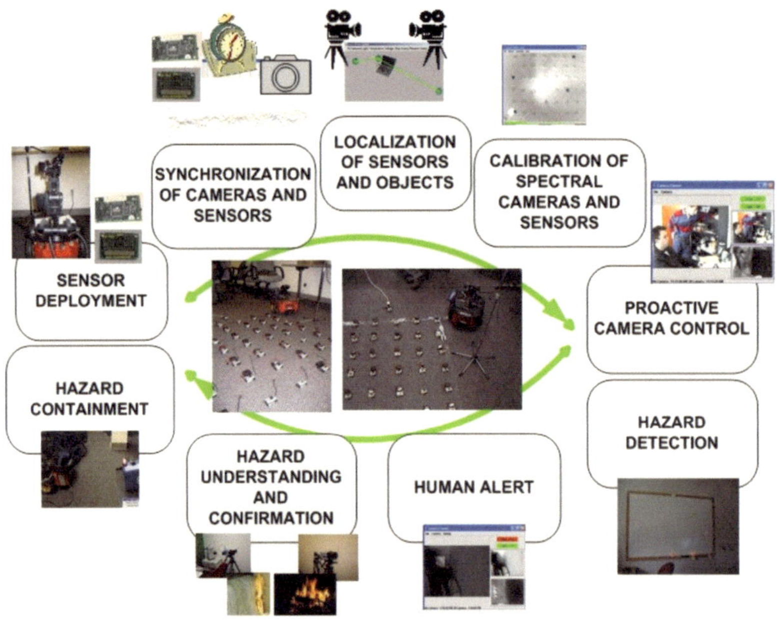

Fig. 1. An overview of several components of the hazard aware spaces (HAS) prototype. The top components represent the setup of HAS while the bottom components correspond to the operation of HAS.

2. Problem description

Research and development of HAS poses several fundamental challenges in the areas of sensing, remote sensor deployment, wireless data acquisition, wireless communication, data integration, distributed signal and image processing, remote and proactive control of sensing, and hazard detection and pattern recognition. It is the understanding of these issues that leads to an optimal real-time system design of HAS. In this chapter, we primarily elaborate on the research themes related to adaptive remote setup of wireless sensor networks. Nonetheless, as one part of the design, we also investigate (a) sensor measurement accuracy and optimal choice of acquisition parameters, (b) information selection for communication bandwidth control, (c) hazard understanding from sensor and image data, (d) human-computer interfaces for human alert, and (e) the use of robotics in HAS application domains. The research themes map into a development of technology components illustrated in Figure 1 and include (1) deployment of point sensors using remote robot control, (2) synchronization of sensors and cameras, (3) localization of sensors and objects, (4) calibration of measurements from sensors and spectral cameras, (5) proactive camera control, (6) hazard detection, (7) human alert, (8) hazard confirmation and understanding, and (9) hazard containment. The rest of this book chapter presents hardware and software for building a prototype HAS system, and theoretical and experimental solutions to the aforementioned technology components.

3. Hardware and software description

In the HAS system design presented, we used the MICA hardware that is manufactured by Crossbow Inc. The MICA hardware consists of (a) 4MHz Atmega 128L processor, (b) 128K bytes Flash, 4K bytes SRAM and 4K bytes of EEPROM, (c) 916MHz radio transceiver with a maximum data rate of 40Kbits/sec, (d) AA battery pack attached to the processor, and (e) plug-in sensor boards like the MTS101CA, connected through a 51-pin expansion connector. For more details, see (Hightower & Borriello), (Hollar).

The MICA sensors are deployed using an intelligent wheeled robot P2DX8 made by ActivMedia Robotics, Amherst, NH, and an on-board computer-processing unit for real-time processing. The robot has a ring of eight forward sonar sensors that can be used for obstacle avoidance and a two-wheel drive plus balancing caster for smooth motion. The robot is connected to a laptop that is either directly cable-connected to the local area network or wirelessly connected with other computers. In order to control the robotic deployment of MICA sensors, we used keyboard-, gesture- and voice-driven interfaces. For the gesture-driven remote control of the robot, we used the IS-300 Pro Precision Motion Tracker by InterSense, with the update rate of 500 Hz, weight of 15 oz, and angular resolution of 0.02 Deg. It measures yaw, pitch and roll using a miniature solid-state integrated inertial instrument "InertiaCube" and these temporal signals serve as inputs to our gesture recognition algorithm. For the voice-driven remote control, we used wireless audio sensors by Audio-technica Corp. To obtain video feedback from the robot, we mounted a pair of wireless miniature color cameras by Samsung on the robot's platform. To find the location of deployed MICA sensors, we equipped indoor hazard aware spaces with RFID tags and mounted an Alien Technology RFID Reader on the robot to find its location in the building based on detected RFID tags with known locations.

The hazard aware space was also equipped with a visible spectrum camera (Network Color Camera SNC-RZ30N PTZ Pan/Tilt/Zoom by Sony, and Canon PowerShot SD100 digital camera) and by a thermal infrared (IR) camera, (the Omega model by Indigo Systems Corporation, Goleta, CA). The thermal IR camera is a long-wavelength (7.5-13.5 microns) uncooled microbolometer camera designed for infrared applications. It is controlled via RS232 serial port and the analog NTSC video output is digitized using a Hauppauge WinTV board. For temperature calibration experiments, we used a regular thermometer used by chemists as the temperature gauge. It is measuring temperature directly in engineering units of degrees Celsius and providing temperature readings in the range [-40°C, 150°C] with a reading uncertainty equal to ±1°C. A set of preliminary experiments to discriminate burning materials was performed with a hyperspectral camera by Opto-Knowledge Systems Inc. This camera is based on a liquid crystal tunable filters (LCTF) technology and operates in two wavelength ranges, such as visible [400nm, 720nm] and near infrared [650nm, 1100nm].

4. Localization of sensors and objects

One of the key aspects of any HAS system is the knowledge of where hazards occur based on sensor locations. The general problem of 3-D information recovery from sensor measurements has been addressed in the past by many researchers in the computer vision, machine vision and signal/image processing communities (Dario, Bergamasco et al.), (Marr), (Priyantha, Balakrishnan et al.), (Wechsler) and in the wireless communication community (Whitehouse), (Whitehouse & Jiang). The motivation for obtaining 3-D information often comes from applications that require object identification, recognition and modeling.

In order to detect hazards, one has to deploy sensors in indoor or outdoor environments. The sensor deployment can be achieved manually by a human or automatically by a robot operating in an autonomous mode. The manual deployment can be accomplished by placing sensors either in-situ or remotely with the help of a robot. Every time a sensor is deployed, the problem of localizing the sensor arises since the sensor readings have to be associated with the location for a hazard to be spatially located.

In general, there are three approaches to the sensor localization problem. First, a person who deploys a sensor records also the location and the unique identification (ID) of a sensor. The ID of a sensor is sent with every measurement to a base station where the ID is converted to a location according to a look-up table prepared by a human. Second, a robot that deploys a sensor uploads its location to the database when the sensor is deployed, the ID of the sensor is then sent with every measurement and the location can be found based on the ID of the sensor. Third, a sensor after being deployed determines its own location by communicating with other sensors or beacons with known locations and transmits the location back with each sensor reading. In all three approaches, the localization information could be defined or found in a relative or absolute coordinate system.

The three approaches above have associated tradeoffs between the reuse of sensors and the cost of accurate localization In order to reduce the cost of the system, reusing sensors is preferred over placing sensors once and not moving them again. However, when sensors are relocated in space over time, the manual localization approaches lead to an increased cost of the labour needed to constantly update look-up tables with the new location of the smart sensors. While the sensor-driven auto-localization saves the cost of manual labour, there is a cost in power/energy spent by the smart sensors to run localization algorithms. For instance, in order to perform acoustic time-of-flight localization, all sensors have to communicate using radio and acoustic signals which are very power-consuming operations. The power expenditure leads to frequent replacements of batteries in smart sensors and hence increased costs of human labour and battery replacements. By analyzing the tradeoffs of multiple sensor localization approaches, we decided to explore two solutions based on the approach described as a robotic deployment of sensors.

Next, we outline the two solutions for localizations; one using stereo and a ‘smart’ sensor network (the MICA sensors) and the other one using Radio Frequency Identification Tags (RFIDs). The first solution aims at accuracy of localization after the sensors have been deployed by combining the results of acoustic time-of-flight ranging and stereo vision algorithms. The second solution aims at power efficiency of localization by deploying RFIDs in indoor environments prior to hazard sensing and recovering the localization information from the locations of RFIDs. The uniqueness of the RFID-based solution lies in the fact that the tags have fixed locations in the environment, and the reader is mounted on a robot and moving in the environment. This is in contrary to the majority of existing RFID-based applications where the RFID reader is static and the RFID tags are dynamic. After identifying dynamic RFID reader location, the MICA sensors for hazard sensing are uploaded with the location information prior to being deployed, and hence the 'smart' sensor batteries can last longer. Before describing these two solutions as the result of our design optimization analyses, we start by introducing the problems and solutions for remotely controlled sensor deployment using a robot.

4.1 Remotely controlled sensor deployment

The remote robot control system consists of three basic software components including (a) acquisition and recognition of control commands from multiple inputs, (b) client-server network communication, and (c) command fusion and execution by a robot and its arm. We used multiple input modalities to generate the control commands for the robot, such as voice and gesture recognition based controls of a robot. For the voice recognition, we considered wired or wireless microphones. In the gesture recognition system, we used wired orientation sensors mounted on human arms. Finally, our prototype remote robot control system included the mouse and keyboard interfaces, as well as a command line interface with files containing scripted command sequences.

The set of gesture commands is based on the US Navy lexicon for navigating aircrafts on the ground (Lementec & Bajcsy), (Urban, Bajcsy et al.). The set of voice and keyboard commands is user defined and maps in our case to the same US Navy lexicon for navigating aircrafts. Fusion of multiple commands is performed by (a) analyzing time delays and (b) assigning different priorities to commands and the clients issuing those commands (Urban & Bajcsy). Consistent and conflicting commands are considered before a selected command is executed by a robot. For an emergency control, a video signal is sent to a monitoring station. The overview of the system with multiple inputs is presented in Figure 2.

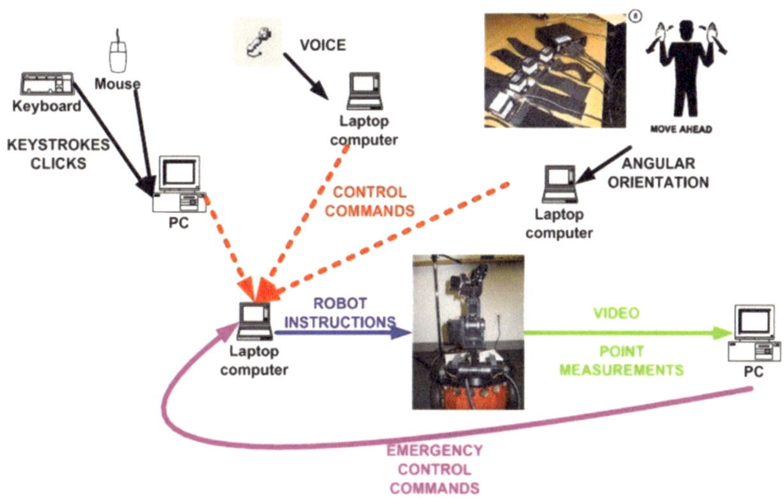

Fig. 2. An overview of a system for remote robot control using sound, gesture and human-computer interface inputs.

First, we implemented an application for the user to control the robot using Human Computer Interfaces (HCI) such as a mouse and a keyboard. A user can use a keyboard and type in commands of his choice and their corresponding parameters.

Second, we developed a template based speech recognition system so that typing can be replaced by more user friendly interface. A database of sound templates is formed by recording commands and extracting Linear Frequency Cepstral Coefficients (LFCCs features) defined by the equation 1, where $i = 0,1,\ldots, P\text{-}1$, Y_k is the audio sample at time k, P is the number of LFCCs features equal to 10, and the set of LFCCs is generated for each K= 256 audio sample points. The audio signal is filtered first using a 4th order high pass Chebyshev filter, to reduce the low frequency background noise. Then, the amplitudes of short length, high amplitude blips, glitches, and spikes are reduced to zero before LFCCs features are extracted. We used the Dynamic Time Warping (DTW) algorithm to match any new audio command to all templates previously created. The template with the shortest DTW error distance is selected unless it is above a pre-defined minimum recognition threshold, in which case the input command is classified as unrecognized.

$$LFCC_i = \sum_{K=0}^{K-1} Y_k \cos\left(\frac{\pi ik}{K}\right) \tag{1}$$

Third, we added a gesture recognition system to accommodate remote control execution in very noisy environments, for instance, an aircraft carrier deck. While there are many approaches to gesture recognition, we chose to research and develop a solution with active sensors IS300 Pro due to our objective to achieve performance robustness and reliability. Given the choice of an active sensor, our approach to the problem of gesture recognition is based on (1) translating arm motion into a temporal sequence of orientation of angles, (2) describing a sequence of orientation angles with its characteristics, (3) building models of gestures in a lexicon using sequence characteristics of orientation angles, and (4) classifying sequences of orientation angles into gesture classes according to the developed gesture models in real time. The basic premise of our approach is an existence of a unique mapping between human gesture represented by arm movements and a temporal sequence of upper arm and forearm orientation angles. Our approach to robust gesture recognition relies on a two-stage classification technique. The first stage characterizes temporal streams of each Euler angle separately. The second stage uses the combination of Euler angle stream characteristics from the first stage to assign gesture labels according to a set of gesture classification models. At any moment each of the 12 streams of angular values from four orientation sensors are labeled as, steady, oscillating or unclassified, and high, medium-high, medium, low-medium and low. These labels are then use to define each gesture model.

Finally, we enabled robot arm control via mouse and keyboard interfaces in order to perform simple loading and unloading operations. For emergency control purposes, we mounted a wireless camera on the platform of a robot to obtain video feedback[1]. In terms of system architecture, the software is designed based on a client-server paradigm as shown in Figure 3. All input devices (microphones, orientation sensors, keyboard and mouse) are attached to multiple computers that represent the clients in the developed system. In our

[1] See a video at http://isda.ncsa.uiuc.edu/gallery.html

laboratory experiments the robot is connected to a laptop using the RS232 connection. This laptop acts as a server and accepts TCP client connections over the network. Each client can issue control commands to the robot by sending commands to the server laptop. The server fuses the commands from all clients, and resolves any conflicts that may occur. After command conflicts are resolved the commands are translated to a set of robot instructions. These instructions are sent to the robot via the RS232 connection, and are then executed.

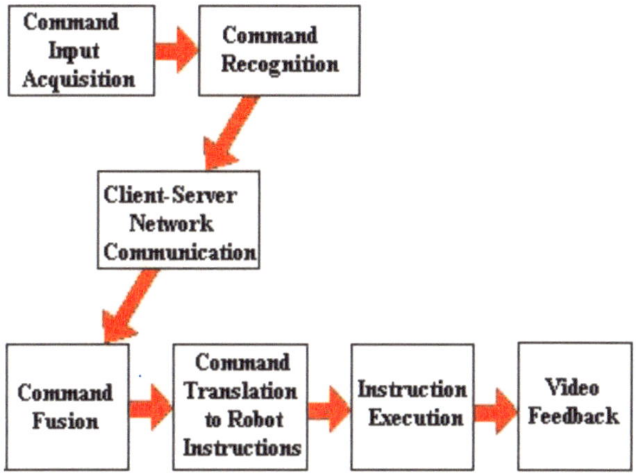

Fig. 3. The signal processing and communication flow of a remote robot control system.

4.2 Location awareness using stereo and MICA sensors

The problem of 3-D information recovery is difficult regardless of whether it addresses static or dynamic object location estimation. In the past, the problem of depth recovery was approached, for example, (a) by vision techniques referred to as shape from cues (Pankanti & Jain) where cues can include stereo, motion, shading, etc., and (b) by communication techniques frequently referred to as location sensing (radio or ultrasound time-of-flight lateration or signal strength analysis (Hightower & Borriello), (Patwari, Ash et al.)). Although the vision and location sensing techniques have been proposed, very few methods are robust and accurate enough to be used in real-time applications. It is well known that many of the depth estimation algorithms are computationally expensive with limited robustness and accuracy in most unconstrained, real-life applications. The need for improved robustness and accuracy of depth estimation motivated our work on stereo and wireless sensor location fusion.

Our approach to the 3D information recovery problem is based on fusing localization data from wireless sensor networks with depth maps obtained through computer vision stereopsis (Scherba & Bajcsy). One could envision performing (1) depth map calibration, (2) sensor location calibration, or (3) depth map and localization fusion. A flowchart depicting the entire process from raw data to calibrated or fused information is shown in Figure 4. We have performed several experiments with synthetic and measured data using the Crossbow MICA2 motes, TinyOS, and Image to Learn (Im2Learn) implementation of the stereo algorithm (Bajcsy, Lee et al.).

Acoustic time-of-flight ranging was implemented according to Figure 5. The first step is to send a message to a ranging endpoint node. The endpoint node, after receiving the message, simultaneously broadcasts a radio ranging message with a 4 kHz chirp. Every node in the

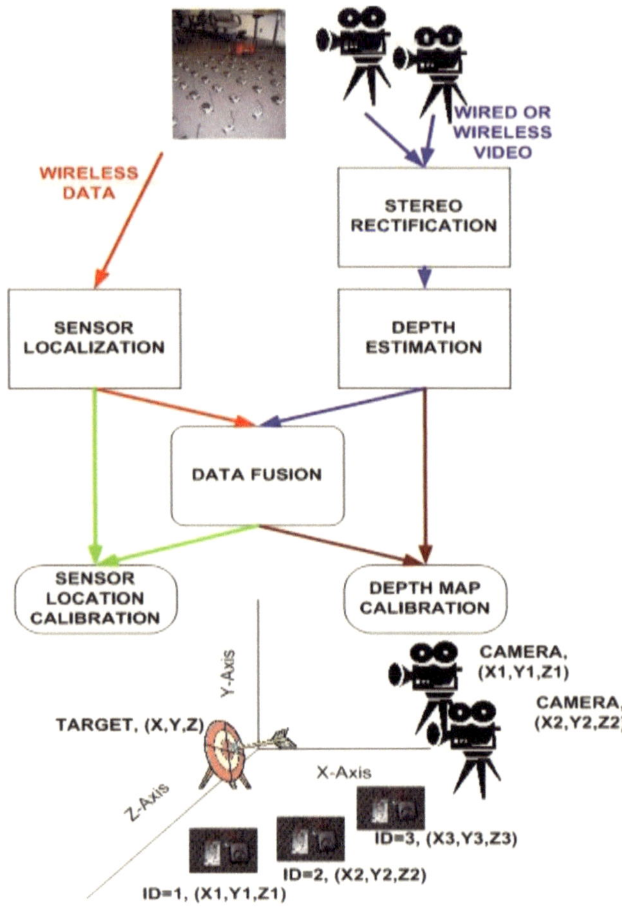

Fig. 4. Flowchart of Sensor Fusion.

network is configured to listen for the radio ranging messages and starts a timer which stops when the audible chirp is heard. A broadcast message announcing the distance between the endpoint and receiving nodes is then sent for all who are interested. Ranging is possible in this setup due to the differential in radio transmission speed (governed by the speed of light, the radio stack, and system-level issues) and the speed of sound in the sensing environment (we use 346.65 m/s for our experiments which corresponds to the speed of sound in air at 25°Celsius). The granularity of the timer on the receiving nodes primarily dictates the uncertainty in the ranging estimates.

Stereopsis is the construction of three-dimensional geometry given multiple views of a scene. The use of stereopsis leads to a depth map that is simply an image of a scene with pixel values given by the depth of each scene point from the camera (the minimum distance from the scene point to the camera plane). We implemented the stereopsis algorithm according to (Hartley). It consists of stereo rectification step (a process which aligns images such that matching points in the resulting images are on the same scanline) followed by pixel matching. In the rectified images, everything can be expressed in terms of disparity that maps inversely to a depth estimate. The stereo algorithm and the acoustic time-of-flight ranging algorithm are described in (Scherba & Bajcsy).

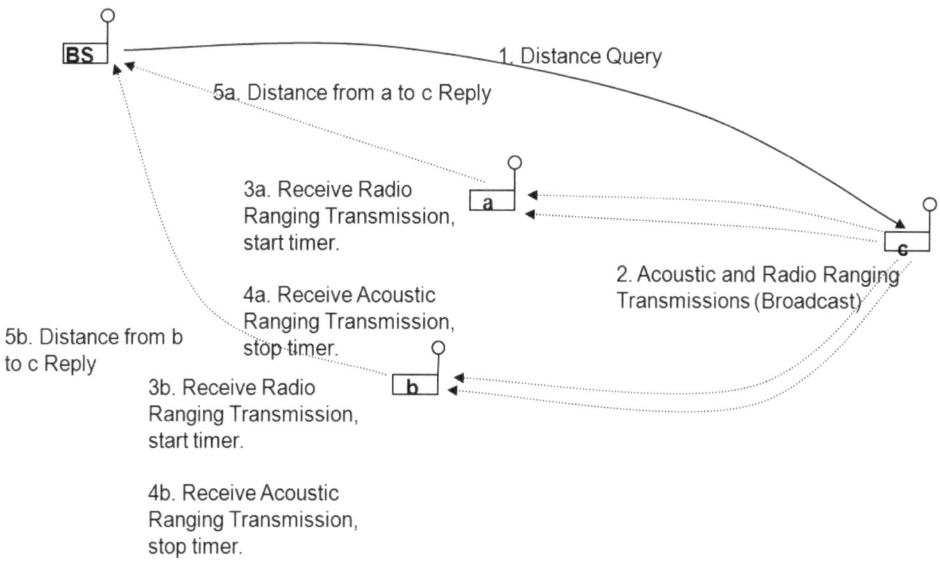

Fig. 5. Acoustic Time-of-Flight Ranging. BS is the base station connected to a computer. The MICA2 motes are labeled a, b and c.

The fusion of sensor localization and stereo depth map results is performed by (a) registering the localization and depth map data, (b) estimating the uncertainty of localization and depth map data as a function depth distance, and (c) fusing the two data by minimizing the uncertainty over the entire depth range. The registration problem is approached by either global optimization or local model-based fitting. The global optimization is achieved by minimizing the difference between depth values and localization values in the least squares sense by solving a non-linear set of equations (number of MICA sensors is equal to the number of equations) using a downhill simplex search. The local model-based fitting approach assumes that a set of *apriori* known sensors is co-planar. Then, the registration is performed by (a) fitting a 3-D surface to a set of *apriori* known co-planar sensor locations, and (b) computing the registration transformation parameters.

In order to fuse the data, the uncertainty of localization and depth map data as a function depth distance is estimated theoretically and verified experimentally. The theoretical uncertainty estimates are derived from a stereo depth disparity equation and from modeling point-point ranging/localization error. Figure 6 shows the determination of a fusion threshold for a particular choice localization and depth map uncertainties. We also developed simulation capabilities for any range of input uncertainties to determine desired fusion thresholds.

To quantify the benefits of fusion, we performed laboratory experiments. The pair of input stereo images is shown in Figure 7 (left) and the MICA sensors were spaced along the depth axis of both cameras. We fused the localization and stereo depth map data sets by minimizing the uncertainty over the entire depth range. The resulting depth map is shown in Figure 7 (right). The accuracy improvements due to fusion for localization and depth map estimation were 95% in our experiments.

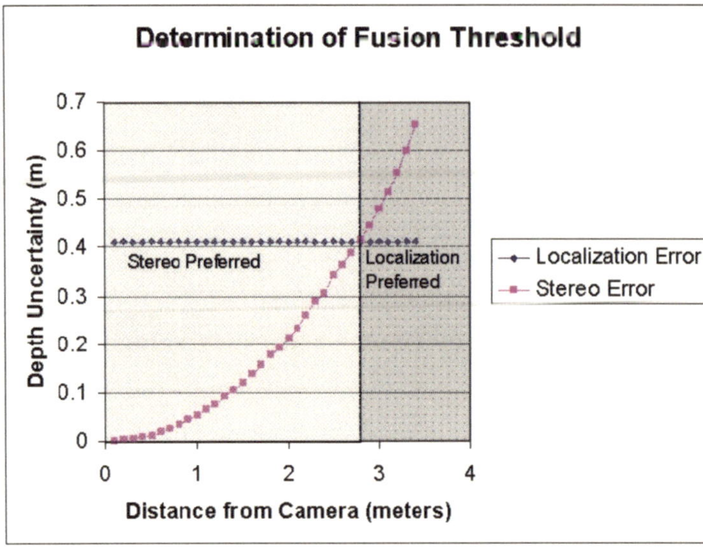

Fig. 6. Fusion decision rule for an image matching error of 4 pixels (stereo error) and a point-to-point ranging error of $\sigma = 0.03m$ (localization error).

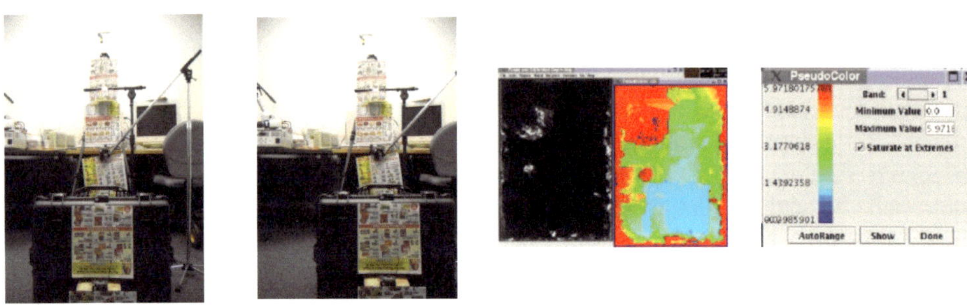

Fig. 7. Left two images show a stereo pair of images taken for quantifying the benefits of fusion in a laboratory setup. To the right from the stereo pair, the images show the resulting depth map after fusing localization and stereo depth map data. Left image is the input depth map before fusion, middle is the pseudo colored depth after fusion, and right is the color legend showing the range of depth values in meters in the pseudo colored depth map.

4.3 Localization with radio frequency identification tags

The aim of this localization approach is to investigate a more power efficient solution by using passive RFID tags compared to the approach using acoustic time-of-flight ranging. Hazard aware spaces might be equipped with passive RFIDs at known fixed locations that could be used for localization of a robot deploying smart MICA sensors. There is a need to explore the cost of deploying RFID tags, their maintenance, robustness and fault-tolerance of reading RFID tags, as well as the accuracy of RFID based localization as a function of the spatial distribution of RFIDs.

In order to evaluate passive RFID technology, we envisioned the following scenarios.

A robot (or person) is moving in a space containing RFID tags, and has a priori knowledge of the passive RFID tag locations. The sensor is an antenna (Alien Reader) that detects tags, and obtains their ID and possibly other information, like temperature, position, etc. The types of RFID tags we used are shown in Figure 8. Our goal here is to use the RFID Reader to detect tags and, given a previously determined map locations of tags, globally localize itself in the space. The location information is uploaded to a central database and then the MICA sensor is deployed by a robotic arm. Future signals transmitted by the MICA sensors will be tagged with the ID of that sensor and the location can be found in the central database. Additional RFID tags on objects in the room will also be used for robot collision avoidance and for hazard material understanding. Thus, the main focus is to incorporate passive RFID tags into the HAS system by (a) building a sensor model for passive RFID tag detection, and (b) creating a suitable localization algorithm.

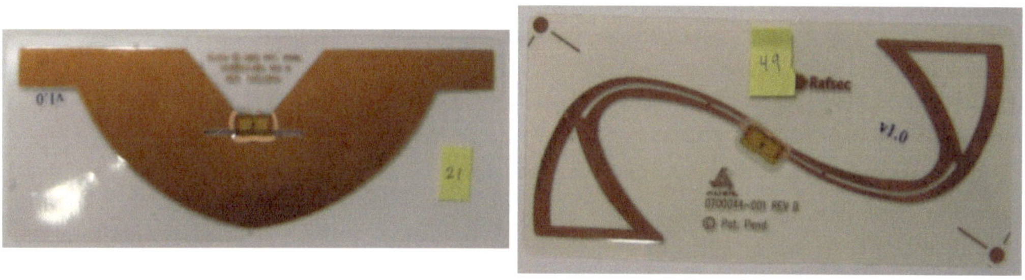

Fig. 8. Tag types used in our experimental analysis.

The key experimental elements to be evaluated during testing can be summarized as follows: (a) Reader and tag geometric configuration, (b) tag detection behavior during occlusion, (c) Reader and tag relative speed, (d) tag density, (e) tag type, (f) and Reader type. Our experimental results show that accurate modeling of the RFID Tag and Reader configuration is the key for Tag detection performance for a given type of Reader and Tag, but once this is done, robust global localization can be achieved. Our results also show that occlusion material between the Reader and Tag is not a very large issue unless that material is metallic, in which case tag detection does not occur, or is severely perturbed.

The process of detecting an RFID tag depends on many factors: antenna footprint, the distance between the antenna and tag, and the relative orientation between the antenna and tag. We have identified so far the robustness of RFID tag detection to be dependent on the following main attributes: (1) geometry of RFID tag versus reader configuration, (2) relative motion of RFID tag versus reader and (3) media properties between RFID tag and its reader (e.g., occlusion material). These dependencies introduce the possibility of false-negative and false-positive RFID tag readings. We have ignored in our RFID tag detection model the non-zero likelihood of RFID tag or reader failure and it could be incorporated after a basic probabilistic RFID tag detection model is developed. For now, we assume that the RFID tag type is not a major contributor to the robustness of RFID tag detection.

4.3.1 Related work

A technology that performs a similar function to the localization task that we wish to accomplish is floor-based sensor systems (Kaddoura, King et al.). Using floor-based sensors such as pressure sensors can be quite useful, but they obviously limit use to an environment with an accessible floor. Also, in order to be low-cost, it is likely that floor-based sensing will only be able to provide 2-D tracking, and not orientation tracking, due to the difficulty of attempting to detect in which direction a person or robot is pointing. Other costs of floor-based sensing involve power usage, wiring, installation of electronics, etc. Also, there are interesting developments in the use of smart sensor technology combined with computer vision for the purpose of tracking objects in a space (El-Zabadani, Helal et al.). The use of passive RFID technology has potential to make simple and robust contributions to such problems, especially when the object being localized is a robot.

Our theoretical framework follows the work of Hahnel et al., (Hahnel, Burgard et al.). The probabilistic sensor model for the RFID antenna was derived for the same technology components (a Pioneer robot and Alien reader/RFID tags) but the geometrical configuration of readers, the number of readers, and the elevation and orientation of RFID tags were different from our experiments (compare Figure 9 and Figure 10). Figure 9 (right) illustrates the angular and range dependency of the RFID tag detection reported in Hahnel et al (Hahnel, Burgard et al.). Also, while Hahnel et al. presented an analysis of the use of augmenting a laser rangefinder with RFID tags during mapping and localization, our objective is to evaluate the strengths and weaknesses of using passive RFID tags as the sole sensor during localization.

Fig. 9. This figure, adopted from (Hahnel, Burgard et al.), illustrates the experimental setup and learning a probabilistic sensor model for the RFID antenna.

4.3.2 RFID tag detection robustness

After preliminary studies of passive RFID technology, we realized the importance of understanding the behavior of RFID tag detection under various conditions, both nominal and off-nominal, for robust performance. Only by examining tag detection behavior will one be able to match the capabilities and limitations of passive RFID technology with their specific application. In our preliminary experiments, we analyzed four main areas: (a) tag detection under varying Tag/Reader geometric configurations, (b) tag detection under varying tag speeds, (c) tag detection during tag occlusion, and (d) tag detection under various tag spatial densities.

Testing tag detection under varying tag/Reader configurations: In this area, we performed a variety of simple experiments, which are summarized below.

- Measurement: lay tag on floor and find average detection range. Result: we found that the average detection range is less than 25 inches.
- Measurement: suspend tag in the air and find average detection range. Result: we found that the average detection range is on the order of 7 feet.

- Measurement: lay tag on floor and find maximum detection range for various angles of the tag with respect to the reader. Result: we found that the maximum detection range varies by approximately 15% depending on the orientation of the tag. This is consistent with the observation that the physical tag is not built with rotational symmetry (i.e. the tag's antenna does have a preferred orientation with respect to the Reader).

Testing tag detection under varying tag speeds: For tag positions in the main detection range of the reader, we found that the Reader detects tag's moving at all practical indoor speeds, with virtually no change in detection rate. We tested this by physically waiving tags, running by the reader, etc. The reader always detected the tag promptly.

Testing tag detection during tag occlusion: We performed simple occlusion experiments which involved completely covering the RFID tag with various materials. We performed these experiments with plastic, wood/paper, and metal. We found that plastic, wood, and paper (up to approximately 2-4 inches thick) have virtually no effect on the nominal detection range, whereas any metal (tinfoil, aluminum sheet metal were specifically tested) covering the tag will completely prevent detection.

Testing Tag Detection with spatially dense tag distribution: Measurement: put roughly 15 tags in the main detection area of the Reader, and see if there is a processing bottle-neck.

Result: We found that when approximately 10 - 15 tags are simultaneously detected, there is a bottle-neck in the Alien Reader processing, and delays of up to 5 seconds occur between reported tag detections (vs. 5-10 ms between reported tag detections when only a few tags are being detected).

To summarize these observations: The dependencies noted above introduce the possibility of both false-negative measurements, in which case a tag is in the expected nominal detection range, but does not get detected, and false-positive RFID tag measurements, in which case the tag is outside the expected nominal detection range, yet still gets detected. In addition to this, it is also noted that depending on the specific tag/Reader configuration, extremely different tag detection behavior can occur. Thus, the tag/Reader configuration is probably the most important factor during the detection process (for a given type of RFID Tag). This complex behavior requires one to spend much effort in building an accurate sensor model, which is discussed in the next section.

4.3.3 Building a sensor model

As will be discussed further later, it is common during probabilistic localization to make use of a sensor model which specifies a likelihood of a sensor measurement given the position of the robot (or person). In this section we present a methodology for developing such a sensor model for use during localization with passive RFID technology. As noted above in discussion regarding tag detection under varying Tag/Reader configurations, it is necessary to build a sensor model which is specifically appropriate for the Tag/Reader geometries and configurations that will occur during deployment of the tracking/localization system. This is because the tag detection behavior changes drastically between different Reader/tag configurations. A simple model will not be sufficient if one's actual implementation differs in any number of fundamental ways, some of which are extremely difficult to foresee before the actual implementation. For example, in our experiments, we mainly worked with RFID Tags distributed in a grid on a carpeted floor, which leads to a very different sensor model than that obtained in (Hahnel, Burgard et al.) (compare Figure 9 and Figure 10). When laid out on a carpeted floor, the maximum read range was roughly 90 cm. In contrast, when tags are held in the air in front of the Reader, the maximum read range is on the order of 200 cm.

This behavior led us to develop the following methodology for building a sensor model. First, one must determine the possible Reader/tag configurations and geometries that will arise in their given implementation. Next, one must fix the Reader's location, and distribute a spatially dense array of tags in the very configurations and geometries that are expected. This array of tags is then used to determine the "footprint" of the Reader, and more specifically, the likelihood of a tag detection given the relative position of the tag and Reader. The sensor model then becomes a manifold of detection likelihoods, which can be used directly in the localization algorithm described below.

In our experimental analysis, we followed this procedure with great success. We distributed tags in front of the Reader (Figure 10 left), and collected statistics on tag detection while varying the angle of the Reader. We then used that information to derive the manifold of detection likelihoods over the space of possible geometries.

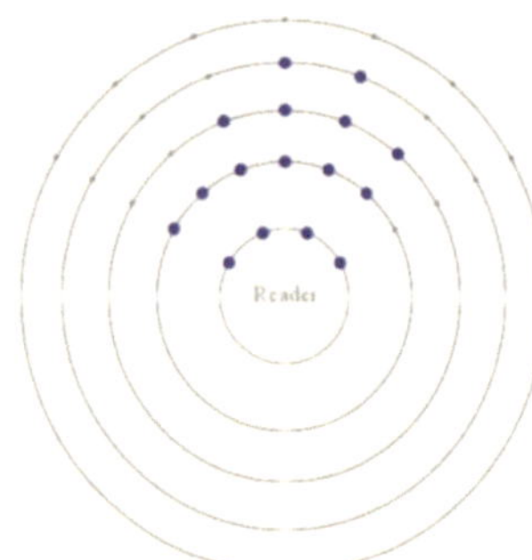

Fig. 10. Left: documentation of setup used to build the sensor model. Right: Image showing nominal detection range when Tags are distributed on the floor and the Reader is mounted on the robot and tilted forward 30 degrees. The 4th "row" has a radius of rough

4.3.4 Probabilistic localization

There are many methods available to perform localization and tracking. We present one probabilistic localization method here, in order to quickly evaluate the accuracy and robustness of using RFID technology as the sole sensing capability during localization. The algorithm we used to perform probabilistic localization, derived from (Choset, Hutchinson et al.), is very simple, but also very computationally expensive. The problem of localization is to estimate the state $x \in X$ of the robot, where X is a discretized state space consisting of possible 2-D positions and orientations. So, for each time step k, we estimate the *posterior probability* $P(x(k) \mid u(0{:}k\text{-}1), y(1{:}k))$ over all possible states, where $y(1{:}k)$ represents the sensor measurements obtained at times 1,..., k, and $u(0{:}k\text{-}1)$ represents motions taken at times 1,..., k - 1. Note that during a given time step, k, it is possible to detect multiple RFID tags, where the likelihood of a tag_m detection is independent of whether or not tag_n was detected (for all m, n such that $m \neq n$). Also note that as background knowledge throughout this analysis, we have a given map of RFID tags.

The key to this probabilistic localization scheme is the recursive Bayesian filtering equation, which specifies how to use the last estimate (the *prior*), in conjunction with current sensor measurements and assumptions about current motions, to calculate a current estimate of the robot's state (the posterior):

$$P(x(k)\,|\,u(0:k-1),y(1:k)) = \eta(k)P(y(k)\,|\,x(k))\Sigma_{x(k-1)\in X}(P(x(k)\,|\,u(k-1),x(k-1))P(x(k-1)\,|\,u(0:k-2)y(1:k-1)) \quad (2)$$

Above, $P(x(k-1)\,|\,u(0:k-2)y(1:k-1))$ represents the prior, or the probability that the robot is at location $x(k-1)$ before motions or sensor measurements are taken into consideration. The term $P(y(k)\,|\,x(k))$ is called the *sensor model*, which represents the likelihood of the measurement $y(k)$ given the robot (or person) is at location $x(k)$; and the term $P(x(k)\,|\,u(k-1),x(k-1)$ is called the *motion model*. In our application, we used a Gaussian distribution centered on $x(k-1)$ as the motion model, and the sensor model is determined as according to the procedures described in the previous section. The value $\eta(k)$ is normalization constant to ensure that $P(x(k)\,|\,u(0:k-1),y(1:k))$ sums up to one over all $x(k)$. We adopted an algorithm from (Choset, Hutchinson et al.), pg. 313, which we used to compute the above posterior at each time step. We initialized the prior to be a uniform distribution across all possible states, which allows us to perform a global localization.
Note that, because we are representing possible states using a discrete grid, the above algorithm leads to $O(N^2)$ computational complexity, where N is the number of points in the grid (Choset, Hutchinson et al.). In the near future, we would like to update the algorithm to improve efficiency to allow larger state spaces and finer time discretization. This can be done by using particle filters and Monte Carlo localization techniques (Choset, Hutchinson et al.),(Dellaert, Fox et al.).

4.3.5 Experimental results

We implemented the approach described above using one Alien Technology RFID Reader and tags of the type shown in Figure 8. We arranged a grid of RFID tags as shown in Figure 11, loaded the Pioneer robot™ platform with the RFID Reader, and moved it in a pre-defined path. Note that in this configuration, the maximum detection range is approximately 90 cm. Ground truth measurements were acquired throughout the experiments by keeping track of the times at which the robot passed certain pre-determined positions. Note that these ground truth measurements include relatively large errors of up to 13 cm.
Our short term goal was to detect at least one RFID every 100 ms while moving the robot through the grid of tags. We found that using a tag spacing of approximately 40 cm allowed us to maintain this detection rate. Figure 12 show an example of the experimental data, and the resulting localization performance. Our localization algorithm quickly arrived at the robots position, and was able to track the movement of the robot very well. Figure 13 shows the localization error throughout one experiment. We believe that much of the variability in the error is due to inaccuracies during ground truth observations. Despite this, we were able to quickly achieve and maintain a fairly smooth position estimate over time to within an average of roughly 20 cm of the actual position.

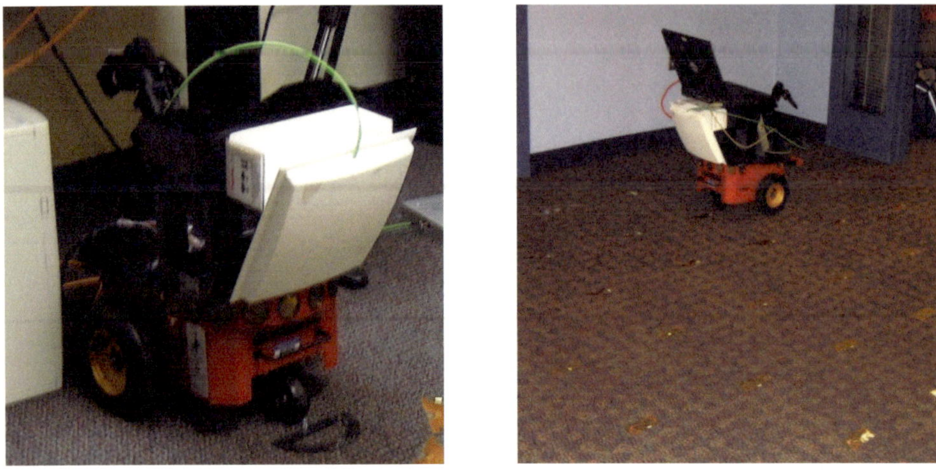

Fig. 11. Experimental setup. The Alien Reader was mounted on the platform of a robot (left). The RFID tags were placed on the floor in a regular grid pattern (right).

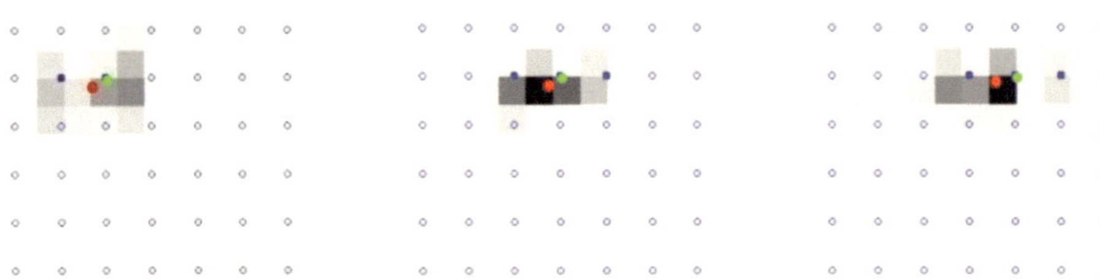

Fig. 12. These images show three time steps in the middle of an experimental run. Open circles represent undetected tags. Filled blue circles represent detected tags. The red circle represents the average robot location, and a green circle represents a ground

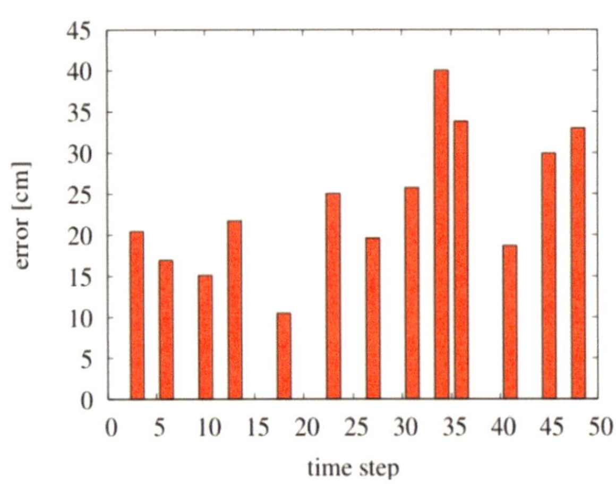

Fig. 13. Plot of 2-D position error over the course of an experiment. Note that our ground truth measurement was extremely noisy, having an accuracy of only +/- 13 cm.

5. Integration of data across disparate sensing systems over time

When integrating data across disparate sensing systems, time will play a crucial point. The readings from different sensing systems need to be time stamped to enable integration. Furthermore, the time stamps need to be with respect to a common clock. It is also important to look at the frequency of updates. When dealing with battery powered devices, like the MICA sensors, we need to consider the tradeoffs between the frequency of sending updates and the power consumption required. In addition, the frequency of sending wireless updates is also correlated with collisions in the sensor network and hence has to be understood.

5.1 Synchronization of sensors and cameras

In order to obtain useful sensor readings, one would like to know when the sensor readings were taken. Thus, our goal was to synchronize cameras with the deployed MICA sensors, or the MICA sensors with cameras. To achieve the synchronization goal we have to collect the data from both cameras and MICA sensors with accurate timestamps. These timestamps combined with the location of each sensor and camera would allow us to associate a MICA sensor value with an image pixel value. Based on our synchronization needs, we designed and developed a temporal calibration technique that follows the schema in Figure 14.
Cameras are attached to a personal computer (PC) and the PC uses its internal clock to set a timestamp for every captured image (the time the first bit of the image is received). If cameras are attached to multiple PCs then the PCs are synchronized using the standard NTP synchronization protocol (Mills). The MICA sensors communicate through an interface board with either the same PC as one of the cameras, or a different PC. In the case of a different PC we assume the camera PCs and the MICA PC are synchronized using the standard NTP synchronization protocol.

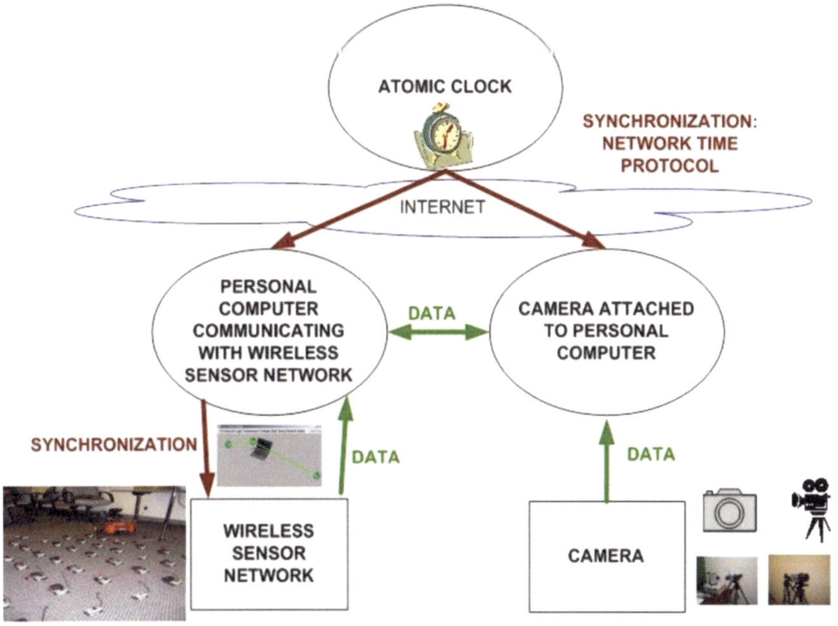

Fig. 14. A schema for time synchronization of MICA sensors and cameras.

The MICA sensors can be viewed as small autonomous PCs. We need to synchronize the time of these small PCs with the PC receiving data. One approach is let the PC receiving data time-stamp the incoming messages. However, if there are some delays in the network, delays due to processing of the incoming messages, or if the MICA sensor buffers multiple readings for their later transmission as a single packet, then the timestamps of the MICA sensor readings would be incorrect. We approached this problem by implementing a simple time synchronization of the MICA sensors with the PC. Since the MICA sensors use a wireless ad-hoc network to send messages back to the base station, we decided to leverage the inherent broadcast of the wireless network to do the time synchronization and not have each of the MICA sensors run the NTP synchronization algorithm.

At the MICA initialization, the PC will send the current time to each of the MICA sensors. A MICA sensor will update this timestamp every 10ms. Each reading on the MICA sensor will be sent back to the PC with the timestamp of the reading. The clocks on the MICA sensors are not as accurate as those of the PC. To prevent too much drift between the MICA sensor and the PC the timestamp on each incoming packet is compared with the current time of the PC, and if the difference between these two is too large (350ms in our case) then the PC will send new time to re-synchronize the MICA sensor.

5.2 Frequency of sensor update

There are several challenges when it comes to (1) acquiring data continuously from wireless sensor networks, (2) dealing with large numbers and high spatial density sensor networks, and (3) performing multi-instrument integration tasks in real time. We investigated experimentally the impact of several communication protocols, spatial sensor arrangements, MICA antenna orientations, presence of other wireless devices, acquisition sampling rates and the dependencies of the number of active MICA sensors on the wireless information loss. Our results were summarized in our past publications (Bajcsy, Kooper et al.), (Saha & Bajcsy), (Scherba and Bajcsy). Next, we only briefly describe the issue related to acquisition sampling rate.

Figure 15 shows how the number of samples received per second decreases with an increasing number of sensors. If there is only one sensor and we sample every 128ms, then we receive almost eight readings per second (1000/128). If we add more sensors then the number of readings decreases because of the collisions in the network. With seven sensors transmitting we obtain the same number of samples per second using 128ms sample rate (and a lot of collisions in the network) or 256ms sample rate. If we would increase the number of sensors then we would continue finding these tradeoff configurations (number of sensors and sample rate configuration assessed by the number of reading lost due to network congestion). From these experimental studies one can make determinations when it is better to switch to a lower sampling rate and still receive the same number of samples per second as at the higher sample rate (while saving battery power). Although unclear from these experiments, it is hypothetically possible that with 128ms sample rate and many sensors in the network one would receive fewer samples per second than with the same number of sensors and 256ms while wasting battery power. Thus, knowing the optimal sampling rate for a given number of active sensors will not only decrease the number of collisions in the network but also save sensor energy. The MICA sensors are battery operated and have a limited supply of energy. Transmission is in most cases the most expensive operation the sensor performs and thus minimizing the number of transmissions will increase the lifetime of the sensors.

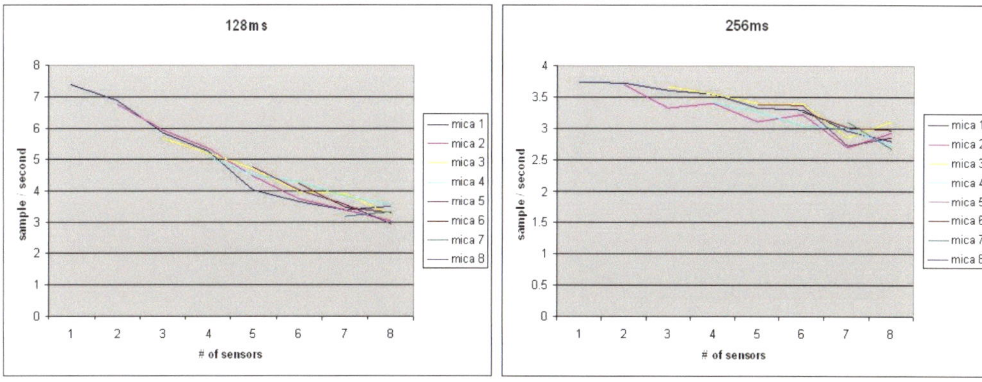

Fig. 15. Samples (sensor readings) received per second as a function of the number of sensors for the temporal sampling rate equal to 128ms (top) and 256 ms (bottom).

6. Calibration of measurements from sensors and spectral cameras

Our objective is to detect and recognize hazards such as fire. We strived to perform continuous wide area monitoring using thermal IR and visible spectrum cameras. Many building already have low resolution visible spectrum cameras to monitor security in the building. Our hazard aware system could use this existing network of cameras to do a low resolution detection of any hazards. Once a hazard has been detected we use the robot to place higher accuracy sensors in the environment to confirm the hazard. In case of a false alarm we want to recalibrate the system to take the new data into account. To be able to achieve this goal we need to integrate the readings from the visible and thermal IR cameras with those from the MICA sensors.

To get accurate hazard detection from raw sensor readings and camera pixel values, for example from the thermal IR camera, one has to convert sensor and camera raw values into engineering units, such as degrees of Celsius or Fahrenheit or Kelvin, otherwise the raw values cannot be used for detection and recognition purposes. This conversion is also denoted as a spectral calibration since temperature (a variable representing thermal wavelength range) could be replaced by any other spectral variable, for instance, a variable that represents visible spectrum, near infrared or radar wavelengths. Figure 16 shows our proposed schema for calibrating raw values.

First, we explored the calibration of thermal IR images using pre-calibrated MICA sensor readings. One can find the need for thermal IR camera calibration in many other areas, for instance, in remote sensing (radiometric and photogrammetry calibration of aerial and satellite imagery), robotics (vegetation detection using near calibrated infrared and red wavelength imagery), astronomy (brightness estimation of stars using thermal IR imaging of the sky) or military (battlefield analyses). We foresee the use of widely distributed and deeply embedded "smart" micro electro-mechanical systems (MEMS) sensors as potential thermal IR calibration gauges for thermal IR cameras in future.

The calibration procedure can be described as follows assuming that all sensors and cameras were synchronized and their locations are known. First, MICA sensors are programmed to sense and send temperature readings over a certain time period. Second, during the same time period, temperature measurements are collected with a thermometer (a calibration gauge). Third, a calibration transformation is established for MICA temperature sensors

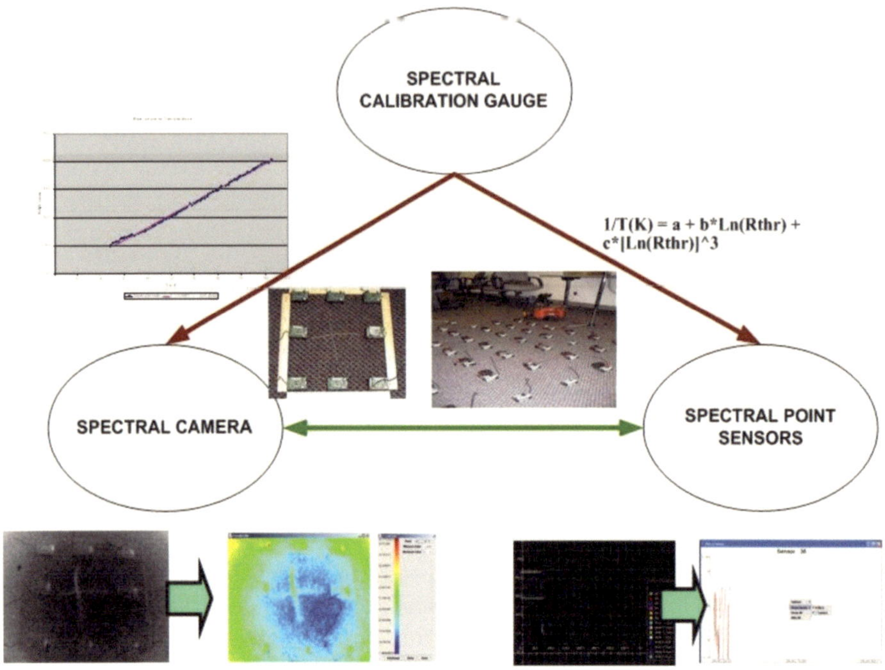

Fig. 16. A calibration schema for MICA sensors and spectral cameras using an appropriate spectral gauge. We used temperature sensors on a MICA sensor board, thermal IR camera and a thermometer as a calibration gauge in our experiments.

using a factory recommended formula and verified with thermometer measurements. Fourth, both thermal IR camera and MICA sensors are initiated to acquire data by broadcasting a RESET signal to MICA sensors and triggering thermal IR camera acquisition. Fifth, MICA sensors transmit every set (packet) of temperature measurements with the state of the internal counter (time stamp) to the base station attached to a personal computer (PC). In meantime, the thermal IR camera acquires data with the time stamp of the CPU clock counting from the RESET signal. Sixth, the MICA raw temperature measurements are received and transformed into degrees Celsius. Seventh, MICA temperature sensor locations in the thermal IR image are identified, and statistics of the transformed MICA temperature measurements and the thermal IR image pixel values at the MICA sensor locations are related to form the final calibration transformation. In this step, if the entire scene viewed by a thermal IR camera is temperature homogeneous then MICA temperature sensor locations in the thermal IR image do not have to be identified and statistics of the thermal IR image can be computed over the entire image.

Second, we explored the calibration of MICA sensor readings using pre-calibrated thermal IR images. This problem turns out to have a great application for large scale sensor deployment scenarios. Calibrating a large number of MICA sensors is a very tedious and time-consuming process since every sensor has to be treated separately. One should be aware that although MICA sensors come with a manufacturer's recommended calibration formula, each sensor has its own hardware characteristics. This is illustrated in Figure 17 for eight MICA sensors in the close proximity of the base station (the PC receiving data). The raw values vary significantly in terms of their amplitude (vertical axis) and slightly over time (horizontal axis).

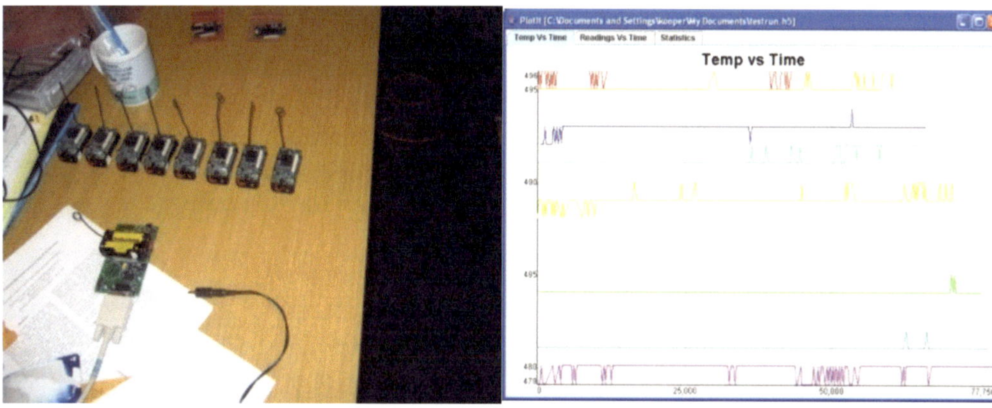

Fig. 17. Left - MICA sensor spatial arrangement. Right - Variations of raw temperature readings obtained from the MICA sensors shown in the left picture.

In this case, the calibration procedure can be described as follows. First, we calibrate a thermal IR camera by acquiring thermal infrared images of a blackbody, such as a cup of hot water, while measuring its temperature with a regular thermometer. Second, both thermal IR camera and MICA sensors are initiated to acquire data by broadcasting a RESET signal to MICA sensors and triggering thermal IR camera acquisition. Third, MICA sensors transmit every set (packet) of temperature measurements with the state of the internal counter (time stamp) to the base station attached to a personal computer (PC). In meantime, the thermal IR camera acquires data with the time stamp of the CPU clock counting from the RESET signal. Fourth, the MICA raw temperature measurements are received and transformed into degrees Celsius. Fifth, MICA temperature sensor locations in the thermal IR image are identified. All calibrated thermal IR image pixel values at the MICA sensor locations are used to form the spatially dependent calibration transformation for MICA sensor readings. The details of the calibration procedures can be found in (Bajcsy, Kooper et al. 2006).

7. Hazard detection, confirmation, understanding and containment

Hazard detection is achieved by comparing calibrated temperatures with a pre-defined hazardous temperature. When a predefined hazardous temperature is exceeded on any sensor, MICA, thermal IR and visible spectrum, the frames are analyzed to detect 2D characteristics (location, shape, color, texture and temporal signatures). We implemented temperature-based thermal IR and color-based visible spectrum image hazard detections from video streams as illustrated in Figure 18. This type of detection guides hazard confirmation step which is performed by deploying more MICA sensors, or by zooming and panning cameras according to the operator's needs.
It is important not only to detect fire hazards, but also to understand what material is burning in order to bring the appropriate extinguishers and to discriminate safe from

dangerous fires, e.g., a couch burning versus wood burning in a fire place. To understand what material is burning, one could analyze hyperspectral images of hazards together with thermal IR images. The assumption of this approach to hazard understanding is that temperature and high dimensional visible spectrum signature of fire flames uniquely defines burning materials. Finally, hazard containment requires studying hazard types, containment methods, hazard accessibility and many other application specific constraints. We have prototyped a demonstration showing that a robot could be used for the purpose of hazard containment when fire hazards in an office building are still relatively small in their extent. It is our belief that constant monitoring of hazards and an immediate hazard containment action could prevent large scale fire hazards and significant financial damages.

Fig. 18. An illustration of color-based visible spectrum image hazard detections from a video stream. Red crosses are placed over candle flames in three frames extracted from a video sequence.

7.1 Proactive camera control and human alert

Once a hazard has been confirmed a human needs to be alerted. We have investigated proactive approaches to camera control, spectral image analysis and human alert mechanisms. We have completed a design phase of a proactive camera control system that can trigger visible spectrum and thermal infrared spectrum cameras based on luminance and temperature sensors mounted on the available MICA sensor boards. Our current design addresses the problems of (a) efficient bandwidth management by proactive camera control (low bandwidth monitoring of hazard awareness spaces with MICA sensors and high bandwidth monitoring of hazard awareness spaces with cameras), (b) hazard understanding (multi-spectral sensing including visible and thermal infrared information) (c) choice of the best spectral modality to capture data about the environment based on the data provided by the MICA sensors and (d) human alert mechanism (image analysis to highlight areas of potential hazard). The proactive camera logic can be described as follows:

- If light is on then visible camera shown.
- If light is off then thermal IR camera shown.
- If temp > thresh & light is on then visible camera & hazard region enhancement are shown.
- If temp > thresh & light is off then thermal IR camera & hazard region enhancement are shown.

Examples of proactive camera control and human hazard alert mechanism are shown in Figure 19. Note in Figure 19 how the system uses the RGB camera when the office light is on, and combines the information about white (hot) regions from the thermal IR camera to highlight the RGB region of interest (the hazard region).

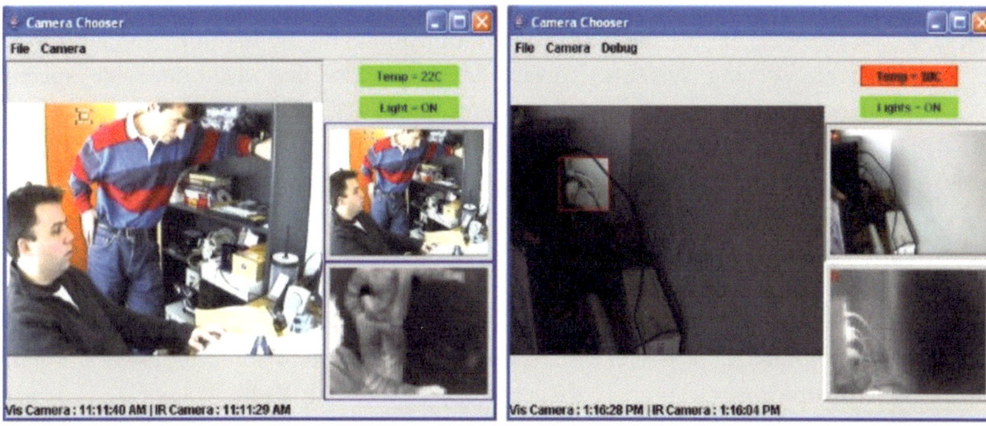

Fig. 19. Left - Proactive camera control. Depending on the temperature and light readings from a wireless network of motes, either a visible spectrum video or a thermal infrared spectrum video is sent to a hazard monitoring station. Right - Illustration of hazard alert by automated region selection and image enhancement.

8. Summary

We presented several research problems and their solutions developed for building hazard aware spaces. We have developed a method for deploying sensors remotely with a robot. To allow us to remotely control the robot we have looked at different modalities of controlling the robot (e.g. voice, gestures, a mouse or a keyboard) and addressed the problems related to fusing multiple simultaneous commands. In multi-sensor and multi-instrument systems like the HAS systems, we have investigated methods how to synchronize, localize and calibrate data. After analyzing the tradeoffs of localization approaches, we explored the effectiveness of using passive RFID technology for localization. We analyzed what elements most effected RFID tag detection robustness, and presented a methodology for building a sensor model for accurate RFID-based localization in addition to the standard acoustic time-of-flight ranging and stereo vision localization methods.

Once the sensors have been configured, we fused the incoming data and maximized our information gain from disparate sensing capabilities. Finally, we showed proactive approaches to camera control, spectral image analysis and human alert mechanisms, and briefly outlined a few challenges in application scenarios similar to the hazard aware space.

9. Acknowledgements

This material is based upon work partially supported by the NAVY STTR N03-T003, TRECC Accelerator Program, and the National Center for Advanced Secure Systems Research (NCASSR). We would like to acknowledge JC Lementec (CHI Systems) and Martin Urban (NCSA) for the contribution to the gesture recognition system, and Miles Johnson for the contribution to the RFID experimentation.

10. References

Abowd, G. D. (1999). "Classroom 2000: An experiment with the instrumentation of a living educational environment." IBM Systems Journal 38 1999: 508-530.

Bajcsy, P., M. Johnson, et al. (2008). Real time localization and 3D depth estimation across disparate sensing systems: Toward hazard aware and tele-immersive spaces. Systems, Signals and Image Processing, 2008. IWSSIP 2008. 15th International Conference on, Bratislava, Slovak Republic.

Bajcsy, P., R. Kooper, et al. (2006). Toward hazard aware spaces: Localization using passive RFID technology, National Center for Supercomputing Applications.

Bajcsy, P., R. Kooper, et al. (2006). Toward Hazard Aware Spaces: Knowing Where, When And What Hazards Occur, NCSA.

Bajcsy, P., S. C. Lee, et al. (2006). "Three-dimensional volume reconstruction of extracellular matrix proteins in uveal melanoma from fluorescent confocal laser scanning microscope images." Journal of Microscopy-Oxford 221: 30-45.

Choset, H. M., S. Hutchinson, et al. (2005). Principles of robot motion: theory, algorithms, and implementation, The MIT Press.

Dario, P., M. Bergamasco, et al. (1992). Multiple Sensing for Dexterous End Effectors. Robots with Redundancy: Design, Sensing and Control. A. K. Bajcsy, Springer Verlag.

Dellaert, F., D. Fox, et al. (1999). Monte carlo localization for mobile robots, Institute of Electrical Engineers Inc (IEEE).

Dey, A. K. (1998). "CyberDesk: A Framework for Providing Self-Integrated Context-Aware Services." Knowledge-Based Systems 11 1998: 3-13.

Dishman, E. (2004). "Inventing Wellness Systems for Aging in Place." Computer: 34-41.

East, E. W. (2004). Facility Area Network Workshop.

El-Zabadani, H., S. Helal, et al. (2006). "PerVision: An integrated pervasive computing/computer vision approach to tracking objects in a self-sensing space." issues 10: 11.

Gupta, P. and P. R. Kumar (2000). "The capacity of wireless networks." IEEE Transactions On Information Theory 46 March 2000(2): 388-404.

Hahnel, D., W. Burgard, et al. Mapping and localization with RFID technology.

Hartley, R. I. (1999). "Theory and Practice of Projective Rectification." International Journal of Computer Vision 35 1999(2): 115-127.

Hightower, J. and G. Borriello (2001). "Location Systems for Ubiquitous Computing." IEEE Computer 34 August 2001(8): 57-66.

Hollar, S. E. (2000). COTS Dust, University of California, Berkeley.

Ilyas, M. and I. Mahgoub (2005). Handbook of Sensor Networks: Compact Wirelss and Wired Sensing Systems. Boca Raton, Florida, CRC Press.

Kaddoura, Y., J. King, et al. (2005). Cost-precision tradeoffs in unencumbered floor-based indoor location tracking, Ios Pr Inc.

Kidd, C. D., R. Orr, et al. (1999). The Aware Home: A Living Laboratory for Ubiquitous Computing Research. Second International Workshop on Cooperative Buildings - CoBuild'99.

Lementec, J. C. and P. Bajcsy (2004). Recognition of Arm Gestures Using Multiple Orientation Sensors: Gesture Classification. 7th International IEEE Conference on Intelligent Transportation Systems.

Mainwaring, A., J. Polastre, et al. (2002). Wireless Sensor Networks for Habitat Monitoring. ACM International Workshop on Wireless Sensor Networks and Applications (WSNA'02). Atlanta, GA.

Marr, D. (1982). Vision. San Francisco, W.H. Freeman.

Mills, D. (1992). Network Time Protocol (Version 3) Specification, Implementation and Analysis, IETF.

Pankanti, S. and A. K. Jain (1995). "Integrating Vision Modules: Stereo, Shading, Grouping, and Line Labeling." PAMI 17 September 1995(9): 831-842.

Patwari, N., J. N. Ash, et al. (2005). "Locating the Nodes: Cooperative localization in wireless sensor networks." IEEE Signal Processing Magazine 22 2005(4): 54-69.

Priyantha, N. B., H. Balakrishnan, et al. (2003). Anchor-Free Distributed Localization in Sensor Networks, Laboratory for Computer Science, MIT.

Rom'an, M., C. Hess, et al. (2002). "Gaia: A Middleware Infrastructure to Enable Active Spaces." IEEE Pervasive Computing: 74-83.

Roush, W., A. M. Goho, et al. (2003). "10 Emerging Technologies." Technology Review 106 2003(1): 33-46.

Saha, S. and P. Bajcsy (2003). System Design Issues for Applications Using Wireless Sensor Networks, NCSA.

Satyanarayanan, M. (2001). "Pervasive Computing: Vision and Challenges." IEEE Personal Communications.

Scherba, D. and P. Bajcsy (2004). Communication Models for Monitoring Applications Using Wireless Sensor Networks, NCSA.

Scherba, D. and P. Bajcsy (2005). Depth Map Calibration by Stereo and Wireless Sensor Network Fusion. The Eighth International Conference on Information Fusion.

Urban, M. and P. Bajcsy (2005). Fusion of Voice, Gesture, and Human-Computer Interface Controls for Remotely Operated Robot. The Eighth International Conference on Information Fusion.

Urban, M., P. Bajcsy, et al. (2004). Recognition of Arm Gestures Using Multiple Orientation Sensors: Repeatability Assessment. the 7th International IEEE Conference on Intelligent Transportation Systems.

Vildjiounaite, E., E. J. Malm, et al. (2003). Smart Things in a Smart Home. the Fifth International Conference on Ubiquitous Computing: 215-216.

Wang, H., D. Estrin, et al. (2003). "Preprocessing in a Tiered Sensor Network for Habitat Monitoring." EURASIP Journal on Applied Signal Processing: 392-401.

Wechsler, H. (1990). Computational Vision. Boston, Academic Press.

Weiser, M. (1993). "Some computer science issues in ubiquitous computing." Communications of the ACM: 75-84.

Whitehouse, K. (2002). The Design of Calamari: an Ad-hoc Localization System for Sensor Networks, University of California, Berkeley.

Whitehouse, K. and X. Jiang (2004). Calamari: A Localization System for Sensor Networks.

18

RFID Context-aware Systems

Jongmyung Choi
Mokpo National University
Korea

1. Introduction

For several decades, RFID technology has been advancing and widening its application fields by penetrating into the traditional industries with the benefits of efficiency, costs saving, and fast processing. As the result, nowadays it is widely used in logistics, inventory, security, supply-chain, manufacturing, and retails industries. For example, Wal-Mart required its top 100 suppliers to attach RFID tags to their goods [1], and logistics and manufacturing companies including Siemense have been adopting RFID rapidly [2]. Experts expect that it will widen faster its application fields thanks to low tag cost, advance in network technology, and various trials of combining ICT (information and communication technologies). With ICT, RFID systems are providing users with smarter services such as world-wide object identification, and world-wide monitoring and tracking objects.

In the ubiquitous computing, context-awareness is one of the hottest issues in the academia and the industries. The basic idea of context-awareness is that the systems are able to recognize not only environmental condition and events but also internal system status, and provide users with smart services that fit on users' intention and their current situation. We call these external and internal conditions and events as context. Therefore, context-aware systems usually have sensing sub-systems that recognize the environmental condition and events and a smart module that determines the current context and the services according to the context. For sensing sub-system, they adopt camera, temperature sensor, GPS, motion sensor, and other sensors. As you might expect, RFID's identifation facility is also an important source of sensing the external condition and events. In fact, there have been a myriad of works that apply RFID to context-aware systems. Some read user data from tag to provide smart services according to user's role [3], and anothers identify user location by reading tags attached to fixed objects to provide services related to the location [4,5]. Still anothers recognize user's intention by reading object identifiers and provide services about the objects [6]. These RFID-based context-aware systems are very diverse and the number of these systems is increasing.

Though RFID has been gaining interests in context-aware systems, little attention has been paid to studies on survey of existing systems or studies on context-awareness in RFID systems in detail. These studies are needed to help system stakeholders understand RFID-based context-aware systems. Therefore, we survey the existing systems and introduce our RFID context-aware system in detail. We have two purposes for this article. The first one is to survey and classify existing RFID context-aware systems. By surveying the existing applications, we identify what kinds of context are used, what kind of context-aware services they provide. We group these applications into five categories according to the data type of tag value: user identifier, location, object, neighborhood, and behavior. We also group context-aware services into five categories: branch, trigger, resource scanning, follow-me, and context recording. The classification of tag value types and context-aware services can help stakeholders to understnad the RFID based context-aware systems. Specially, these are helpful to analysts and designer with understanding and modleing context and services. After survey, we define some concepts on context-awareness including context attribute, context, and other concepts in measurable terms, and we also introduce some desing issues on context-aware system.

The second purpose is to show how we apply RFID to our museum exhibition guide system, named MyGuide. This system is able to provide user specific information on the exhibit to users. For example, when a user does not have knowledge about the exhibit and he/she speaks Korean, it explains the exhibit with plain Korean. It can also provide multimedia data to help the user to understant the exhibit when he/she has a multimedia-enabled device. For this context-aware service, it adopts four context information: location, background knowledge level, device facility, and language. RFID is used to locate users in the musemum, and the system gets other information from user registration on the web. After considering context modeling issues such as context extensibility, inference, simplicity, and implementation, we draw two contexts -TEXT and MULTIMEDIA - from four context attributes. For software architecture, we adopt ECAM (event-context-action-model) architecture. The architecture is suitable for RFID context-aware systems, because it supports event-drivenness and context-awareness. MyGuide prototype system is implemented in Win32 platform with Java language, 13.56 MHz RFID readers, card type tags, Oracle DBMS, and Apple's Darwin Streaming Server. MyGuide shows how to model context from multiple context information including RFID and how to provide user-specific contents according to the context.

This article consists of five sections. In Section 2, we describe about RFID system: its elements and various existing systems. After this, we define some concepts on context-awareness such as context attribute and context and explain context-aware features in RFID systems in Section 3. In Section 4, we introduce our RFID system, MyGuide, which provides exhibition guide according to user's location and other user's information and preference. Finally, we reveal our conclusions in Section 5.

2. RFID system overview

2.1 RFID system elements

RFID is an identifying technology that uses wireless radio frequency, and it is widely used in various fields because of its merits: efficiency, costs saving, fast recognition, multiple recognition, and long recognition distance. In general, an RFID system is basically made up of three elements as shown in table 1: readers, tags, and servers. Table 1 shows the three elements and their classification.

Tags keep a small amount of data such as identifier in their internal memory, and they are classified according to their memory types: read-only and read-write. Most of tags have mobility because they can be carried with users; however, some others are immovable because they are attached to fixed objects. Mobile tags have close relationship with users because they move with users, but immovable tags are related with fixed objects or locations.

Elements	Criteria	Classification		Description
Tag	Power Source	Active		Long recognition distance
		Passive		Short recognition distance
	Memory	Read Only		Unchangeable value (ex. Identifier)
		Read/Write		Changeable value (ex. State, money)
Reader	Mobility	Stationary		Attached to fixed location
		Mobile	Mobile	Connected to wireless network, PDA type
			mRFID	Connected to CDMA or GSM, Cell phone type
Server	Service	Primary		Primary server for the services
		Secondary		Auxilary server

Table 1. RFID Elements Classification

RFID readers read data from tags and pass the data to the primary server. They can be classified into two groups: mRFID [8,9] and stationary RFID according to whether they are connected to a mobile communication network including CDMA and GSM. The former is connected to mobile communication networks, and the later is connected to wire-networks. The mobile RFID has two advantages: mobility and connectivity with the network. It is mobile because it is attached to a user's hand-held device such as a PDA or mobile phone. It is connected to the network anytime and anywhere because it uses CDMA or GSM network. Therefore, users can use it any places.

There are two kinds of servers: primary server and secondary server. The primary server is the server that receives the data passed from readers and processes it according to its business logic to meet users' requests. In stationary RFID systems, computer systems usually play the role of the primary server. However, in mobile RFID systems, hand-held devices usually play the role of the primary server. The secondary server is the server that helps the primary server by processing requests from it. ONS (Object Naming Server) in EPC (Electronic Product Code) systems [10] is one example of a secondary server. As RFID systems are getting complicated, the importance of the secondary servers increases. In addition, a variety of secondary servers come out. In mobile RFID systems, a service broker [8] is also an example of the secondary servers.

Typical RFID systems are connected to each of three elements shown in Fig. 1. In the figure, T_1 and T_2 mean RFID tags, and R represents RFID readers: fixed reader or mobile reader. P is a primary server and S is secondary servers for the system services.

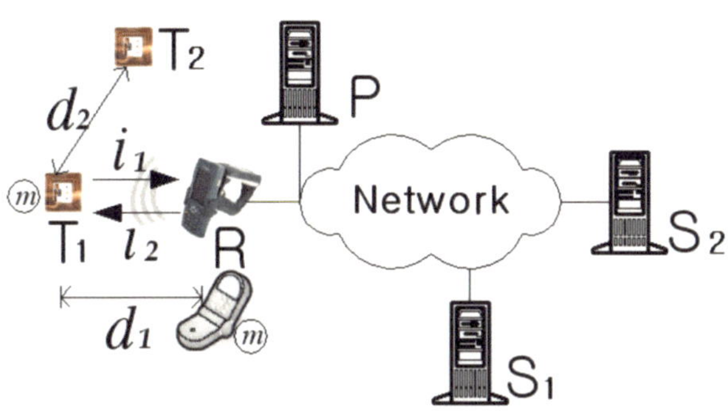

Fig. 1. RFID System Model

As illustrated in Fig. 1, tags have mobility in stationary RFID systems; however, readers have mobility in mobile RFID systems. In both cases, readers can access the tag data within the distance d_1. When readers recognize the tag data i_1, they transmit the data to the primary server P over the network. Then primary server P configures the services, and if it needs support from the secondary servers S_1 and S_2, it transmits the requests to them. Finally, the primary server P processes the data and provides the services to the users. In some cases, tags are writable, and readers may write data i_2, into tags. Furthermore, some applications, such as smart box [7], use several tags at the same time, and they should stay within a constant distance d_2.

2.2 Various RFID systems

We survey the existing RFID systems, and identify various types of them. During the survey, we find out that most of tags keep identifer valus, but the meaning of them are all different. Therefore, we classify the tag values into five groups according to their meaning or types: user, object, location, neighborhood, and behavior. Table 2 shows the five types of tag values, their meaning, and their usage patterns in the RFID systems.

Tag value	Meaning	Usages Patterns
User	A person who owns the tag	User's ID, role, preference, etc
Object	An object that a tag is attached	Object ID, object = action[28], monitoring
Location	The location where a tag or a reader is	Location based services, Follow-me, etc
Neighborhood	The relationship among the objects to which tags are attached	Cooperation among objects, Relationship among objects, monitoring, etc
Behavior	User's behavior or intention	Monitoring

Table 2. Tag values in RFID Systems

2.2.1 User

One of the most popular areas of RFID applications is identifying users. These systems identify users for their own purposes; access control, security, payment, and managing people. With these systems, users keep their own tags and there are stationary readers at some location such as gates, doors, or other places. The systems identify users by reading the values stored in the tags and matching them to their databases. Fig. 2 shows a user identifying system. By approaching the user's tag to the reader, the reader scan the value in the tag and the system determine who the user is or whether to allow him/her to access to the cetifacte data or places.

User information in the tag can be combined with other data in the server system and provide more sophisticated services. RFID applications considering user's preference or tastes usually use this information plus his/her preference information together for determining user specific services. Another example is the role-based services, in which user information is used to determine the role of the system. This kind of applications are closely related to the security, authorization, and authentication. However, other areas can also apply this role based service to their services. For example, Bravo [3] adopted two types of roles - student and teacher - in his application. If a teacher approaches, the system shows the lecture notes. However, if a student approaches, it shows his/her homework.

Fig. 2. User Identification

2.2.2 Object

A RFID tag may contain information - usually identifier - about an object which it is attached to. By reading the information, RFID applications can identify objects, get information about the objects, and keep track of them. These applications are used for logistics, inventory, and retail. In logistics, applications classify and deliver goods using RFID tags attached to them, and monitor to keep track of the goods while delivery. In inventory, they use tags to check which goods are stocked in and which items are out of stocks. In retail, companies can figure out which goods are sold. Retail industries enjoy the benefits of fast processing and preventing shoplifting.

Object identifiers in tags are the gateway which connects the physical world and the virtual world. Users can get information about the object and control the object from the virtual world. The EPC(Electronic Product Code) is the standardized object code developed by EPCglobal [10]. Using EPC, RFID applications send requests to EPCIS(EPC Information Service) which manages the information about objects and provides the services for the inquiries. ONS(Object Name Service) provide the location of data about the object which is indentified by the object identifier. EPC Discovery Service is the search service for the data location for the objects.

We find that some interesting researches which treat the object as the users' intention. It means that reading object identifier triggers an action which is closely related to the object with users' intention. For exmaple, during shopping a shopper lifts an item, it means that he/she has interests on the item. Therefore, the shopping assistant application may show information about the item or suggest another items that match the picked item. Fig. 3 shows the Smart Shopping Assistant system researched by Michael Schneider [11].

Another example is Trevor Pering [6]'s Elope system. The system identifies objects and provides a services which are related to the purpose of the objects. For example, users scan the tag on the remote controller for presentation, and the system allows users to use presentation. Fig. 4 shows the architecture of Elope system.

These RFID systems consider objects as the user intention, and this view changes the user interface in the computing environment from WIMP(Windows, Icons, Menu, and Pointing device) to object = action paradigm. Ailisto [12] argues that object=action paradigm can

Fig. 3. Smart Shopping Assistant

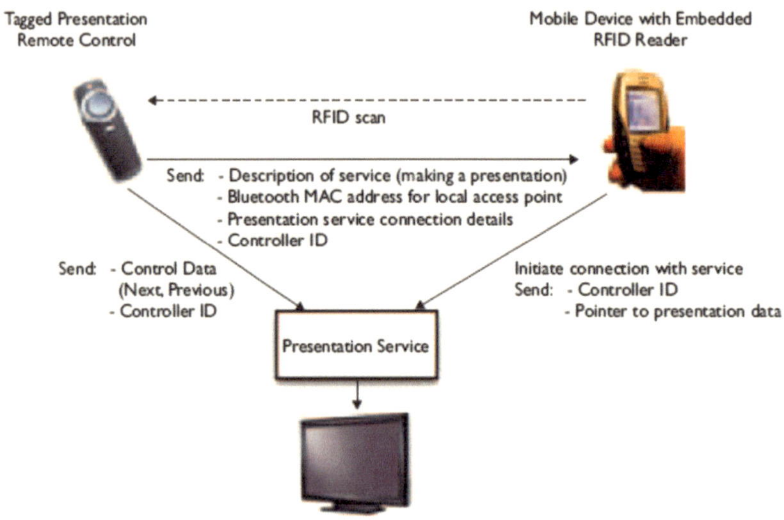

Fig. 4. Elopse System

change user interface paradigm of desktop computer by touching objects with mobile RFID readers. The typical example is the business card with RFID tag [13]. When a user touches the card with his/her cell phone equipped with RFID reader, the phone calls the person automatically.

2.2.3 Location

In context-aware systems, location is very important context information. Location based services (known as LBS) are good examples of context-aware services using location as context information. Searching for nearest restaurant from the user's location and weather service according to users' location are typical examples of LBS. There are two types for location system: one for absolute location with latitude and longitude and the other for symbolic location with relative location to the base location. There are several technologies for determining location inclusing GPS, RF singnal length, triangulation, and RFID. RFID is good for determing indoor location because it is easy to implement. Lionel's [14] does researches on determining a user's location and movement in the building. Lionel uses active tags and determines a user's location by calculating users' location with reference tags and a tracking tag.

There are RFID systems that use location information. Michael Crawford [15] developed the internet-based cooperating system. This system provides user's physical location for cooperation with people spread around the world. The location information helps the participants understand each other. For example, they know others' local time and know who participates and who leaves. There are serveral tour guide system that utilize RFID to locate users. Sherry [4] introduces an exhibition guide system using RFID. The main feature of this system is its bookmarking function. It means that visitors can bookmark any exhibit that draws his/her interests, and then he/she can make web pages with the information about the exhibits. Jani [5] also introduces an RFID-based system, mTag, which provides information to the pre-specified devices when the user carrying a tag approaches the RFID reader.

2.2.4 Neighborhood

Neighborhood is similar to the symbolic location, but it has unique features that it uses multiple RFID tags within a limited scope for cooperation of them. The systems determines their services according to the combination of tags within the scope. This neighborhood is captured from the symbolic location information. There are several interesting systems using neighborhood as context information. The representative system is Lampe's Smart Tool Box [7]. This system notifies users when he/she left tools outside tool box. In the system, the tool box embeds a RFID reader and tags are attached to each tool. Fig. 6 shows Smart Tool Box.

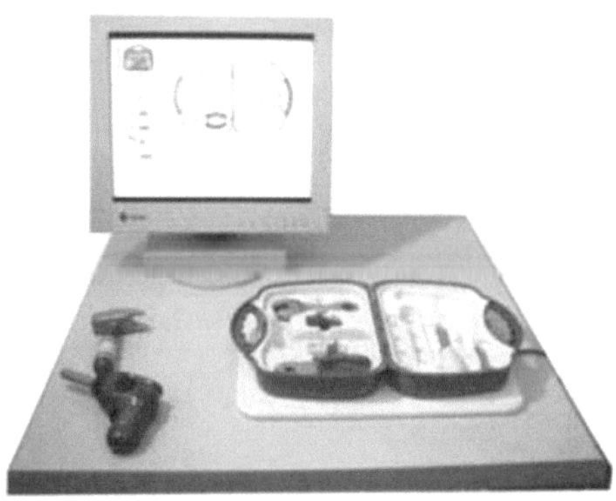

Fig. 6. Smart Tool Box

The neighborhood information can be used with user information. Bravo's classroom system [3] provides services based on the neighborhood information and user ID. When teacher moves to the blackboard, the system shows lecture note. However when a student moves to the blackboard, it shows the answer that the student makes.

2.2.5 Behavior

RFID can be used to identify user's behavior [16]. Context-aware systems should be able to figure out user's intention in order to provide suitable services according to context. User's behavior is a good hint for figuring out his/her intention. There are several methods to identify user's behavior: camera and active badge. However these methods are good for identifying short term behavior, but it is not suitable for long term behavior. Joshua [16] created a glove equipped with RFID reader, which reads RFID tags attached to things. This system can figure out what the user is doing. Fig. 7 shows i-Glove system, which is used to figure out user's behavior in a hospital.

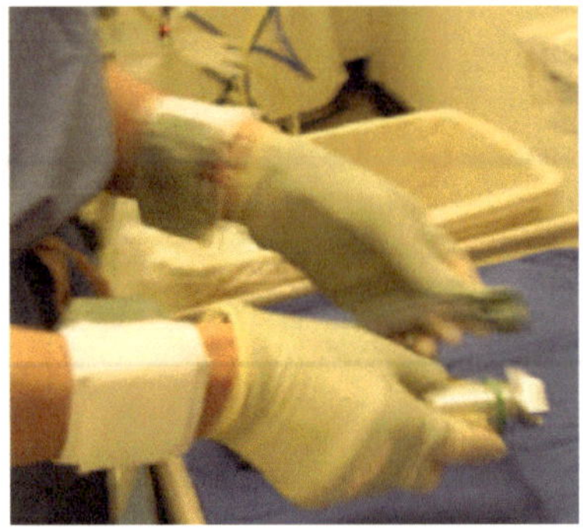

Fig. 7. iGlove System

3. Context-aware RFID systems

3.1 Definitions and terms of context-awareness

Context-aware systems are able to choose and provide the most suitable service from the having the same semantics but different implementations to meet the changing contexts. In general, context-aware systems have two kinds of services: normal business services and context-aware services. Context-aware services are the business services whose contents and formats are determined by the context. That is, context-aware services are business plus context-aware features and the latter is our main concern. Context logic has a close relationship with the implicit interaction between users and external environment, and it also has a relationship with business logic because it triggers or determines the business services. Fig. 8 shows the relationship among business logic, context logic, and user interface.

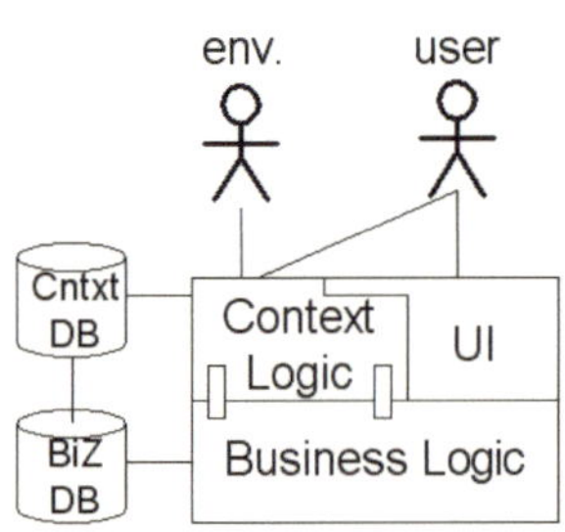

Fig. 8. Conceptual Layer of Context-aware Systems

In the context-aware systems, context is the most important issue, but there is no commonly accepted definition on it. Shilit and Theimer [17] define the context as location, identies, and objects. Hull [18] describes context as the aspect of the current situation. Dey and Abowd [19] define context as "any information that can be used to characterize the situation of entities that are considered relevant to the interaction between a user and an application, inclusing the user and the application themselves."

In this Section, we define context in more concrete and measurable terms by extending our previous work [20]. First, we define the most basic piece of information of context, and we call it context attribute. A context attribute is a semantic data type of value sensed from the external world via a sensor or an internal state value stored in a variable. Therefore context attribute is defined as Def. 1.

Def. 1. Context Attribute

Context attribute, a_i, is a sematic data type of an atomic value sensed via a single sensor or a sematic data type of a value stored in a variable.

Temperature, location, and time are good examples of context attribute. Temperature and location are semantic data types of external values sensed by sensor subsystems, but time is a semantic data type measured internally. We use the term of "semantic data type" in order to distinguish it from the meaningless data type in the programming language. For example, 75 is represented as "int" data type in the programming language, but it is meaningless in the context-aware systems because we do not know what it means. Therefore, the value should have the semantic data type such as temperature, pressure, or other concrete meaning to be used as context information.

Since context attribute is data type, there are possible values set for each context attribute. We call it context attribute value, and it is defined as Def. 2. For example, if a context attribute, H, is humidity type, and the current humidity is 60, then the 60 is the context attribute value for H.

Def. 2. Context Attribute Value

A context attribute value, v, is an element of the possible value set of context attribute, a.

Some system use one context attribute for their context, but others adopt multiple context attributes for their contexts. Therefore, we need to identify what kinds of context attributes are used in the systems, and we define the set as Def. 3.

Def. 3. Context Attribute Set

Context attribute set, A, is the set of context attributes used in a context-aware system.

Context is an abstract and single concept in human mind, but we need context attribute values to determine what the current context is. For example, when we think about "talking" context, it is an abstract and simple concept but we have to check context attribute values whether their values are in the range of "talking" context. The more complex thing is that there can be multiple sets of context attribute for a context. Let's think about "talking" context. A computer system is able to notice the "talking" situation by sensing the speech via a noise sensor, however another system may use a camera and recognize talkers. Therefore, we have to define abstract context and concrete context.

Def. 4. Abstract Context

Abstract context type, AC, is an abstract and simple concept for the situation.

Concrete context template is a tuple of context attribute for an abstract context. We can define it as shown Def. 5.

Def. 5. Concrete Context Template

Concrete context template, T, is a tuple of the finite number of context attribute: $T = <a_0, a_1, \ldots>$.

Context instance is an instance of abstract context, and it consists of context attribute values. In general, context instance is determined by checking attribute values or inferencing from attribute values. For example, we can determine "talking" context if nosy frequency value is within audio frequency, and nosy pattern is in human voice pattern. Therefore, context instance is determined by checking attribute values as defined in Def. 6.

Def. 6. Context Instance

Context instnace, c, is an instnace of context, and it is determined by evaluation of $c = f<v_0, v_1, \ldots>$, where f is the evaluation function which interpretes attribute value tuple and generate a context instnace.

Most context-aware systems have several situations in their lifecycle, and we call the set of contexts in a system the context set.

Def. 7. Context Set

Context set, C, is the set of contexts used in a context-aware system for its life time.

In context modeling, we determine the context set, concrete context template, and evaluation function algorithm f. During context modeling, there are context-related issues to be considered carefully: context extensibility, dynamic re-configurability, dynamic adaptation, inference, simplicity, and ease of implementation [21].

- Context extensibility: Is it easy to extend context? Is able to add more contexts or context attributes?
- Dynamic re-configurability: Is it possible to re-configure the system after extending contexts?
- Dynamic adaptation: Is the system possible to adapt itself to the change of context dynamically?
- Inference: Does the system need inference facility for generating context from context attribute values? Is the inference correct and efficient?
- Simplicity: Is it easy and simple to model context?
- Ease of implementation: Is it easy to implement context model?

We can classify context-aware systems according to extensibility of context into two types: closed and open. The systems of the first type have the fixed context set; i.e. the contexts of the system are fixed at design time and never change. In other words, the context-aware services are also determined at design time.

Def. 8. Closed

A context-aware system is called 'closed' when its context set C is fixed and determined at design time.

On the other hand, the systems of the second type use open context set. These systems allow new context and new context attributes to be added at run time. If the context attribute set changes at run time, the system is open context attribute and it should support the ability to handle it.

Def. 9. Open Context

Context-aware system is called 'open context' when its context set C is not fixed at design time. At runtime, there is a time t that meets the condition; $|C|_t = n$ and $|C|_{t+1} = m$, where $n \geq 1$, $m \geq 1$, and $m \neq n$.

Def. 10. Open Context Attribute

Context-aware system is called 'open context attribute' when its context attribute set A is not fixed at design time. At runtime, there is a time t that meets the condition; $|A|_t = n$ and $|A|_{t+1} = m$, where $n \geq 1$, $m \geq 1$, and $m \neq n$.

Context-aware services are another important issue to be considered. They influence system usage pattern, service scenarios. As the result, it also influence on systems architecture, design pattern, and development platform. We survey the existing context-aware systems, and classify context-aware services into five types according to the role of context in the services [20]. The five types are as follows:

- Branch: context determines a service among several ones, its contents, or its output format.
- Trigger: change to context triggers the execution of a service.
- Resource Scanning: the system scans the external resources and provides services that are combined with external resources. Context plays the role of medium of cooperation.
- Follow-me: the system keeps user's context, follows him/her, and provides services with the help of surrounding context. Context plays the role of the medium that connects the past and the present.
- Context Recording: the system records and saves the current context for later retrieval.

The service types listed in the above are not exclusive to each other, so that a service may be included in one or more groups. For example, a context-aware service named "A" is triggered on the change of context, and the contents of the service are determined by the context. Then the service is included in both Branch type and Trigger type.

3.2 Features of RFID systems for context-aware services

RFID systems can get user's intention by noticing the change of a tag value or understand the surrounding conditions by reading tags attached to objects, humans, or locations. In general, RFID has five interesting features compared to other sensors. First, the value that it reads is clear and explicit. Other sensors may have unclear values because of jitter or noise. RFID systems can read multiple data concurrently at higher speeds than other sensors. Furthermore, in many cases, tag values are categorized and registered in a database. It means that we do not need to apply complex algorithm such as statistics, probability, or filtering to get tag values as context attribute values.

Second, to perceive environmental values, RFID needs at least two elements: a tag and a reader. Therefore, when the system reads a tag value, it means that the system has at least two values: the tag value and the reader's ID. In most cases, the reader's ID plays the role of location identifier. However, in mobile RFID systems, the reader can provide the location of the user and the user identifier because it is combined with the user's mobile phone.

Third, a tag value may have a variety of meanings based on its data type. For example, the data from a tag may mean a user identifier or a location identifier according to the meaning of the data. In this case, the user identifier and the location identifier are significantly different when determining the contents of services.

Fourth, RFID is easily combined with other sensors. RFID cannot sense the physical environment such as temperature and humidity by itself, but it is able to fuse with other sensors and to perceive the physical environment. Even though the system adopts multiple sensor types, RFID will play the main role in the system because in most case, RFID triggers the services in the system.

Fifth, RFID system is event-driven. RFID system starts its operations on the event of reading tags. Therefore, the system architecture should be designed to meet event-driven features, and implementation should be done in event-driven programming.

3.3 Analysis of context-aware RFID systems

3.3.1 Analysis

There is no clear cut boundary of what RFID systems are context-aware and what are traditional. Therefore, we classify them according to whether they have smart services or not. If it provide smart services with the help of RFID, we consider it as a context-aware system. With this criteria, we summarize the existing context-aware RFID systems, and show the context and the service types in Table 3. The six types of context are user (U), object (O), location (L), neighborhood (N), and behavior (B). In addition, the service types are Branch (S), Trigger (T), Resource Scanning (C), Follow-me (F), and Context Recording(R).

Systems	Features	Context Attribute	Service Types	Context Extensibility
Bravo's [3]	Provide services according to a user's role	U	T, S	Closed
Elope [6]	Provide services with the paradigm of object = action	O	T	Closed
City Tour Guide [4,5]	Provide services based on the user's location	L	T, S	Closed
Refrigerato [29]	Provide services based on the objects in neighborhood	N	S, C	Closed
iGlove [16]	Recognize user's activity or record user's activity	B	R	Closed
RFID Game[30]	Provide users with game score or rules based on the internal status	O	S	Closed
Experience Blogging [31]	Record user's experience and context, and share them with others	L, O	R	Closed

Table 3. Context-aware Services

3.3.2 Context attribute model for RFID systems

From the five data types of tags, we get the following information: Who (User), What (Object), Where (Location), With Whom/What (Neighborhood), and Why (Behavior). With tags and readers, we get answers to the 5W questions. When combined with internal status, the context information will be much richer because it can provide time and extra information. With tag, reader, and internal status, we can model the context of RFID systems.

Our context models are of two types according to the reader's types: stationary and mobile. We separate them because the context information in each reader is different. Fig. 9 shows the context model for stationary RFID systems. In these systems, the reader has identification information. And it provides location information that is about absolute position or about symbolic position. Therefore, tag values may be used for user identification or object identification. The context may need time and purpose information, and these are maintained as internal values within the system. Context also needs extra sensory data for environmental information.

The context model for mobile RFID systems is a little different from stationary RFID systems because the mobile RFID reader provides both location information and user identification. In mobile RFID systems, the reader is attached to the cell phone or PDA. Fig. 10 illustrates the context model for mobile RFID systems.

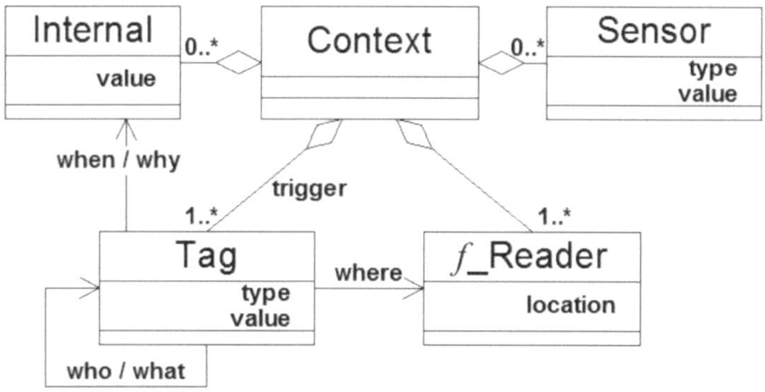

Fig. 9. Context Model for Fixed RFID Systems

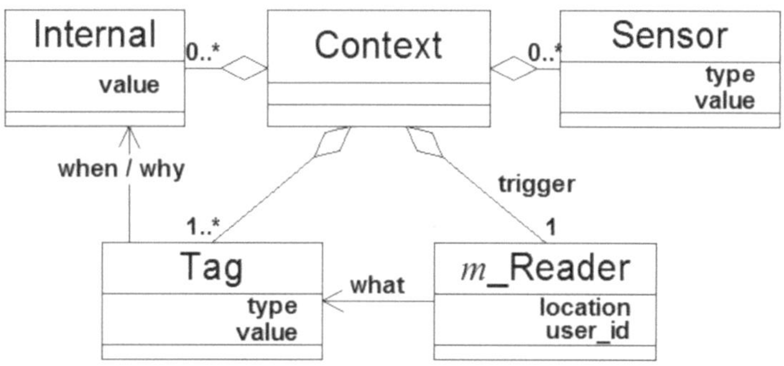

Fig. 10. Context Model for Mobile RFID Systems

3.3.3 System context trasition model

In general, RFID systems start their operation based on event-driven approach, and they have internal states. Their states changes according to the events. Similarly, the context-aware RFID systems have their current context and have all possible contexts as their context set. During the service life time, the context changes from one to another and the services also change according to the change of context. Therefore, we can represent the transition of context with context transition diagram. The context transition diagram is similar to the state transition diagram and it shows the change of contexts in the system. Fig. 11 shows the example of the context transition diagram. In the figure, state "Idle" and "Calling" are business logic, and "Guiding" is a context-aware service. In the context-aware service, there are state transition between "Multimedia Guide" and "Text Guide."

3.3.4 System architecture

Software architecture is important enough to determine the system's framework and design. Most of RFID systems trigger their services on the event of RFID tag reading, so they are considered event driven. With this feature, architects determine the event driven architecture for context-aware architecture, there are event-driven architecture with context information. We propose ECAM (Event, Control, Action, Model) pattern [22] for RFID context-aware systems because it supports event-driven and context-aware. This architecture is the extension of the ECA (Event, Control, Action) [23] pattern by adding Model for context information.

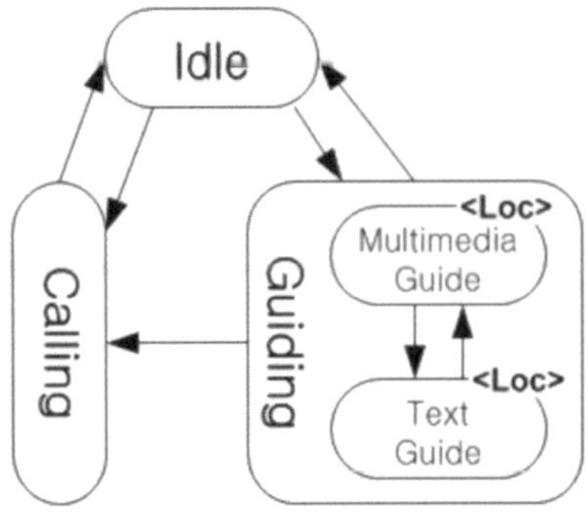

Fig. 11. An Example of Context Transition

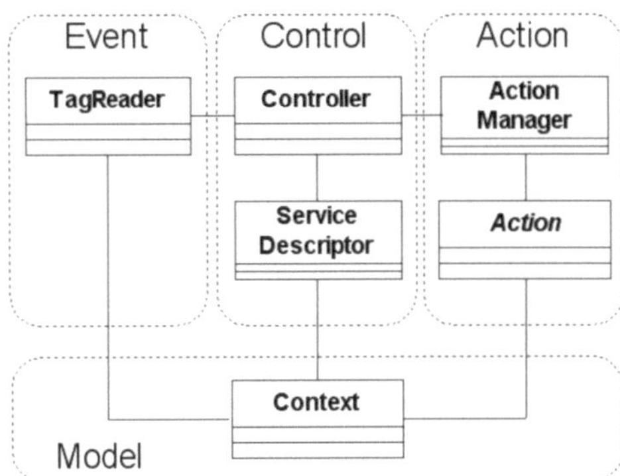

Fig. 12. ECAM Architecture

In ECAM pattern as shown in Fig. 12, TagReader reads the RF tag and generates a READ event. The event is transferred to the Controller in Control, and Context in Model. Context gathers context information from multiple sources and determines the current context of the system. Therefore, it should implement the context evaluation function defined in Def. 6. Controller determines the service contents based on the Service Descriptor and the current context determined by Context. Then the Controller transfers the information about the service to ActionManager. ActionManager provides the service.

4. Case study of a context-aware RFID system: MyGuide

4.1 System overview and requirements

There have been several RFID-based and context-aware tour guide systems because it is a interesting research topic for the academia and the tourism industry. We also decide to design and implement a RFID-based and context-aware guide system, named MyGuide [24], but we have to add more context information into our system to meet the system requirements. The main goal of our system is to provide the most suitable information to the visitors in the most suitable format. The following scenarion shows the basic usage senario and functional requirements of the system.

Scenario

When a visitor arrives at the museum, he/she registers his/her information (background knowledge, language) and his/her mobile phone information (facilities, phone number), and receives an RF tag.

After registration, the visitor walks around the exhibits, finds an interesting exhibit, and accesses the RF tag to the reader attached to the exhibit. Then the system sends a message to the visitor. The message contains the information about the exhibit, or the URL for the multimedia information about the exhibit.

From the scenario, we draw the sequence diagram as shown in Fig. 13. We figure out RFID reader, database for context information, context-server, and contents server from the scenario. The RFID reader sends the tag value to the context server, and it identifies the visitor. After that, it also can access context information stored in the context database using the tag value as a key. The contents server keeps the explanatory contents about the exhibits, and provides suitable contents to the visitor.

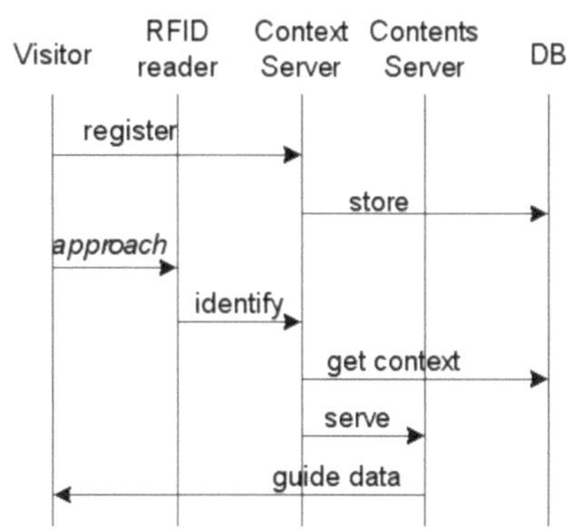

Fig. 13. Sequence Diagram for Scenario

From the basic scenario and the sequence diagram, we determine the overall architecture of the system. Fig. 14 shows the overall architecture of MyGuide system. When a visitor approached to the exhibit, the context server determines the context, and contents server provides guide information through the network.

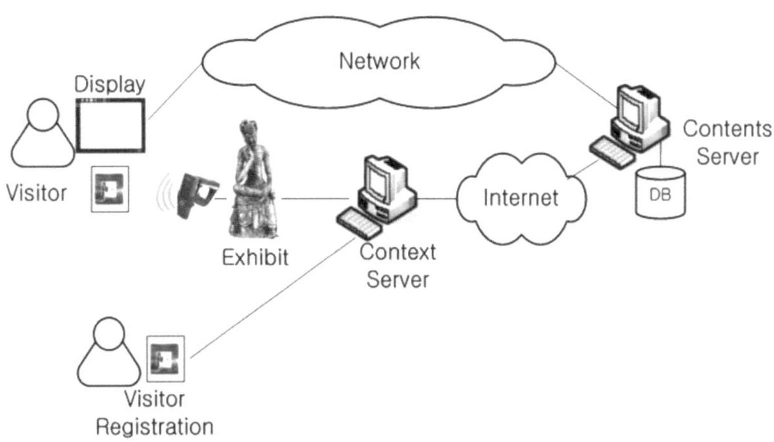

Fig. 14. Overall Architecture of MyGuide System

From the goal and the scenario of MyGuide, we are able to draw the functional and non-functional requirements. There are two basic functional requirements: "register users" and "provide information.

- (F1): register user information. This information is used to determine user-customized guide information.
- (F2): provide a user with information about the exhibition next to the user in user-customized format and with user-customized contents considering user's background knowledge of the exhibition.

Besides the two basic functional reqruiements, there are at least six non-functional requirements as follows. Most of these requriements are related to the usuability issues, but we exclude performance, throughput, and security requirements because we intend to focus into RFID and context-awareness in our system.

- (R1) Customized Services: Existing guide systems provide the same services to all visitors without considering the visitors' background knowledge or age. This can feel unsatisfactory to the visitors. Children and beginners don't understand the explanation with jargon, and experts are not satisfactory with an explanation that contains only general information. Therefore, our system considers the visitors' background and provides appropriate information according to their profiles.
- (R2) Easy Use: The end-system or terminal should be easy to use, and the visitors should feel comfortable using the technology.
- (R3) Personal Usage: The visitor should be able to get and control the services at any time when he/she wants in the museum.
- (R4) Multimedia Services: Our system should be able to deliver multimedia data to the visitors. Multimedia content is very effective to explain information to a visitor whether he/she is a child or an adult, or a beginner or an expert. On the other hand, if the terminal does not support multimedia service, it should be able to provide text data at least.
- (R5) Less Cost: Our system should be implemented with the least costs and efforts because of limited budgets. Therefore, we prefer to use the existing technologies without developing new technologies.
- (R6) Multiple Languages: One of the interesting requirements from stakeholders is for the system to support multiple languages as the number of visitors from other countries increases. Therefore, our system should support guide contents in multiple languages.

In order to meet the basic requirements, we have to determine our fundamental plarform and technologies. Since there are several altanative platforms and technologies, we have to evaluate them and choose one of them. Table 4 shows the altanative technologies and the evaluation result.

●: 5, ○: 3

Technologies / Requirements	Location			Mobile Device		Messaging Protocol	
	RFID	mRFID	Triangulation	Cell Phone	PDA	WAP	J2ME
F2	●	●	○	●	●	●	●
R1	●	●	●	●	○	N/A	N/A
R2	●	●	○	●	○	N/A	N/A
R3	●	●	●	●	●	N/A	N/A
R4	N/A	N/A	N/A	●	●	●	●
R5	○	○	○	●	○	●	●
R6	N/A	N/A	N/A	●	●	N/A	N/A

Table 4. Evaluation of Possible Technologies

After the evaluation, we choose three basic technologies: RFID, mobile phone, and context-awareness. There are several reasons why we choose these technologies as follows;

- RFID: An RF tag is small enough to carry with the visitor and it is natural and easy to use. It is a stable technology because it has been used in various fields for a long time. Furthermore, there are ample system components and software modules for RFID systems, so we can reduce costs and efforts by using the existing components instead of developing them from the scratch.
- Cell Phone: In order to support mobility, we need mobile devices including PDA, cell phone, and other dedicated mobile terminals. From several kinds of devices, we choose mobile phone based on the following merits. First, most of the new products support multimedia display, and even older ones support at least a text message service without adding any hardware components or installing any software modules. Second, we can save cost and effort in developing end-systems for providing guide service. Third, the mobile phones are easy to carry and people tend to keep their phones with them. Fourth, they are user friendly. It is their own phone, and they use it every day. Fifth, a mobile phone is always connected to a CDMA or GSM network at any place. We don't need to consider a situation in which the end-system is disconnected.

4.2 Context modeling

In the usage scenario of MyGuide, we assume two things. First, the system database has a table that maps the exhibit and the RFID reader's identifiers. Therefore, we can get the visitor's location when a reader reads his/her RF tag. Second, the visitor registers his/her information with an RF tag value.

There are countless environmental elements around the context-aware system. However we cannot consider every contextual element, so that we limit the number of contextual elements in our system to the information that is necessary to meet the requirements. We identify four kinds of context attributes that affects the service of our system.

- The visitor's location
- The visitor's background knowledge
- The facilities of the visitor's mobile phone
- The visitor's preferred language

For each context attribute, it has its own value set. The visitor's location is a primary context attribute. The system can determine which exhibit draws a visitor's attention according to his/her location. We use RF tags and RFID readers to identify the visitor's location. When the visitor with an RF tag approaches to an exhibit, the reader attached to the exhibit identifies his/her tag value. In the perspective of guide information, the location determines the subject of data to be served.

Considering the visitor's background knowledge is the most valuable feature of our system. The more we divide the level of the background knowledge, the more the service will be satisfactory. However, we should consider development costs and the maintenance problem of contents, so that we group the background knowledge into three levels.

- Beginning: basic information for children and those unfamiliar with the subject
- Normal: general information for ordinary adults and middle/high-school students
- Advanced: advanced information for experts or college students

The facilities of the visitor's mobile phone determine the media type of information. Some phones support multimedia data display, but others may have the only minimum functions

for mobile phone. Therefore our system should consider the facilities of the visitor's mobile phone. We therefore categorize mobile phones into two types: Text or Multimedia.
The visitor's preferred language is also an important context attribute to be considered. As the world becomes smaller, we need to serve the visitors from other countries in museum. We consider total three languages: Korean, English, and Chinese.
MyGuide adopts four context attributes for its services. The notable feature of them is that they are orthogonal and independent of each other. For example, the visitor's location has nothing to do with his/her preferred language, and the subject and the language of the contents have nothing in common. The four types of context information can be represented orthogonally in the four-dimensional coordinate as shown in Fig. 15.

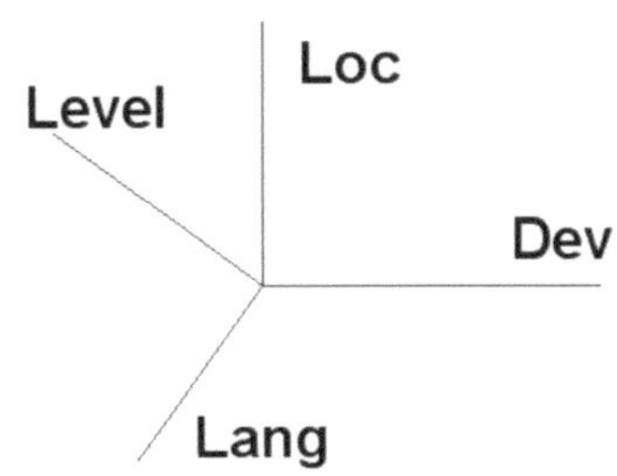

Fig. 15. Context Information Space of MyGuide System

According to our definition, a context may be represented as a dot in the coordinate system in Fig. 15. However, in MyGuide system, we have too many contexts. Let C be the set of all possible contexts in MyGuide system. Then a context c ($\in$ C) can be represented as a tuple: c = <loc, level, dev, lang>, where loc $\in$ Loc, level $\in$ Level, dev $\in$ Dev, and lang $\in$ Lang. We can imagine that Loc = { x | x is an exhibit. }, Dev = {multimedia, text}, Lang = {Korean, English, Chinese}, and Level = {beginning, normal, advanced}. Then we can calculate the number of contexts in MyGuide as follows:

$$| C | = | Loc | \times | Dev | \times | Lang | \times | Level |$$

If we define context as the tuple of context attribute values, this system has too many contexts. Therefore, we need to categorize them according to commonality and variability. In MyGuide system, media type determines the metadata and service server, so we categorize context information according to media type. Furthermore, we choose closed context model because we will not extends our context model later. Considering these issues, we model context as shown in Fig. 16. The notation used in the figure follows our prior studies [20]. The notation is similar to class diagram. According to this notation, a context may be implemented in multiple ways, so that it is represented as an interface. Similar contexts have common features, and the common features can be generated as context type. MyGuide has two contexts: Multimedia and Text and four context attribute type: location, backgounf knowledge level, mobile phone type, and language
The system maps the four context attributes to one of the two contexts after the following context reduce algorith. In the algorithm, retrieve(T, s, v) means that it retrieves data from the database meeting the condition. As you see the algorithm, you will notice that the media type determines the context.

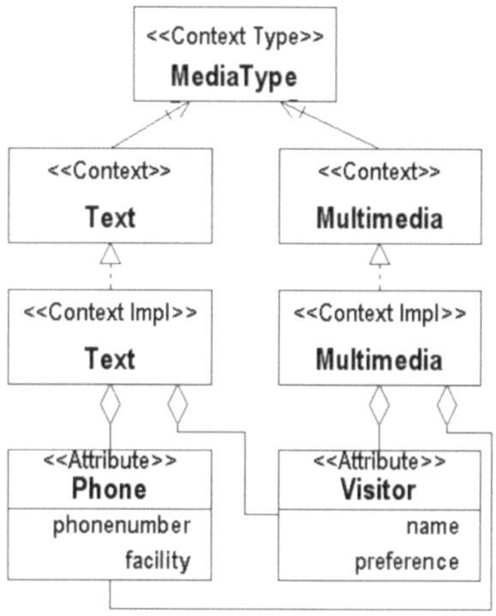

Fig. 16. MyGuide Context Model

```
Algorithm. Determine_context ( TagId, ReaderId )
  v = retrieve (Tag, tagID == TagId, "visitorID")
  (pnum, mcode) = retrieve (Phone, visitorID == v, "phonenumber, mcode")
  type = retrieve ( MediaType, mcode == mcode, "desc")
  if ( type == MULTIMEDIA )
    new MultimediaContext
  else if ( type == TEXT )
    new TextContext
  end-if
```

4.3 Implementation of prototype system

We implemented MyGuide prototype system on the Microsoft Window platform. We chose the Java language for software implementation, and MIDP [25] for mobile phone programming. We used 13.56 MHz RFID readers and card-type RF tags. We used GNU java serial communication library [26] to connect to the RFID readers. For multimedia serv ice, we adopted Darwin Streaming Server from Apple [27]. We managed the context information and the guide data on the exhibits in Oracle 10g DBMS. Fig. 17 shows the architecture of the prototype system and its components.

MyGuide starts services on READ event from RFID reader. The visitor's profile and his/her mobile phone information are stored in DBMS. Fig. 18 shows the E-R diagram of the prototype system. Contents entity means the guide data for an exhibit. For an exhibit, there are several Contents records to provide visitors with suitable data. Visit entity means the trace data of a visitor's access to the exhibits.

The class diagram of MyGuide is shown in Fig. 19. TagManager gets the values stored in the RF tag and the reader's identifier, and generates READ event. On the event, the Controller takes the control and coordinates the cooperation between CntxtReducer and MsgSender. CntxtReducer infers the current context from the values of contextual elements. It uses DbFacade for retrieving contextual data stored in DBMS. Controller determines the contents of service using the context, and asks MsgSender to generate Msg for the service and send it to the visitor. Msg consists of the explanation of the exhibit and the metadata.

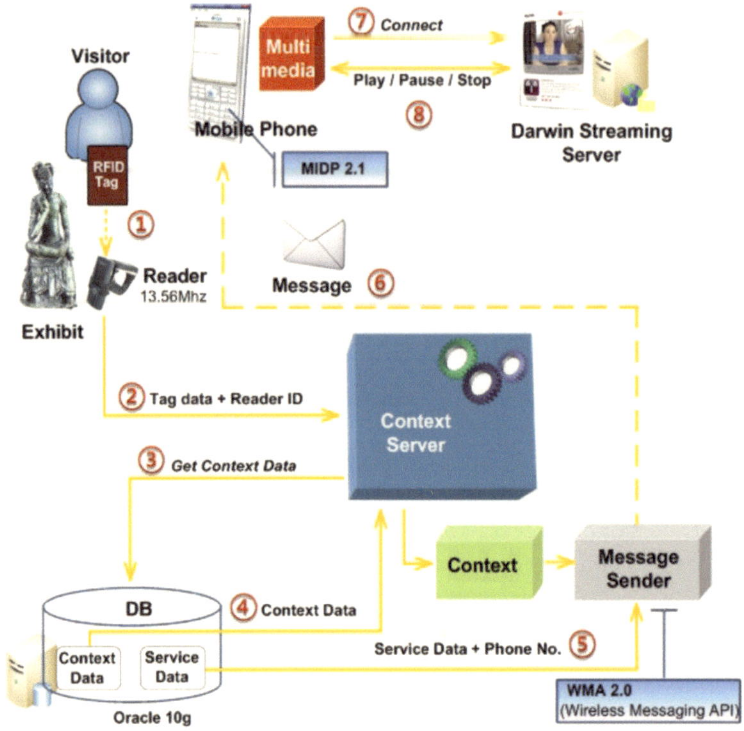

Fig. 17. Architecture of MyGuide Prototype System

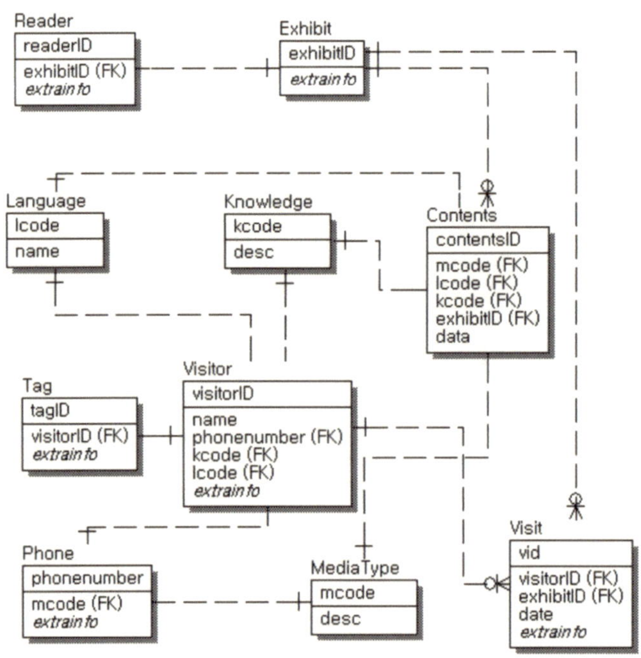

Fig. 18. ER Diagram for MyGuide Prototype System

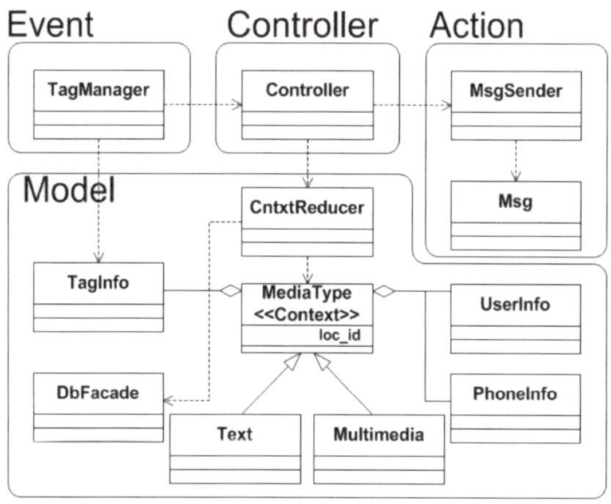

Fig. 19. Class Diagram of MyGuide Prototype System

Fig. 20 show the running examples of MyGuide prototype system. Fig. 20 (a) shows the text service at beginning level in Korean. Fig. 20 (b) shows the multimedia service at beginning level in Korean.

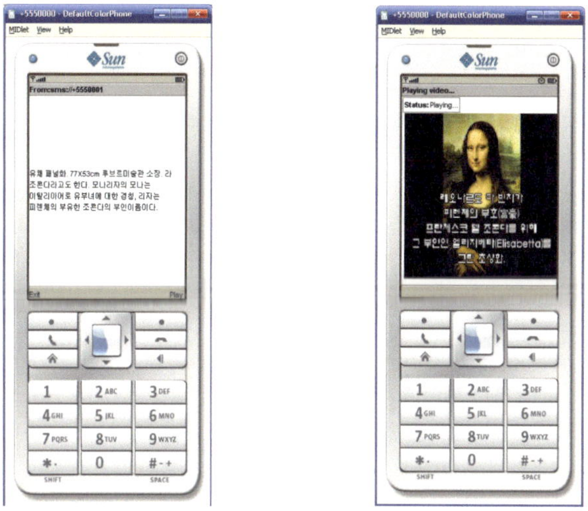

(a)Text Service (b)Multimedia Service

Fig. 20. Running Examples

5. Conclusions

There have been many attempts to adapt RFID technology to context-aware applications. However, until now they have paid little attention to survey of existing systems and to analyze them. In this article, we surveyed the existing context-aware RFID systems, and identified interesting issues: the meaning of tag values, their usage patterns, and context-aware services.

We also introduced a mobile guide system for exhibitions and museums. It considers four kinds of context information: location, background knowledge, media type, and language.

Based on the context information, the system provides the most suitable services to the visitor. For example, when a Korean visitor is an expert on the exhibit, and he/she has a mobile phone that supports multimedia, the system provides the multimedia type expert information in Korean. Furthermore, we introduced some interesting issues when developing context-aware systems: system requirements, context information, context modeling, software architecture, and its prototype system. We drew system requirements from stakeholders during the informal interview. From the requirements, we chose three basic technologies for the system: RFID, mobile phone, and context-awareness. We also found out the four types of context information. Because the context data are independent each other, there are too many possible contexts. Therefore, we model context from context information and determine two kinds of contexts. We chose ECAM architecture pattern for our system because it is event-driven and it can handle multiple kinds of context information. Our prototype system is implemented with 13.56MHz RFID reader, Darwin Streaming server, and MIDP APIs.

6. References

[1] Ron Weinstein, "RFID: A Technical Overview and Its Application to the Enterprise," IT Pro, May/Jun, IEEE, pp. 27-33, 2005.

[2] Peter Harrop, "Rapid adoption of RFID in manufacturing and logistics sectors," Manufacturing and Logistics IT, Aug. 2007, available at http://www.logisticsit.com/absolutenm/templates/article-critical.aspx?articleid=3190&zoneid=31

[3] Bravo J., Hervas R, Chavira G, and Nava S., "Modeling Contexts by RFID-Sensor Fusion," Proc. of Pervasive Computing and Communications Workshops, pp. 30-34, 2006.

[4] Sherry Hsi and Holly Fait, "RFID Enhances Visitors' Museum Experience at the Exploratorium," Comm. of the ACM, Vol. 48, No. 9, Sep. pp. 60-65, 2005.

[5] Jani Korhonen et al., "mTag - Architecture for Discovering Location Specific Mobile Web Services Using RFID and Its Evaluation with Two Case Studies," Proc. of AICT-ICIW, pp. 191-200, 2006.

[6] Trevor Pering, Rafael Ballagas, and Roy Want, "Spontaneous Marriages of Mobile Devices and Interactive Spaces," Comm. of the ACM, Vol. 48, No. 9, pp. 53-59, Sep., 2005.

[7] Matthias Lampe and Martin Strassner, "The Potential of RFID for Moveable Asset Management," Proc. of Workshop on Ubiquitous Commerce at Ubicomp, 2003.

[8] Magnus Holmqvist and Gunnar Stefansson, "Mobile RFID, A Case from Volvo on Innovation in SCM," Proc. of Hawaii International Conf. on System Sciences, 2006.

[9] Penttila K., Pere N., Sioni M., Sydanheimo L., and Kivikoski M., "Use and interface definition of mobile RFID reader integrated in a smart phone," Proc. of the 9th International Symposium on Consumer Electronics, pp. 353- 358, 2005.

[10] EPCglobal, http://www.epcglobalinc.org/.

[11] Michael Schneider, "A Smart Shopping Assistant utilising Adaptive Plan Recognition," Workshop on Adaptivity and User Modeling in Interactive Software Systems, pp. 331-334, 2003.

[12] Heikki Ailisto, Johan Plomp, Lauri Pohjanheimo, and Esko Strommer, "A Physical Selection Paradigm for Ubiquitous Computing," LNCS 2875, Springer, pp. 372-383, 2003.

[13] Lauri Pohjanheimo, Heikki Keränen, and Heikki Ailisto, "Implementing TouchMe Paradigm with a Mobile Phone," Proc. of Joint sOc-EUSAI Conf., pp. 87-92, 2005.

[14] Lionel M. Ni, Yunhao Liu, Yiu Cho Lau, and Abhishek P. Patil, "LANDMARC: Indoor Location Sensing Using Active RFID," Wireless Networks, Springer, Vol. 10, No. 6, pp. 701-710, Nov., 2004.

[15] Michael Crawford, et al., "RFID Enabled Awareness of Participant's Context in eMeetings," Proc. of Pervasive Technology Applied Real-World Experiences with RFID and Sensor Networks, 2006.

[16] Joshua R. Smith, et al, "RFID-based Techniques for Human-Activity Detection," Comm. of the ACM, Vol. 48, No. 9, pp. 39-44, Sep., 2005.

[17] Schilit B. and Theimer M, "Disseminating active map information to mobile hosts," IEEE Network, Vol. 8, No. 5, pp. 22-32, 1994.

[18] Hull R, Neaves P, and Bedford Roberts J., "Towards situated computing," Proc. Of the International Symposium on Wearable Computers, 1997.

[19] Dey A. K and Abowd G. D, "Towards a better understanding of context and context-awareness," Proc. Of the Workshop on the What, Who, Where, When and How of Context-Awareness, ACM Press, 2000.

[20] Jongmyung Choi, "Context-Driven Requirements Analysis," Proc. of ICCSA 2007, LNCS 4707, Springer, pp. 739-748, 2007.

[21] Jongmyung Choi, "Context: From Birth to Design," Proc. of ALPIT, IEEE, 2008.

[22] Jongmyung Choi et al., "Context Model for RFID Systems," Proc. of 2nd International Conference on Ubiquitous Information Technologies and Applications, 2007.

[23] P. Dockhorn Costa, L. Ferreira Pires, and M. van Sinderen, "Architectural Patterns for Context-Aware Services Platforms," Proc. of the Second International Workshop on Ubiquitous Computing, 2005.

[24] Jongmyung Choi and Jong-bae Moon, "MyGuide: A Mobile Context-aware Exhibit Guide System," Proc. of ICCSA, Springer, LNCS, 2008.

[25] Mobile Information Device Profile (MIDP), available at http://java.sun.com/products/midp/.

[26] RXTX, available at http://www.rxtx.org/

[27] Open Source Streaming Server, available at http://developer.apple.com/opensource/server/streaming/.

[28] Matthai Philipose, et. al., "Inferring Activities from Interactions with Objects", IEEE Pervasive Computing, pp. 50-57, 2004.

[29] M. Levinson, "All-in-One Appliance - THE REFRIGERATOR", CIO Magazine, Feb., 2003.

[30] Christian Floerkemeier and Friedmann Mattern, "Smart Playing Cards - Enhancing the Gaming Experience with RFID," Proc. of the Third International Workshop on Pervasive Gaming Applications, 2006.
[31] Yun-Maw Cheng, Wai Yu, and Tzu-Chuan Chou, "Life is Shareable: Blogging Life Experience with RFID Embedded Mobile Phones," MobileHCI, pp. 295-298, 2005.

19

Beyond Identification: Representing Real World Objects and Actors in Radio Frequency Identification

Krishan Sabaragamu Koralalage, Noriaki Yoshiura, and Takaomi Shigehara
Saitama University
Japan

1. Introduction

Due to the improvement of technology, the way human beings do things are changing day by day. Radio frequency (RF) identification has been acting a big role in recent technological developments. Although there are three types of Radio Frequency IDentification tags (RFID): active, passive and semi-passive, almost all the tags are being used to store only a unique number in addition to the processing information. For a particular application, all the relevant details of the tag have to be taken from the database system. There are several applications which are not viable under this infrastructure. Intelligent Transportation System (ITS) including railway, is one of them. ITSs can be improved using the characteristics and the behaviours of RF tagging systems. However currently available RF tags are not suitable for such applications, mainly as it does not provide role base access control mechanisms and have no ability to stand alone to enable offline communications. Thus there is a necessity to look into this area of research.

After extensive survey and hardships, a novel RF tag architecture called Object Tag (OTag), to represent the real world objects and actors in RF tags were designed with common communicational protocols. The OTag can provide access control mechanism and have abilities of being stand alone, being interoperable, being self-describing, and also being plug-and-playable. Then we defined common communicational protocols to communicate with reader/writer and tag. Finally we derived possible applications using our architecture. Our architecture can be used in two ways: one is just as a real world object and second is as an actor or agent. Agent is a combination of a reader/writer, processing unit and OTag. These three units together compose the actions of an agent that can work according to predefined instructions. Following sections describes how proposed technology work in real world by taking ITS as an example by deriving novel applications to support safe, reliable, comfortable, and productive social life in eco-friendly manner.

1.1 Motivation

The first motivation to develop novel architecture for ITSs was raised because there is no proper system to support emergency vehicles such as ambulance, fire brigade, and rescue vehicles which work on saving vital seconds of invaluable human life(Al-Khateeb et al., 2008). The main reason for that is the lack of proper communication among three actors:

users, vehicles, and infrastructure including vehicle to vehicle in ITSs. Here the passengers and the pedestrians are referred as users. Therefore there is a necessity to create a common platform to communicate among three actors to solve the above issue.

If it is possible to develop such a common platform, prioritizing emergency vehicles, changing traffic lights adaptively, prior informing of phase changing on traffic lights, understanding vehicle movements, detecting collisions before entering to intersections or merging points, enforcing rules and regulations such as speed enforcements, assisting elderly driving, understanding accidents, managing traffic congestions, vehicle registrations and ownership transference, can be automated to the maximum possible level (Lee et al., 2008).

RFID Technology can be considered as one of the best candidate technologies to develop a common platform to communicate among users, vehicles and infrastructure including vehicle to vehicle. Although there are a few applications in ITS like electronic toll collection (ETC) using RFID tags, those tags cannot be used for this purpose because they are unable to provide interoperability, ability to be stand alone, ability to be self describing, ability to be plug and play with role base accessing mechanisms.

Unlike current ETC systems, if such a platform is developed, there will be many readers in different systems reading the tag information in a given time. Each reader should be allowed to access only the authorized data depending on their rights. But this cannot be achieved using the existing or the previously proposed tag architectures. Instead it is necessary to have several tags to allow being read by different systems. Additionally, attaching several tags for different systems to work on vehicles are a burden for users as well as for managing authorities. Therefore existing tags cannot be used to develop a common communication platform. On the other hand, such a platform can only be built with one tag, if the access control to the stored data can be provided depending on the roles or the access privileges. Then, it will be possible to subscribe many services or allow different reader to read authorized portions of the tag and thereby the cost and management can be improved. Therefore, using one tag with above features and secure communication protocols, will lead to innovate novel era of ITSs enabling ultimate infrastructure support for autonomous driving in future. Nevertheless, there are no major weaknesses of current applications of ITS.

1.2 Goal and objectives

The goal of our research is to develop a novel RF Tag that can represent real world object and their attributes and behaviours. Then create autonomous agents with those objects to improve existing navigation systems and traffic management systems, enhance electronic fee collection system, assist safe driving, increase efficiency in infrastructure management and commercial vehicle operations and support public transport and emergency vehicle operations by enabling extensive communication among the user(pedestrian/passenger), vehicle and infrastructure of ITS as shown in figure 1.

To achieve above mentioned goals we set following three main objectives: first is to design a novel radio frequency tag architecture with the ability of being intelligent, stand alone, self-describing, and plug & playable while enabling role based access controlling mechanisms to improve the ITSs. Second is to define security and privacy enhanced application layer protocols to communicate among user, vehicle and infrastructure including vehicle to vehicle. The third objective is to derive electronic vehicle registration and ownership transferring mechanism to improve the flow of tagged vehicles throughout their lifecycle.

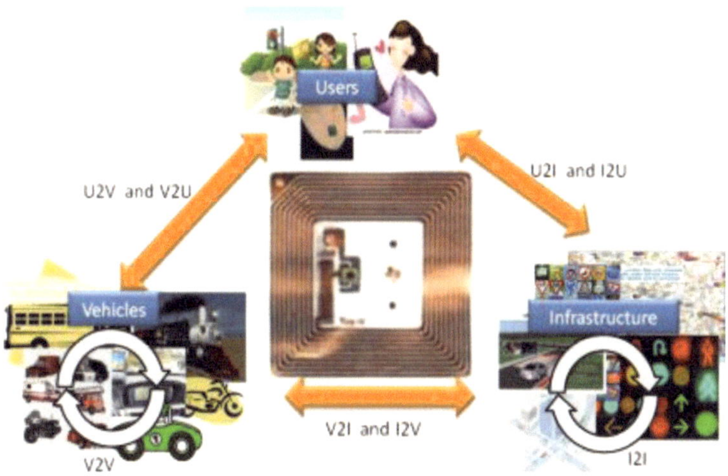

Fig. 1. Illustrate the focused communication among user, vehicle and infrastructure.

1.3 Organization of this chapter

The main aim of this chapter is to introduce the above architecture and communication protocols considering case studies on vehicle to vehicle and vehicle to infrastructure communication.

The remainder of this chapter is organized as follows. Section 2 describes the literature review on RFID and ITS. The proposed architecture and communication protocols are explained in Section 3 and Section 4 respectively. Applications of the proposed architecture in ITSs are described in Section 5 while brief discussion is made in Section 6 addressing sample cases with OTag architecture and finally the Section 7 concludes the chapter with future works and remarks.

2. Related work

As the characteristics and behaviours of RFID tags help to build the novel ITS applications RFID became a suitable candidate technology. Currently, there are RFID systems used in toll collection, parking management, transport payments and logistics management systems by using conventional RFID technology. However, when the capability of RF communication is properly analyzed, it can be seen that there are more possibilities beyond that.

After considering the characteristics and behaviours of RF communication, it is possible to build rich ITS applications that improve the infrastructure by ensuring the safety, security, comfort-ability, productivity, and interoperability in eco-friendly manner. Existing applications are using conventional RFID tags and store only a unique number and the data related to the tag is taken from the proprietary databases. Furthermore interoperability and plug and playable features are not supported.

Though the RF tag can play a big role to achieve safety, security, productivity and comfortability in ITSs, current usage of them are not up to the maximum possibility due to many reasons. They are mainly used for electronic toll collections and proprietary parking management systems. If a vehicle is to use both of these applications, vehicle should be equipped with two RFID tags to work with two different systems.

2.1 Tags used in automated toll collection system

Electronic toll collection (ETC), an adaptation of military "identification friend or foe" technology, aims to eliminate the delay on toll roads by collecting tolls electronically. It is thus a technological implementation of a road pricing concept. It determines whether the cars passing are enrolled in the program, alerts enforcers for those that are not, and electronically debits the accounts of registered car owners without requiring them to stop (Hae et al., 2004).

Enforcement is accomplished by a combination of a camera which takes a picture of the car and a radio frequency keyed computer which searches for a driver's window/bumper mounted transponder to verify and collect payment. The system sends a notice and a fine to cars that pass through without having an active account or paying a toll (Isamu et al., 2005).

Norway has been the world's pioneer in the widespread implementation of this technology. ETC was first introduced in Bergen, in 1986, operating together with traditional tollbooths. In 1991, Trondheim introduced the world's first use of completely unaided full-speed electronic tolling. Norway now has 25 toll roads operating with electronic fee collection (EFC), as the Norwegian technology is called AutoPASS. In 1995, Portugal became the first country to apply a single, universal system to all tolls in the country, the Via Verde, which can also be used in parking lots and gas stations. The United States is another country with widespread use of ETC in several states, though many U.S. toll roads maintain the option of manual collection (Hae et al., 2004).

Electronic toll collection systems rely on four major components: automated vehicle identification, automated vehicle classification, transaction processing, and violation enforcement (Isamu et al., 2005).

2.2 Automated vehicle identification

Automated vehicle identification (AVI) is the process of determining the identity of a vehicle subject to tolls. The majority of toll facilities record the passage of vehicles through a limited number of toll gates. At such facilities, the task is then to identify the vehicle in the gate area.

Some early AVI systems used barcodes affixed to each vehicle, to be read optically at the toll booth. Optical systems proved to have poor reading reliability, especially when faced with inclement weather and dirty vehicles. Most current AVI systems rely on radio-frequency identification, where an antenna at the toll gate communicates with a transponder on the vehicle via Dedicated Short Range Communications (DSRC). RFID tags have proven to have excellent accuracy, and can be read at highway speeds (Keiichi et al., 2001).

2.3 Automated vehicle classification

Automated vehicle classification is closely related to automated vehicle identification (AVI). Most toll facilities charge different rates for different types of vehicles, making it necessary to distinguish the vehicles passing through the toll facilities.

The simplest method is to store the vehicle class in the customer record, and use the AVI data to look up the vehicle class. This is low-cost, but limits user flexibility, in such cases as the automobile owner who occasionally tows a trailer. More complex systems use a variety of sensors. Inductive sensors embedded in the road surface can determine the gaps between vehicles, to provide basic information on the presence of a vehicle. Treadles permit counting the number of axles as a vehicle passes over them and, with offset-treadle installations, also detect dual-tire vehicles. Light-curtain laser profilers record the shape of the vehicle, which can help distinguishing trucks and trailers (Keiichi et al., 2001).

2.4 Transaction processing

Transaction processing deals with maintaining customer accounts, posting toll transactions and customer payments to the accounts, and handling customer inquiries. The transaction processing component of some systems is referred to as a "customer service centre". In many respects, the transaction processing function resembles banking, and several toll agencies have contracted out transaction processing to a bank.

Customer accounts may be post paid, where toll transactions are periodically billed to the customer, or prepaid, where the customer funds a balance in the account which is then depleted as toll transactions occur. The prepaid system is more common, as the small amounts of most tolls make pursuit of uncollected debts uneconomic. Most post paid accounts deal with this issue by requiring a security deposit, effectively rendering the account a prepaid one (Keiichi et al., 2001).

2.5 Electronic congestion pricing

Congestion pricing is a system of surcharging users of a transport network in periods of peak demand to reduce traffic congestion. Examples include road pricing, and higher peak charges for utilities, public transport and slots in canals and airports. This variable pricing strategy regulates demand, making it possible to manage congestion without increasing supply. At the same time, users will be forced to pay for the negative externalities they create, making them conscious of the costs they impose upon each other when consuming during the peak demand, and more aware of their impact on the environment (Lionel et al., 2004).

The application on urban roads is limited to a small number of cities, including London, Stockholm, Singapore and Milan, as well as a few smaller towns. Four general types of systems are in use; a cordon area around a city centre, with charges for passing the cordon line; area wide congestion pricing, which charges for being inside an area; a city centre toll ring, with toll collection surrounding the city; and corridor or single facility congestion pricing, where access to a lane or a facility is priced (Lionel et al., 2004).

Congestion pricing or urban toll schemes were implemented to enter the downtown area using ETC technology and/or cameras and video recognition technology to get the plate numbers in several cities around the world.

2.6 Violation enforcement

A violation enforcement system (VES) is useful in reducing unpaid tolls, as an unmanned toll gate otherwise represents a tempting target for toll evasion. Several methods can be used to deter toll violators.

Automatic number plate recognition, while rarely used as the primary vehicle identification method, is more commonly used in violation enforcement. In the VES context, the number of images collected is much smaller than in the AVI context. This makes manual review, with its greater accuracy over fully automated methods, practical. However, many jurisdictions require legislative action to permit this type of enforcement, as the number plate identifies only the vehicle, not its operator, and many traffic enforcement regulations require identifying the operator in order to issue an infraction (Keiichi et al., 2001).

2.7 Tags used in automated parking management system

Automated parking management systems offer great benefits to owners, operators, and patrons. Some of the main benefits include reduced cash handling and improved back-office operations, high scalability, automatic data capture and detailed reporting, improved traffic flow at peak hours, improved customer service, cash-free convenience, and also provinces to arrange special privileges for VIP customers such as volume discounts, coupons, and other discounts for daily parkers (Keiichi et al., 2001).

RFID enabled automated parking access control systems eliminating customers' need to fumble for change, swipe cards, or punch numbers into a keypad. Vehicles can move smoothly through controlled entrances, and more parkers can be accommodated, thereby increasing revenues. There are no cards to read, tickets to read and sort, and the whole system is a convenient, hands-free way to ensure easy vehicle parking (Lionel et al., 2004).

TransCore is one of the pioneers to provide parking access control systems which offer using RFID technology. They uses proven eGo® and Amtech®-brand RFID technology to deliver reliable, automated parking access control systems solutions. TransCore's parking control systems facilitate airport security parking and parking at universities, corporations, hospitals, gated communities, and downtown parking facilities (Lionel et al., 2004).

ActiveWave is another company who provide such systems using RFID technology. In their system, surveillance cameras or video recorders can be triggered whenever a vehicle enters or exits the controlled area (Lionel et al., 2004).

Although there are number of companies who provide RFID based parking management systems, almost all the systems are proprietary, and provide no interoperability. Some of them are using passive RFID systems. As described in Chapter 1, there is no stand alone, self-describing, plug and playable, interoperable and access controlling mechanisms built into those tags. Hence, all the existing and previously proposed RFID tags are not suitable for the real requirements that are demanded by ITSs.

3. New architecture: OTag

Object tag (OTag) is a radio frequency (RF) tag which can represent real world objects such as vehicles. Considering the main concepts of Object Oriented Architecture, RF tag was designed with its own attributes and methods to stand alone by itself. The tag can be developed as active, passive or semi-passive depending on the application requirement. The attributes and the methods of the OTag may differ according to expected characteristics and behaviors of the real world objects. In other words the attributes and the methods are defined according to object class. For instance, the attributes of a vehicle OTag differ from the attributes of a road symbol OTag. Each OTag can be considered as an instance of an object class. If a vehicle is the object class, tags attached to all vehicles are considered as instances of vehicle object class like in object oriented concepts. Furthermore, each attribute of the OTag has got access modifiers to manage the access to its information based on roles. They are categorized as public, friendly, protected and private. With above features OTag is capable of being stand alone, interoperable, self-describing, and also plug-and-playable. Interoperability is provided by developing common communicational protocols to communicate between tag and interrogator (reader/writer). Howeer the access modifiers used here are not exactly the same as in object oriented concepts.

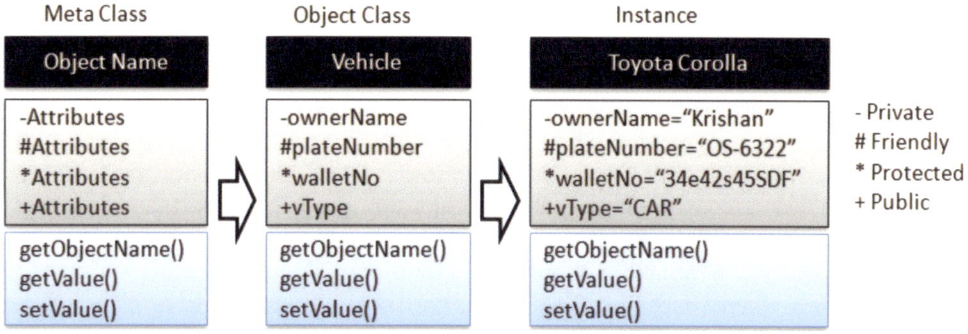

Fig. 2. Illustrate the Instantiation Process of Object to Instance.

OTag architecture can be used in two ways: one is just as a real world object and the other is as an agent. For example when it is necessary to enable communication with vehicle, attaching vehicle OTag instance to the vehicle enables the communication with vehicle. If vehicle need to act autonomously, vehicle will be installed with an agent which is a combination of a reader/writer, processing unit and OTag as shown in figure 3. These three units together compose the actions of an agent that can work according to predefined instructions (Krishan et al., 2009).

Fig. 3. Illustrate the composition units of OTag agent.

Access modifiers can be considered as roles and one of the four access modifiers is assigned to each attribute. With the help of those modifiers, OTag acts four roles: public, friendly, protected and private. Public means no security and any reader can read any attribute value of public area. Private, Protected, and Friendly attributes of the OTag need secure access. Furthermore the writing permission is also available for these areas, but that is controlled by both keys and memory types. For instance write once memory is used for some attributes in friendly area to prevent updating fixed attribute values whereas rewritable memory is used for dynamically changing attribute values according to the behaviors of object class. Details of each role definition are described in later paragraphs.

Figure 2 represents the Object, Class and Instance structures of the OTag. This is the process of instantiation. Any object in the real world should contain its own characteristics and behaviors. In other words each object should have attributes and methods to access those attributes. Tag object has such attributes and methods with access modifiers. When defining a class from object, the class name will be given. Therefore, OTag gets its class name such as vehicle which is a class level attribute, plus implementations of get and set methods that are defined at the time of fresh OTag class creations. In other words, when a vehicle class OTag is created, it will consist of vehicle attributes and vehicle method implementations inside the tag. Such a tag becomes an instance after attaching it to a vehicle. Such as attaching to the vehicle OTag into a vehicle and populating its attribute values make the proper instance of vehicle OTag. Similarly other objects, classes, and instances can be defined.

Once the OTag instance is created, that instance of OTag becomes self-describing and stand alone using the features of object oriented architecture and radio frequency technology. Furthermore it has a common interface to communicate and thereby interoperability is guaranteed.

Plug and playability is a capability of automatically recognizing, configuring and resolving resource conflicts when attached or detached a hardware component to or from a system when running or ceased without user intervention by notifying its availability to the relevant actors. Thus it provides ease of use, maintenance and scalability while supporting the phase base implementation which is worth when installing OTag at many of places. To achieve plug and playability, tag must be self-describing, stand alone and have common communication protocols to communicate with any actor in a system.

3.1 Logical structure and roles of OTag

Role base accessing methods are implemented using access modifiers, memory types and encryption algorithms. As described in the above paragraphs, four roles of vehicle OTag are defined here.

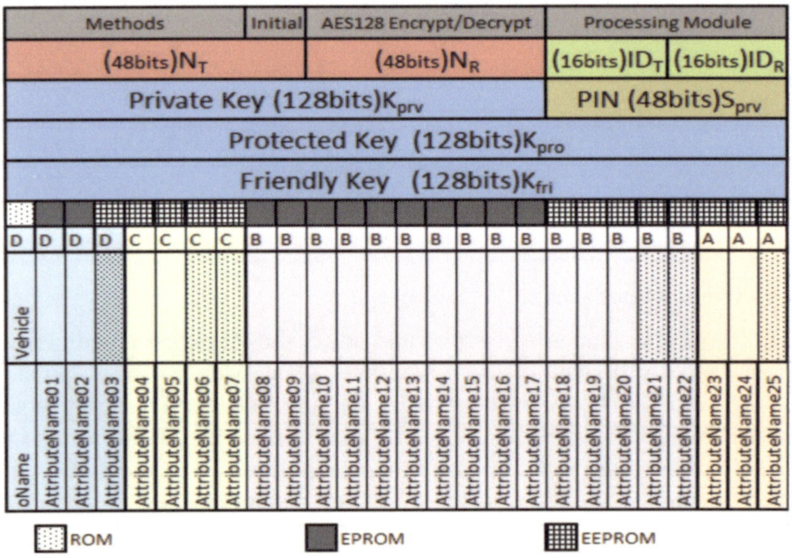

Fig. 4. Illustrate the Logical Structure of an OTag Class.

Memory types used here are ROM, EPROM, and EEPROM. OTag controls writing permissions according to the process flow by using these three types of memories. ROM is used for read only data and they are created by time of fresh tag manufacturing. EPROM is used for one time writing data and they are all about product information. EEPROM is used for all the rewritable data and those data may be changed in any stage of the product lifecycle by relevant authorities. For example, if the attribute values belong to the owner may change after some time of usage. Such as address changing after changing the residence.

Public modifier allows reading and writing of any attribute value by any interrogator who has a common communication interface. Private role is granted to the vehicle owner. He has the permission to read/write any attribute of his own vehicle other than the protected area which is used by one government authority such as department of motor vehicle registration. If the owner needs to change some attribute value in friendly area, he must use the key of friendly area. Thus an owner has two keys with him: private key and friendly key in addition to a PIN to provide secure transference of ownership of the vehicle.
Protected modifier controls the area used by the government authority in vehicle tag. After first registration of the vehicle, rewriting or updating of attribute values in this area can only be done with the protected key. Vehicle's license plate number, inspection expire date and etc are available in this area.
Friendly modifier allows several services to be catered in effective manner. The owner of the vehicle must register with friendly key to allow reading/writing of such area to use the service render by the service provider. For instance ETC or automated parking collection facility can be provided after prior registration with relevant authorities. Emergency vehicles like ambulances, fire brigade, rescue, and police petrol use this area for traffic signal prioritizations.
Figure 4 illustrates the clear image of the memory types, access modifiers, and respective keys of each role. Encryption algorithm used hers is AES-128 stream cipher. Therefore each key size is fixed to 128bits. No secured information is transmitted without sufficient authentication.
OTag contains methods, data "Initial", 128-AES algorithm and processing module. OTag generates nonce N_I whereas the reader generates three nonce values N_T, ID_I, and ID_T to carry out proper mutual authentication. In addition that three role keys and PIN are stored to ensure the security of data.
OTag can store its own attribute and methods inside the Tag. Therefore, there is no need to run a database lookup to retrieve detail information from a database. Thus it provides stand alone capability. Unlike conventional RFID tag, OTag can also control and manage role base accessing depending on four roles: public, friendly, protected and private. Furthermore, the OTag itself contains the necessary information to describe its characteristics and behaviours. Thus OTag can represent any object and when combined with interrogator it can also act as an agent.

4. Communication protocols

OTag has five main protocols: non-secured reading, secured reading, secured writing, key updating and transferring tag ownership. Transferring ownership of OTag is a combination of above protocols. It is only used in authorized centres since specific readers will carry out this task. Section 5.1 describes the usage of ownership transference in detail.
Protocol notations are as follows.

K_{prv} - *Private Key (128bits)*
S_{prv} - *Shared Key or PIN (48bits)*
K_{pro} - *Protected Key (128bits)*
K_{frn} - *Friendly Key (128bits)*
PIN - *Personal Identification Number (48bits)*
N_I - *Nonce generated by Interrogator (40bits)*

N_T - *Nonce generated by Tag (40bits)*
ID_I - *Interrogator generated ID (16bits)*
ID_T - *Interrogator Generated ID (16bits)*
Initial - *Publicly defined initial message (16bits)*
R - *Response value – Attribute Value, or Successful/Failed [1/0]*
{M}K - *Message "M" encrypted by Key "K" using AES128 steam cipher algorithm*

4.1 Non-secured reading protocol

As figure 4 illustrates, non-secured reading protocol is used for public reading. Any reader can query the public attribute values by passing the attribute name to the OTag. For that attribute names of the given object class must be available for the reader to query the attribute value. For instance, attribute name list of vehicle class should be available with the interested readers. Interested readers can query "Type" attribute value from OTag. Then the OTag will answer with value "CAR".

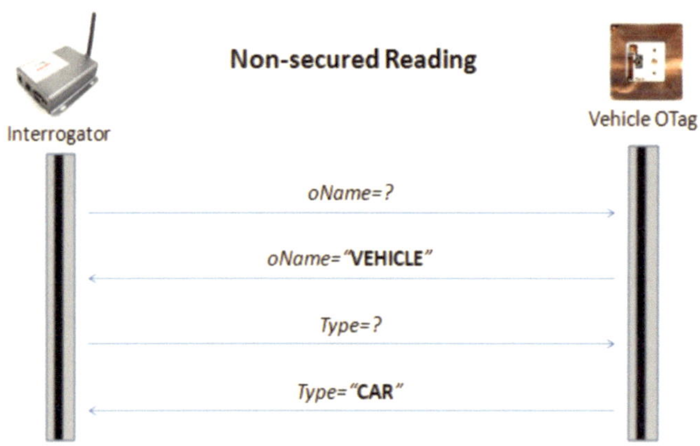

Fig. 4. Illustrate the public reading protocol.

4.2 Private, protected and friendly reading

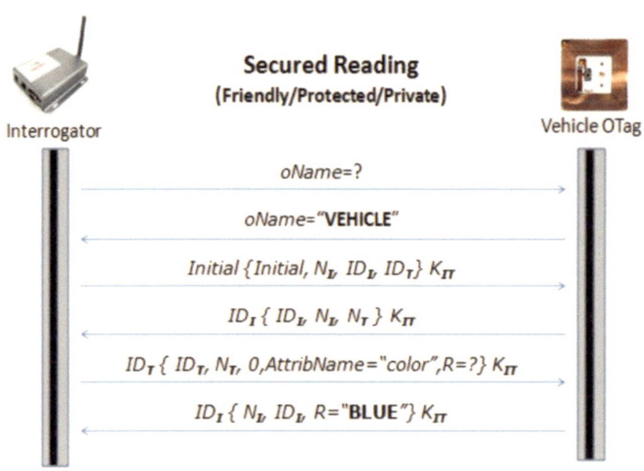

Fig. 5. Illustrate the secure reading protocol with private, protected and friendly role keys.

Each role attribute values should be managed securely. Reading is allowed only after successful mutual authentication. Here the K_{IT} denotes the Encryption key used in each party. For instance, K_{IT} denotes the private key in private reading where as K_{IT} denotes the protected key in protected reading. Reading is only granted for authorized parties and they can only read the data which belong to them. To ensure security of each reading messages are encrypted with authorization key and send as shown in Figure 5.

4.3 Private, protected and friendly writing

Same as in the secure reading, the secure writing needs successful mutual authentication. Attribute name and attribute value pair should be passed for each writings. After successful writing, interrogator will be acknowledged as shown in Figure 6.

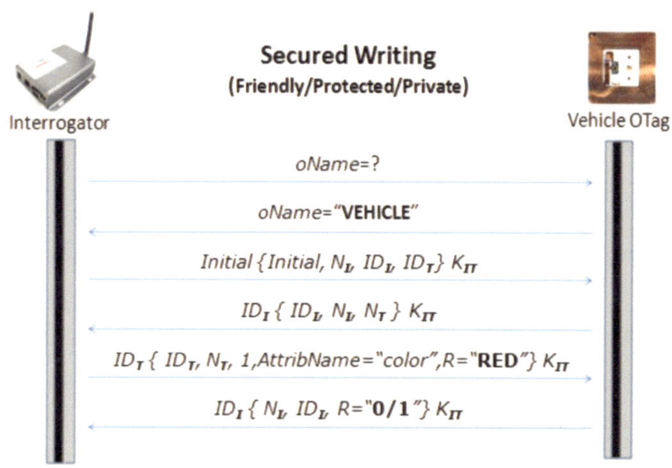

Fig. 6. Illustrate the secured writing protocol with private, protected and friendly role keys.

4.4 Secure key updating

Unlike secure writing as shown in Figure 7, it is necessary to confirm the keys to be updated before actual key updating because the communication will be no longer possible if a wrong key is set. Therefore, there are two more passes in this protocol other than secure writing as shown in Figure 6.

4.5 Transferring the ownership

This is also a secured writing process. Before transferring the ownership, personal data must be deleted. Therefore all the private data are deleted using secured writing protocol. After successful deletion, key updating will be carried out. If the new owner and the old owner are present, keys will be updated on the spot. Otherwise keys will be converted to temporary secrets and passed to the authorized agents to handle ownership transferring process. This process is carried out by a specific reader place in authorized centres.

Once the ownership is transferred the predecessors cannot read the tag as they are unable to provide new keys to get the access to the area now belonging to successors. Therefore, predecessors' and successors' security and privacy are ensured.

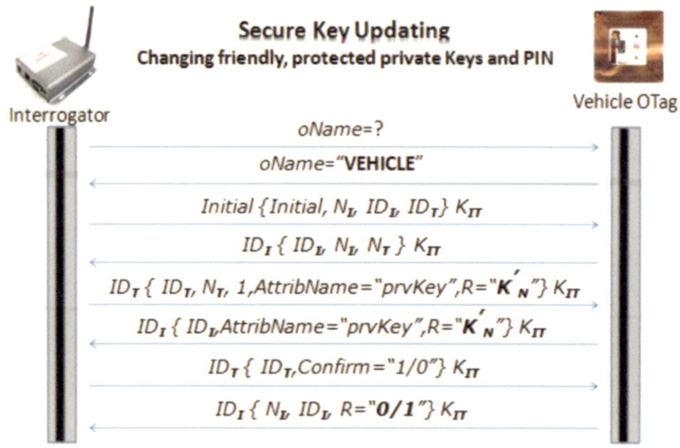

Fig. 7. Illustrate the role key update protocol for private, protected and friendly role keys.

5. Possibilities and applications with OTag

As explained in Section 3, OTag inherits all the characteristics and behaviours of RF communication. Additionally, OTag is intelligent, stand alone, self describing, plug and playable, and interoperable with role base accessing system that allows to communicate with different actors. Here the OTag agent is embedded into a vehicle.

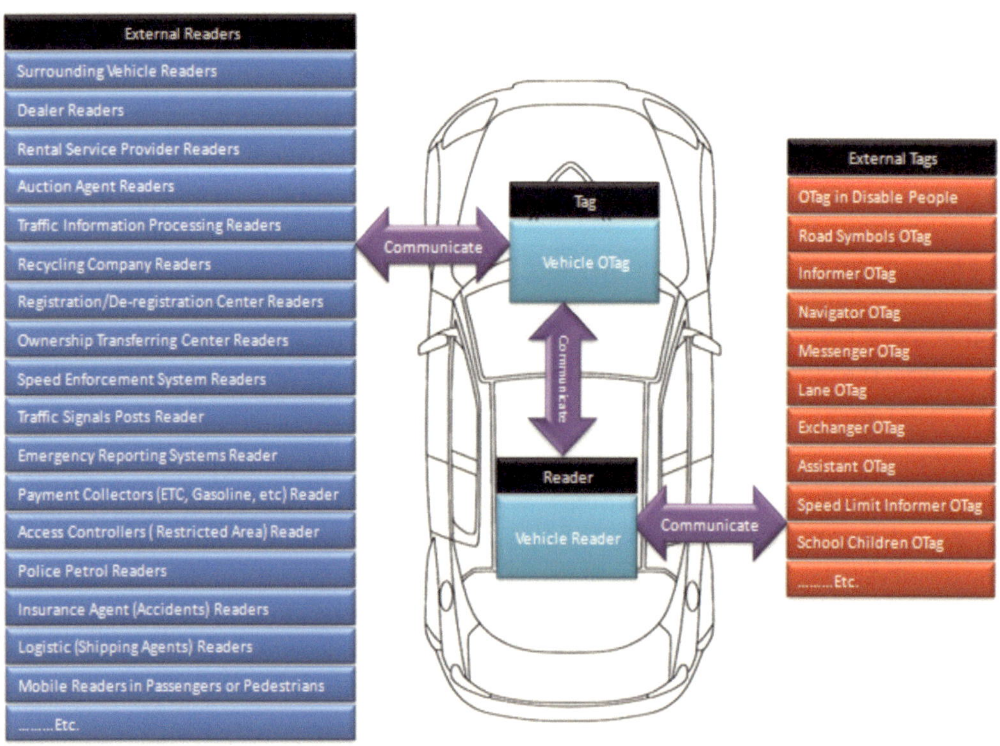

Fig. 8. Illustrate the actors in ITS as readers and tags.

OTag can be used to represent most of the agents in the ITS system. Each Object that interacts with ITS system can be modelled as an OTag. For instance, the main object in ITS system is the vehicle. Other than that there may be traffic signals, road symbols, road guides, disable assistants, information providers, rules and regulation controllers, etc. As OTag can be used to create an interactive agent for each task, efficient and effective infrastructure can be built to support ITSs.
Figure 8 illustrates the different actors in ITS, that are interested in communicating with the main actor vehicle. External Readers can read OTag in vehicles and the readers in vehicle can read infrastructure and passenger or pedestrian OTags.

5.1 OTag in vehicle

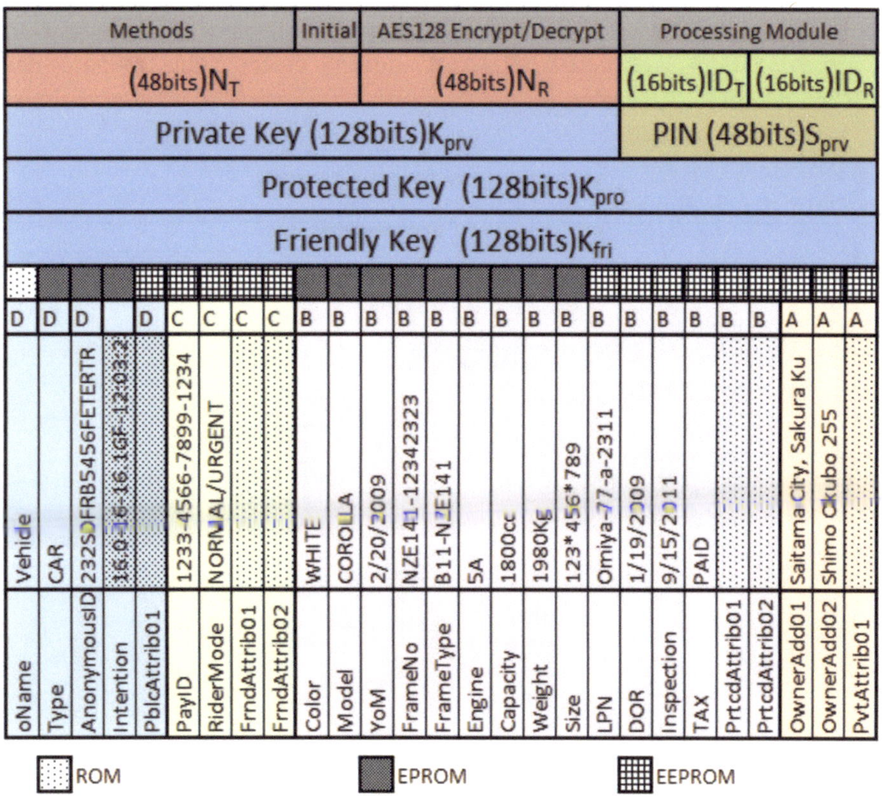

Fig. 9. Illustrate the Logical Structure of an OTag instance in Vehicle.

Imagine that each vehicle has got a RF reader and an active OTag in it. This arrangement makes vehicle an intelligent interface. Therefore, OTag enables large array of applications to improve ITSs.
Figure 9 represents the logical structure of an OTag in vehicle. Four areas for four roles are marked with A, B, C and D as described in section 3. Information stored in the public area can be read by any reader whereas the friendly, protected and private areas are secured with reading and writing permission. Each area is used by different agents throughout the lifecycle of the vehicle. Four areas of the vehicle OTag can be used as follows.

Public area stores the object class name, type, anonymous ID, intention and customizable attribute called PblAttrib01. Information stored in this area can be used to understand object and its movements. Identifying those public information leads to applications like collision avoidance, traffic signal management, real time traffic information, obstacle identification, road congestion management, road rule enforcements, disaster messaging systems, driving assistant systems, etc.

Friendly area stores the payID or walletID, riderMode, two customizable attributes named as frndAttrib01 and frndAttrib02. Information stored in this can be used to subscribe variety of services provided by companies. Any registered service provider can use this information to provide the comfortable service to the vehicle user. Electronic fee collection systems like toll collection and parking can use walletID whereas riderMode can be used to prioritize emergency vehicles.

Protected area stores the colour, model, manufactured year, frame number and type, engine, capacity, weight, size, license plate number, date of first registration, inspection validity, and tax payment status. Additionally it also has got two customizable attributes named as prtcdAttrib01 and prtcdAttrib02.

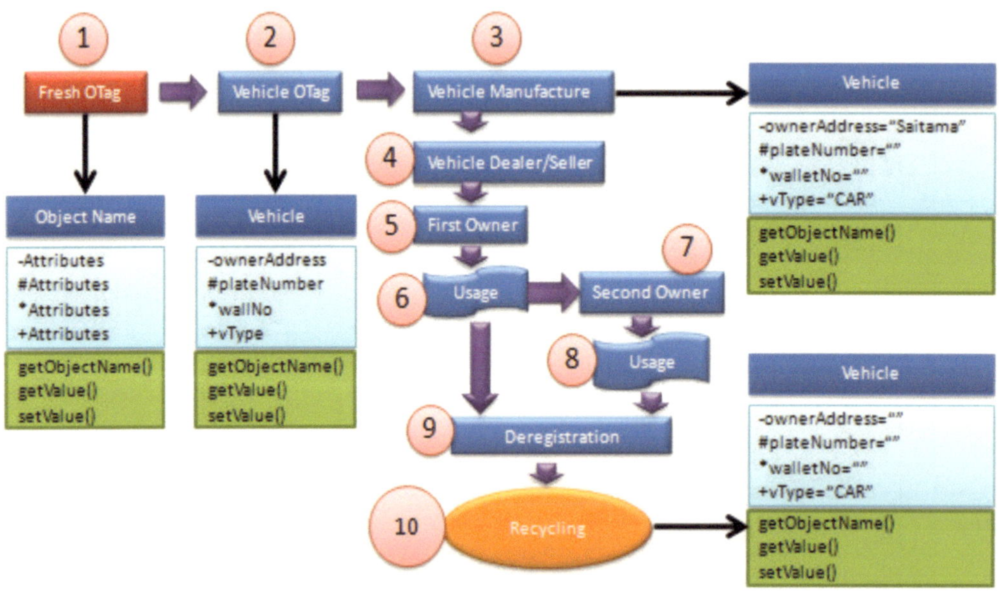

Fig. 10. Illustrate the flow for the OTag embedded vehicle lifecycle.

Information stored in this area is devoted to the vehicle governing authority. Mostly this is the government service. Therefore, after first registration the data stored in this area can only be manipulated by government authority such as department of motor vehicle registration. Additionally, protected information can be read only by the owner or the granted authorized persons in police. Inspection validity, insurance, tax, etc. helps to identifying the current status of the vehicle. Recognizing illegal, fake, clone, stolen or altered vehicles, issuing fines for inspection expired vehicles, verification of tax payments, management of carrying garage, temporary and brand new vehicles are some of the main applications using this area.

Private area stores owner name, address and one customizable attribute named as pvtAttrib01. Information stored in this area is to prove the ownership of the vehicle. No one can read or write data into this area without the permission of the owner. When the vehicle is sold ownership information will be changed.

5.2 Tag flow

As shown in Figure 10, step 1 creates fresh OTags according to the object class. Each class has got different attributes as of the real world object. In this step, the tags contain the attribute names "oName" and "anonymousID". Then the OTag instance for vehicle class is created in step 2. It contains the attribute names relevant to vehicle class. Only the values of attribute names "oName" and "anonymous ID" are filled in vehicle OTag before attaching to a vehicle. The value of "oName" is "VEHICLE" and that of "anonymous ID" is a random unique number. Next the OTag is passed to the vehicle manufacturer.

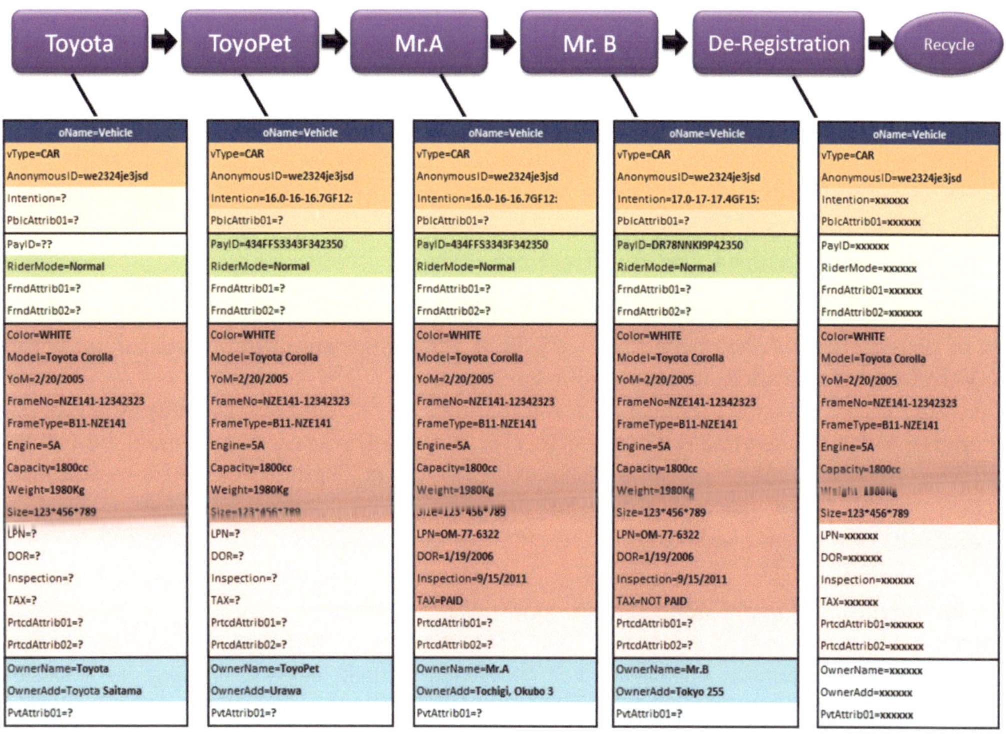

Fig. 11. Illustrate the sample flow for the vehicle OTag throughout its lifecycle.

Step 3 when vehicle manufacturers receive the vehicle OTag instances with empty attribute values except "oName" and "anonymousID". Additionally the three role keys and the PIN are set to their secrets. All the relevant instance attributes will be fed to the OTag. For example Toyota may request 1000 vehicle OTags to attached in brand new vehicles and then feed the instance attribute values such as vehicle type, frame number, engine etc as shown in Figure 12. Then the vehicles are passed to the dealers or sellers after transferring the ownership to the dealers.

In step 4 vehicles in dealers possession contains all information fed by the manufacture but the ownership information, role keys and PIN are changed to dealers' secrets. When a customer bought a brand new vehicle as in step 5, first registration will be carried out. By that time the protected role key will be handed over to the vehicle governing authority while friendly, and private role keys with PIN will be changed to the customers' secrets as shown in Figure 11. Additionally the ownership information will also be changed to his own details. Since the customer has the friendly key to allow secure communication between desired service provider and his own vehicle, services like electronic toll collection, parking payments, gasoline payments, etc. can be subscribed easily.

Step 6 and step 8 emphasize the usage of the vehicle. During the usage period customer may extend the inspection period, pay taxes, etc. Then the relevant OTag attribute values may be updated as shown in Figure 11. When the customer needs to sell the vehicle to someone, ownership information should be transferred by visiting the authorized centres' readers.

In step 9 if the vehicle may be no longer needed, cancellation of registration can be done in the same centre and pass the vehicle for recycling as shown in step 10. By the time of de-registration only the minimum required information will be kept and other information including personal data will be deleted to protect security and privacy of users as shown in de-registration tag information in Figure 11. Then the recycling company also can use RF communication to improve their process of gathering information on recycling units.

Note that each user's ability of writing is also controlled by using three memory types: ROM, EPROM and EEPROM.

6. Sample applications and discussions

After having installed the infrastructure and vehicle with OTags and interrogators, whole lot of possibilities will be opened up. Vehicle to Vehicle communication and Infrastructure to Vehicle communication are focused for sample cases.

Detecting vehicle to vehicle collisions at intersections, at merging and demerging and when changing lanes has a vital importance in ITS. If one vehicle can understand how other vehicles' move, the collisions can be detected. Once the collision is detected, precautionary action can be taken. Though the prior detection is very important, guaranteeing the avoidance is not an easy task but the consequences due to collisions can be reduced.

6.1 Identifying and expressing the intension of the vehicle

Identifying the current position and intended action including driving direction is very important to avoid fatal accidents. For that each vehicle should express its intended movements at least within a given block or time slot. Following paragraphs explain how a vehicle can express its intension to other vehicles or infrastructure in detail.

There is OTag class called Informer who can explain the next target, current position, current route and next crossing lane, etc. to a requester vehicle. These tags are positioned in the centre line of the road. It can be read by the vehicles running towards the both directions.

As shown in Figure 12. Informer OTag contains the next target information in the form of attribute name and value. Once the immediate previous 16.1 Informer OTag position is passed by a vehicle, intended action "16.1GF" of that vehicle is sent to the 16.2 Informer OTag via the vehicle reader requesting the next target information. Then the Informer OTag will respond with the next target information to the vehicle. If there is no previous Informer, the reader in the vehicle requests yourPosition attribute of first Informer OTag to understand the current position. Informer OTag can be of type passive.

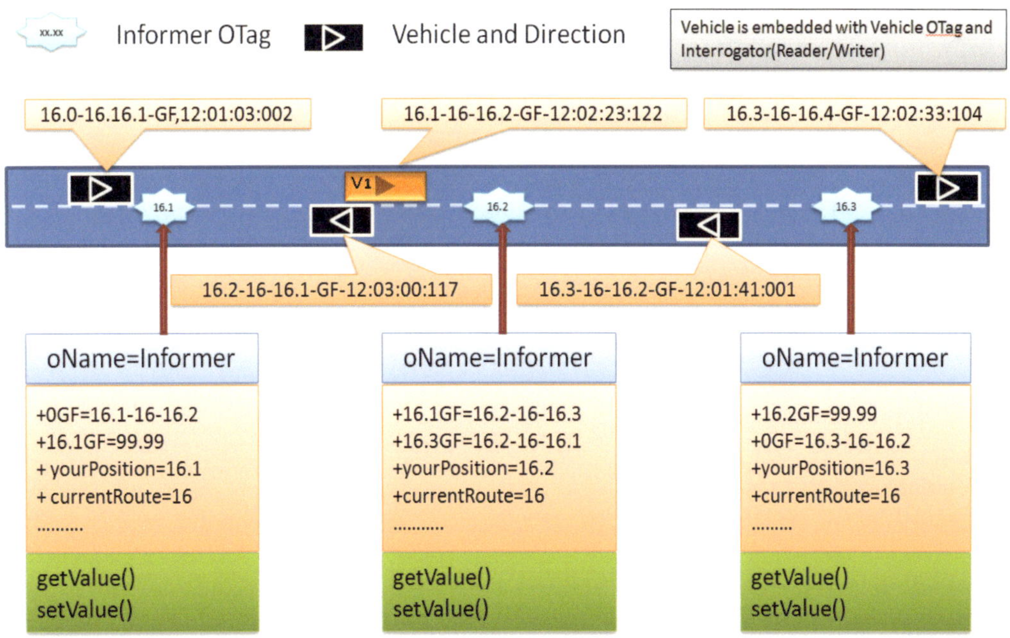

Fig. 12. Illustrate the informer instances of OTags and positions on the roads with moving vehicles.

As described, when a vehicle passes the very first Informer, the interrogator in the vehicle gets to know the next immediate target information. After that, the interrogator writes that information with intended action, running speed, and expected reach time of current immediate target and lane number if it exists, to the public area of the vehicle OTag. If there is no lane number, or previous position information before the Informer, the predefined not applicable dummy values will be used to compose the intention attribute. Once the vehicle OTag is filled with intention, it will start expressing the intended movements as shown in Figure 12. This process continues until the no Informer tag is found. Whenever changes happened to the values relevant to the parameters used to compose above message, recalculation is done and vehicle OTag starts retransmitting the new message. For instance, if the speed is changed the recalculation is done and after that the retransmission starts. A target message interpretation is represented in Figure 13.

If a moving vehicle could not pass any Informer tag, the vehicle OTag will transmit only the available information with predefined dummy values. In this situation, the vehicle that is lacking main information must take care of its mistakes, and only simple warning about surrounding vehicle can be given to such vehicles. This can happen if a vehicle enters to the road between two Informer tags.

Here the intended action is divided to seven categories: Turn Left (TL), Turn Right (TR), Go Forward (GF), Go Backward (GB), U-Turn (UT), Hazard Stop (HS), and Emergency Accident (EA). Intended movements are not expressed when a vehicle is in parking mode.

The message in Figure 13 is interpreted as a vehicle is moving over the route number 16 by passing the Informer position 16.0 and heading forward (GF) to 16.1. Current estimated reach time to the 16.1 is 12:02:23:122 and the vehicle is running at a speed of 60kmph in lane 0. Here the lane 0 means road has only one lane for one side.

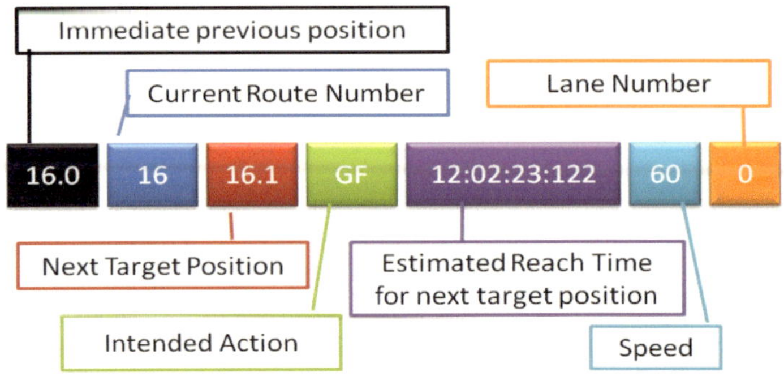

Fig. 13. Target message expressed by moving vehicle.

6.2 Vehicle talking to vehicle

In this section we explain two main cases to avoid collisions when crossing an intersection and changing lanes. For both of them it is necessary to understand the movements of the surrounding vehicles. Our solution considers six types of vehicle to vehicle collisions. They are the collisions that may occur when crossing the intersection, when changing the lanes,

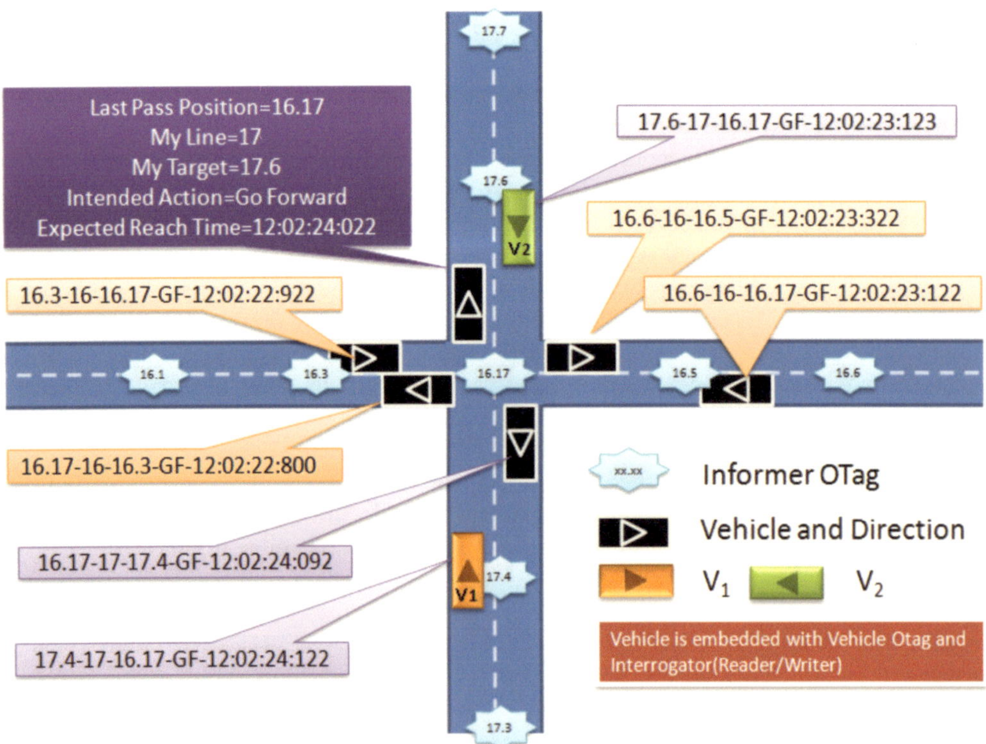

Fig. 14. Illustrate Detecting collision in intersection with 8 vehicles from 4 sides going forward.

when merging and demerging, when running face to face, when running to front vehicle and when backing up. Though it is possible to solve each of them with the method proposed in following section, only crossing intersections and merging lanes are discussed here.
To identify the vehicle movements, following assumptions were made.

1. Infrastructure Informer OTags have to be placed according the following diagram.
2. Each Informer OTag know its position, and immediate Informer position.
3. Vehicle should be equipped with interrogator (Reader/Writer) and Vehicle OTag.

6.3 How to detect the collisions at intersection

Consider the route number of the vertical route as 17 and horizontal route as 16 as shown in Figure 14. Suppose that the vehicle V1 is moving towards the intersection passing the 17.4 position in the route 17. V1 explains others about his intended movements in his message as shown in the same figure. This message means that the V1 is heading forward (GF) his next immediate target 16.17 by passing the 17.4 over the route 17 and he is supposed to reach the target by 12:02:24:122.
Similarly, in his assistant screen there will be an image similar to Figure 14 explaining how the other vehicles are coming towards the intersection and also leaving the intersection by interpreting their intended movement messages. Since real time calculation is done and also every change is calculated once change occurs to any of the parameters used to create the message, reach time of each vehicle can be estimated accurately. If two or more vehicles are reaching a same target by the exact time, there is very high possibility to collide them. Therefore once the reader in the vehicle could understand such situation, the driver can be warned or asked to take precautionary steps to avoid the predicted collision.
In case of turning vehicles, the message explains their intensions by setting the composed value with TL or TR. Suppose that the vehicle V1 is going straight and the vehicle V2 is going to take a V1's left, V1 must understand the possible collision and take precautionary steps. If V2 is going to take the turn to right side of the V1, no collision will be happened. Thus quick decision can be taken and smooth, safe and fast crossing of intersection can be realized with OTag.

6.4 How to detect the collision when merging lanes

Figure 15 illustrates the movements of merging points. There are three main types of actions which lead to collisions: merging, demerging and turning to right or left. Each movement should be properly identified and drivers are informed possible collision.
Consider the route number of the road in Figure 15 as 16 and it has got two lanes for one direction and one merging point as illustrated. Informer OTag (16.3) should also been placed at exact line of the merging point shown in the Figure 15. The number of the entering lane is considered as 100. One Informer OTag (16.2) must be installed in the entering lane (100) before reaching the merging point and two lane tags must be also installed in lane 110 and 111 to recognize the lanes as shown in the Figure 15. When lane tag is passed the vehicles get to know their own lanes via their readers. Entering lane is one way and one lane, thus the Informer OTag 16.2 contains the lane number as 100 with next target (16.3) information. Therefore, there is no need to place another lane tag in the entering lane.
Suppose that the vehicle V1 is running straight forward over the route 16 while the vehicle V2 is coming towards the merging point to merge with lane 110. V1 expresses his intended

movement as 16.0-16-16.3-GF-12:02:23:082-110-95 which means that V1 is heading forward (GF) over the route 16 passing the 16.0 Informer and targeting the 16.3 Informer. Also V1 can reach the target by 12:02:23:082 as he is moving at a speed of 95Kmph in lane 110 whereas V2 express the message that can be interpreted as V2 is heading forward (GF) over route 16 lane 100 passing 16.2 targeting 16.3 at a speed of 40Kmph and his estimated reach time is 12:02:23:122 by the intention "16.2-16-16.3-GF-12:02:23:122-100-40".

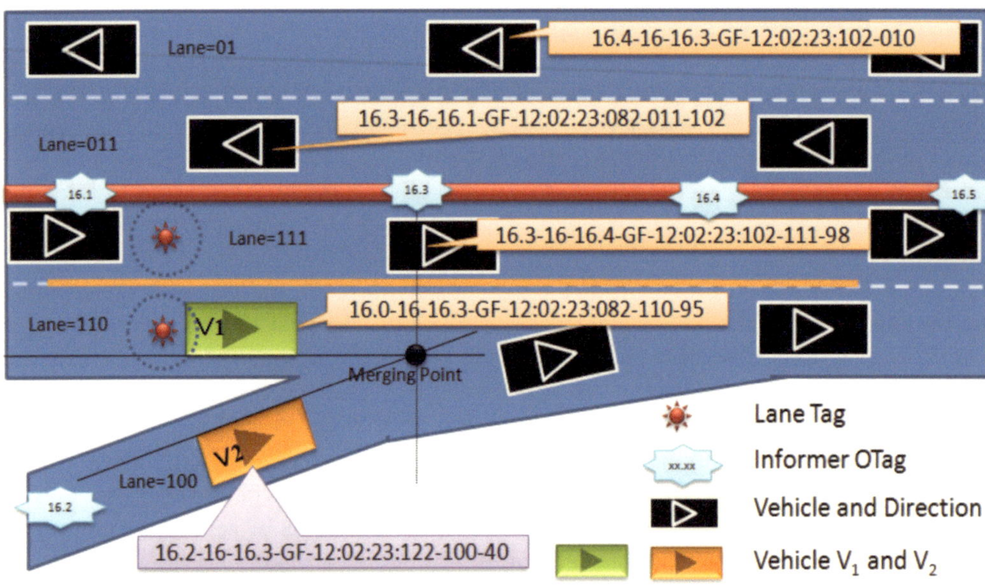

Fig. 15. Illustrate detection of collisions when merging lanes.

Since the estimated time is so close, both of them will be informed by each other's screen that the drivers of V1 and V2 are heading to a possible collision. Then they can take precautionary actions while V2 must know that it has the lowest priority in this scenario. Similarly, all the vehicles driving towards the merging point can be informed possible collisions. Then the drivers can decide their next steps or autonomous system may tackle it in future.

Installation of these tags can be done in essential points where accidents occur frequently. Specially, in toll roads, accident rates are high when merging and demerging. Though it is possible to make safe and smooth merging with two lane tags and four or five Informer OTags, implementing same method in lane changing throughout the road needs to have more lane tags installed over the road which is costly at the moment. Therefore the essential points can be identified and installed.

6.5 Infrastructure talking to vehicle

Infrastructure will be consisting of tags and readers. The following cases are described assuming that there is a reader in the traffic signal post and it knows the friendly key of four emergency vehicle types: ambulance, fire brigade, rescue and police patrol.

6.6 How to prioritize emergency vehicles

One main application we consider is changing the traffic signal in intersection to green when the emergency vehicle is reaching the traffic signal post. Secure identification of emergency vehicles such as ambulance, police petrol, fire brigades and rescue vehicles should be in place to prevent frauds. Otherwise some vehicles may act as an emergency vehicle and try to get the advantage of signal prioritizing silently.

Currently all the emergency vehicles are identified by the tempo and tone of the siren hone. When the siren signal is heard, everybody has to stop and arrange the road to pass the emergency vehicle. This is always not successful; as there are long queues in congested situations which rarely could allow any movements before the traffic signal changes to green. Additionally in the intersections such system cannot prevent accidents because someone may enter to the intersection according to the traffic light guidance without understanding that an emergency vehicle is heading towards the intersection. With OTag, it is possible to change the signal for emergency vehicles' path and allow smooth traffic. By this way time taken to reach the destination by emergency vehicles can be reduced and thereby lives can be saved greatly.

There is a reader in the Intersection traffic signal post. It has the friendly key of the emergency vehicles. When the OTag in the emergency starts to say that I am an emergency vehicle, reader validate the vehicle and change the colour light to green to allowing emergency vehicle to pass smoothly. The changing logic will work as shown in Figure 16.

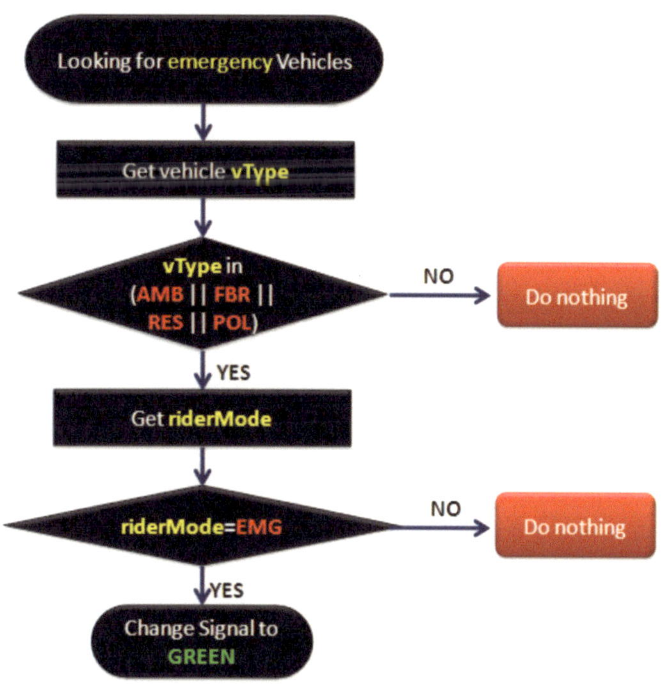

Fig. 16. Illustrate decision flow chart of changing colour lights according to the emergency vehicle request.

Emergency vehicles can change the status from NORMAL to URGENT depending on the patients condition. Then the OTag will describe its status to readers in the traffic signal. Once the reader understands the vehicle type, reader authenticates the vehicle and the green signal is issued for emergency vehicle to cross the intersection smoothly with others.
On the other hand drivers in the normal vehicles are informed that the emergency vehicle is approaching using vehicle to vehicle communication described in above section. Thus they can cooperatively help to pass the emergency vehicle without any tension, delay or accident.

6.7 How to enhance intersection traffic

Varying traffic volumes during the peak hours and midday makes it very difficult to enhance traffic signals. Similarly the areas that experience heavy traffic congestion, needed traffic signal timing improvements to implement effective traffic flow as well as air quality and fuel consumptions. Currently several methods are being used to detect and count the vehicle coming towards the intersection. Several systems are capable of monitoring the traffic arrivals and adjusting timings based on the detected inputs. Traffic detectors may range from metal detectors, infrared readers, image detectors, etc. Metal detectors are the most popular in use though they provide minimum information. Image detection devices exhibit numerous problems including degradation during bad weather and lighting.
Consider that there is a reader who has a reading distance of 100meters radius is installed in the traffic signal post as shown in Figure 17. All the vehicles are embedded with OTag and

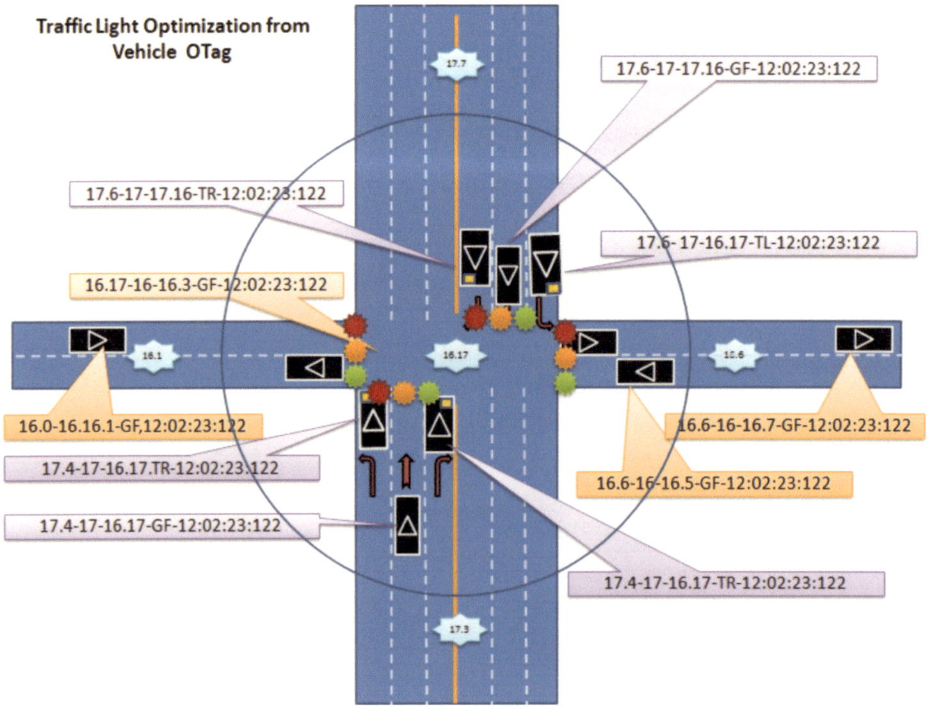

Fig. 17. Optimization through adaptive reading via Vehicle OTag and traffic signal agent.

they are expressing intended movements. By reading this message the reader in the traffic signal post understands the number of vehicles, in its range, their intended directions and the number of vehicle types including the availability of emergency vehicle or common transportation units such as bus, etc.
Depending on the policy of the country, traffic optimizing algorithm can be implemented considering above knowledge. Unlike in other conventional intelligent traffic systems, an extra knowledge can be mined and that knowledge can be used to optimize the traffic signal adaptively because OTag can provide detail information to take a better decision. Additionally if a tag is installed in the signal and set to describe starting and ending phases of transition to incoming vehicles, vehicle stopping can be made smooth and collision due to misinterpretations can be minimized.

6.8 How to prevent collisions in railroad crossings

In railway crossings, collision between vehicles and locomotives can be avoided using the similar arrangement. In this case the priority is given to the locomotive, so the vehicle passing the crossing must take the precautionary steps or its system might take such action automatically if driver neglect to act accordingly. Since the same arrangement is proposed for vehicles and railways, interrogator in the vehicle can understand that there will be a possible collision with moving locomotive if it proceeds without any precautionary action as shown in figure 18.

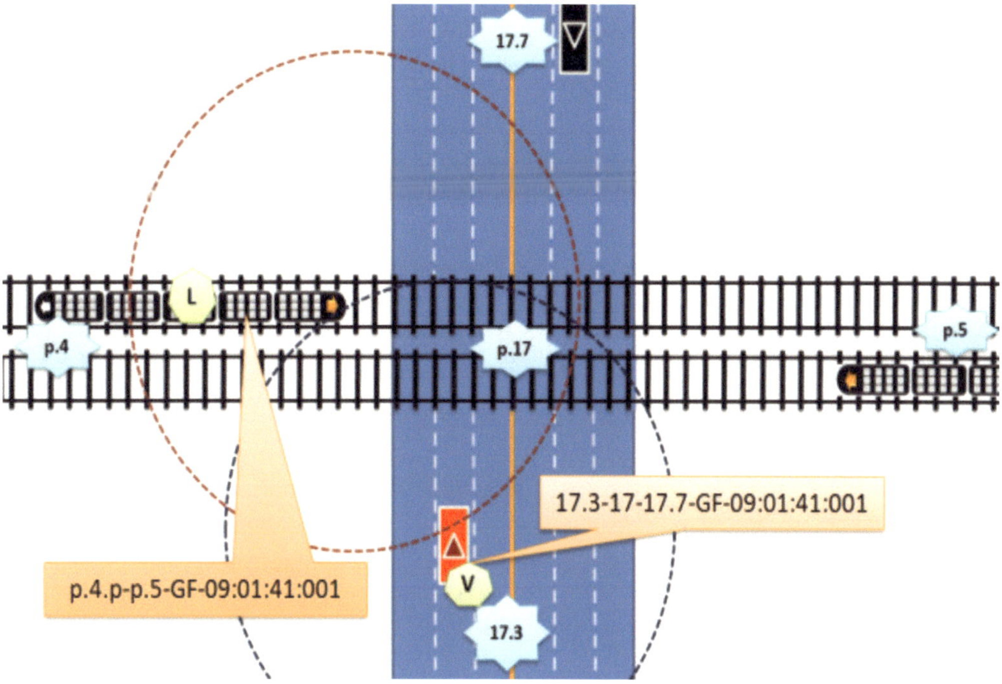

Fig. 18. Collision detection between vehicles and locomotives in rail road crossings.

Here the locomotive L is heading forward to pass the crossing and detect that there is a possible collision ahead but locomotive will not take any action unless and until IA issues any instructions to stop since the locomotive is given the first priority. A vehicle V heading forward towards the railway crossing detect that there will be a possible collision with passing locomotive and ask the driver to take precautionary steps to avoid accidents or activate automated breaking system. Thus collisions can be avoided or consequences due to the collision can be reduced.

6.9 Benefits and possibilities

Since the OTag's rich infrastructure supports intelligent, stand alone, self-describing, plug and playable, interoperable features with role based access controlling mechanism, main benefits like time savings, improved throughput, reduced crashes and fatalities, cost avoidance, increased customer satisfaction, and energy and environment friendly technology can be achieved while providing the safe, secure, productive and comfortable life in eco-friendly manner.

6.10 Other possibilities

As traffic signs provide the driver various information for safe and efficient navigation, one of the important areas is representing traffic signs on OTag. Then the automatic recognition of traffic signs can be provided to support for automated driving or driver assistance systems. Therefore the reader in vehicle can understand the road sign and explain the driver or control the vehicle according to the instructions. OTag can also be used to control the speed of a vehicle according to the road condition or climate changes. It also could be used to control the maximum allowed speed as well as the minimum speed. Similarly, there are scenarios where one vehicle trying to enter a single lane without seeing the coming vehicle due to building etc. Such situations can be mitigated without much infrastructure facilities and thereby eliminate the unnecessary movements of vehicles creating traffic congestions.

Not only that but also vehicles can be registered electronically, transferring ownerships can be automated, uninsured vehicle restrictions and taxations can be enforced without human intervention while many more innovative applications can be developed with this system as each chip has got user memory area for such applications including supporting for future autonomous driving infrastructure systems.

7. Concluding remarks

OTag is introduced to improve the existing intelligent transportation systems enabling large array of novel applications. The architecture of the OTag, communication protocols, capabilities, and usage are being explained. Collision detection in intersections and merging points, traffic signal prioritizing for emergency vehicles and adaptive traffic signal improving systems are discussed under vehicle to vehicle and vehicle to infrastructure communication.

As a conclusion, if the OTag can be implemented and used in vehicles and infrastructure, the power of the identification technology beyond the unique number creates a platform to communicate among user, vehicle and infrastructure assuring the safety, security, comfortability and productivity in eco-friendly manner.

In future we will release the other infrastructure tag designs by complying with OTag using the same protocols proving the interoperability, self-describing ability, ability to be stand alone, and plug and playability with role base access control mechanisms.

8. References

Al-Khateeb A. S., et al., "Dynamic Traffic Light Sequence Algorithm Using RFID Khalid" Journal of Computer Science 4 (7): 517-524, 2008 ISSN 1549-3636 c 2008 Science Publications.

Hae D. C,"Using RFID for Accurate Positioning",Samsung Electronics Co., LTD, Presented at GNSS 2004, The 2004 International Symposium on GNSS/GPS.Sydney, Australia 6?8 December 2004.

Isamu Y., et al., "Secure Active RFID Tag System " Ubicomp2005 Workshops.

Ishida Y.,"Strategic Railway Systems Using ADS Technologies in Japan", Proceedings of the Sixth International Symposium on ADSous Decentralized Systems (ISADS'03) IEEE, pp.1-5, 2003.

Jun L., et al., "The Security and Privacy of Smart Vehicles" Joint work with Srdjan Capkun

Kitahara F., Kera K., Bekki K.,"Autonomous Decentralized Traffic Management System", Proceedings of ADS2000 International Workshop, pp.87-91, 2000.

Keiichi M, et al."Development of Wireless Communication Technology for ETC System", Mitsubishi Heavy Industries, Ltd.Technical Review Vol.38 No.3 (Oct. 2001)

Krishan S. K, M. R. Selim, J. Miura, Y. Goto, and J. Cheng, "POP Method: An Approach to Enhance the Security and Privacy of RFID Systems Used in Product Lifecycle with an Anonymous Ownership Transferring Mechanism", In Proc. SAC, ACM Press, 2007, pp. 270-275.

Krishan S. K. and Yoshiura N., "OTag: Architecture to Represent Real World Objects in RF Tags to Improve Future Intelligent Transportation Systems", International Journal of Convergence Information Technology(JCIT), Vol. 4, No. 2, pp. 30- 48, June 2009..

Lee L. T. and Tsang K.F., "An active RFID system for railway vehicle identification and positioning", International Conference on Railway Engeering(ICRE 2008), pp.1-4, 2008.

Lionel M. Ni and Yunhao L, "LANDMARC: Indoor Location Sensing Using Active RFID". Wireless Networks 10, 701?710, 2004 Kluwer Academic Publishers. Manufactured in The Netherlands.

Matthias S, et al., ERTICO ? ITS Europe, Tanja Kessel, EICT, "Preventive and Active Safety Applications Integrated Project". eSafety for road and air transport report.

Samuel S. Gould, "How Auto-Identification and Data Collection (AIDC) is Transforming in the RFID" World Manufacturing Advantage Conference May 21, 2008.

Song Z. and Z. Qiu, "The application of UHF RFID technology in mine locomotive positioning system", IT in Medicine and Education 2008(ITME2008) IEEE, pp.1079-1082, 2008.

Takashi K, J Nishiyama, H Sugahara, T Okada, F Yamato and M Matsumoto, "A Railway Signal Control System by Optical LAN and Design Simplification", Journal of Networks (JNW) VoL3 NO7, pp.8-15, 2008.

Takenori W., K. Bekki and Y. Yokosuka, "Leading-Edge Solutions for Next-Generation Railway Systems" Journal of Hitachi Hyoron Vol87No9, pp,699-704, 2005.

Yoshimichi S and Koji M,"Development and Evaluation of In-vehicle Signing System Utilizing RFID tags as Digital Traffic Signs" International Journal of ITS Research, Vol. 4, No.1, December 2006.